Bio-hydrogen Energy and Sustainable Development

Dr. Deepak Vyas

Associate Professor, Department of Botany,
Dr. Hari Singh Gour Vishwavidyalaya, Sagar, M.P.

Dr. Rajan Kumar Gupta

Associate Professor, Department of Botany,
Govt. P.G. College, Rishikesh, UK.

2016

Daya Publishing House®

A Division of

Astral International Pvt. Ltd.

New Delhi – 110 002

Cataloging in Publication Data--DK
Courtesy: D.K. Agencies (P) Ltd. <docinfo@dkagencies.com>

Vyas, Deepak, 1964- author.
Bio-hydrogen energy and sustainable development / Dr. Deepak Vyas, Dr. Rajan Kumar Gupta.
       pages cm
       Includes bibliographical references and index.

ISBN 978-93-5124-362-5 (International Edition)

1. Hydrogen as fuel. 2. Biomass energy. 3. Sustainable development. I. Gupta, Rajan Kumar, 1963- author. II. Title.

TP359.H8V93 2016   DDC 665.81   23

Published by   :   **Daya Publishing House®**
                            A Division of
                            Astral International Pvt. Ltd.
                            – ISO 9001:2008 Certified Company –
                            4760-61/23, Ansari Road, Darya Ganj
                            New Delhi-110 002
                            Ph. 011-43549197, 23278134
                            E-mail: info@astralint.com
                            Website: www.astralint.com

Laser Typesetting   :   **Classic Computer Services,** Delhi - 110 035

Printed at   :   **Sanat Printers**

DEPARTMENT OF CHEMISTRY
DOCTOR HARISINGH GOUR UNIVERSITY, SAGAR (M.P.) 470003 INDIA
(A Central University)

Dr. A.P. Mishra
D.Phil, D.Sc.(Alld.), FIC, MNA Sc.
Professor of Chemistry
(Sectional President 102nd ISC)

'A' Grade

Res. : C-51, University Campus
Sagar(M.P.), 470 003 (India)
☎(R) : 07582-265661
(O) : 07582-265265
(M) : 09425425938
e mail : apmishrasagar@gmail.com

Foreword

The current fossil fuel-based generation of energy has led to large-scale industrial development. However, the reliance on fossil fuels leads to the significant depletion of natural resources of buried combustible geological deposits and to negative effects on the global climate with emissions of greenhouse gases. Hence, enough efforts have to be made to shift from fossil fuels to non-polluting and renewable energy sources. One potential alternative may be bio-hydrogen (H_2), a clean energy source with high catalytic yields; on the combustion of H_2, H_2O is the only major by-product. In recent decades, the attractive and renewable characteristics of H_2 led us to develop a variety of biological routes for the production of H_2. Based on the mode of H_2 generation, the biological routes for H_2 production are categorized into four groups: photobiological fermentation, anaerobic fermentation, enzymatic and microbial electrolysis, and a combination of these processes. Thus, this book primarily focuses on the evaluation of the biological routes for the production of H_2. Attempt has also been made to assess the efficiency and feasibility of these bioprocesses with respect to the factors that affect operations, and we delineate the limitations. Besides above alternative options such as bioaugmentation, multiple process integration and microbial electrolysis for improving process efficiency have also been discussed to address industrial-level applications.

A.P. Mishra

Preface

Due to the continuously increasing high energy demands of society and the finite nature of fossil fuels, alternative energy sources are becoming exceedingly important. Hydrogen is a promising alternative fuel because of its clean, renewable and high energy content Given the current situation it is logical to aim at energy saving and more efficient conversion of fossil fuels. In addition, "CO_2-free" operation is required. The objective of this "clean fossil" strategy is to enable utilisation of fossil fuels in a societally responsible manner.

A substantial contribution of renewable energy sources (solar, wind, hydropower, biomass and geothermal energy) to the energy economy is expected in the long term. This requires development of an integrated energy system comprising innovative and highly efficient energy conversion technologies. This enormous challenge calls for a creative approach, while a concomitant modification of the current energy structures is expected.

The environmental impact to a hydrogen economy would be the ideal of sustainability, but only when conditions were appropriate. And this is not the case. To stop issuing between 70 per cent and 80 per cent of CO_2 into the atmosphere, it would mean the possibility of regeneration in a relatively small space of time, as well as, improve air quality in cities, especially the most polluted as Mexico. Even with only operate worldwide with hydrogen cars, only the issuer would reduce greenhouse gases by 20 per cent. These values refer to hydrogen fuel with no carbon. Not taken into account, the hydrogen obtained from natural gas. If the application is done in optimal conditions, the hydrogen would be a sustainable energy vector.

Political action of this magnitude, in a globalized world where oil is currency, and the big multinationals of oil and oil producing countries are strangling the world economies,. However,the economy of the hydrogen would be near to a social justice, prosperity and equality. But,Looking in to the present scenario developed countries hold the key, because the hydrogen economy requires a very high technological level which lacks the most disadvantaged countries and oil producing countries. The incursion of the hydrogen economy in society will produce an unprecedented social change and energy; we might even attend a social revolution. This application will not only transform the concept of energy also ways of life.

When we refer to the social effects produced by something, it should define what is going to be evaluated, then proceed to the analysis of the facts. This requires a life cycle assessment and impact of technology, becoming part a series of elements such as economic, environmental, social, health, risk, human needs, sensitivity, development objectives of society and political impacts decision making.

Therefore, taking these issues on the horizon, and emphasizing that the implementation of technology can never make the risks significantly outweigh the potential social, as any damage to the citizenship for their implementation, and undermining their quality of life instead of getting beneficial technologies reported. Reflect on the effects of the hydrogen economy.

Depending on what the source of production of hydrogen vary substantially the price. Given this fact we must ask two questions: If the question is to eliminate energy dependence on fossil fuels, or reduce the emission of CO_2 into the atmosphere. According to the election, the effects on the social impact vary substantially. If the intention is to reduce emissions of greenhouse gases as stated in previous sections, the current method of producing hydrogen via natural gas is the most profitable. Companies, universities and research centers continue to explore different components to reduce costs, making it competitive and produce fewer emissions of CO_2.

Another way is through acquiring hydrogen from biomass. The energetic use of the biomass is the gasification, which allows obtaining gas of synthesis (CO + H_2). The synthesis gas obtained can be used as direct fuel, as well as H_2 source or chemical feedstock to make other fuels. The production of hydrogen by gasification of biomass is an interesting option, it has the advantage over the conventional (steam methane reforming of water) to use a waste and not a chemical feedstock. It arises as a hope to the energy consumption, though it issues CO_2 there diminishes the dependence of the fossil fuels. It has an important disadvantage, his corrosive effect reducing considerably the life of the fuel cells and pipelines of transport.

On the other hand, if you wanted to eliminate the dependence on fossil fuels and emit no CO_2, the process is quite different. To achieve carbon-free hydrogen the most common is water, and electrolysis treatment. This method requires a lot of energy to break water molecule and split it into hydrogen and oxygen. The power supply either through renewable energy or nuclear energy is higher than that after the hydrogen is exploited.

Hydrogen is a valuable gas as a clean energy source and as feedstock for some industries. Therefore, demand on hydrogen production has increased considerably in recent years. Electrolysis of water, steam reforming of hydrocarbons and auto-thermal processes are well-known methods for hydrogen gas production, but not cost-effective due to high energy requirements. Biological production of hydrogen gas has significant advantages over chemical methods. The major biological processes utilized for hydrogen gas production are bio-photolysis of water by algae, dark and photo-fermentation of organic materials, usually carbohydrates by bacteria. Sequential dark and photo-fermentation process is a rather new approach for bio-hydrogen production. One of the major problems in dark and photo-fermentative hydrogen production is the raw material cost. Carbohydrate rich, nitrogen deficient solid wastes such as cellulose and starch containing agricultural and food industry wastes and some food industry wastewaters such as cheese whey, olive mill and bakers yeast industry wastewaters can be used for hydrogen production by using suitable bio-process technologies. Utilization of aforementioned wastes for hydrogen production provides inexpensive energy generation with simultaneous waste treatment. This book summarizes bio-hydrogen production from Algae, Bacteria, Cyanobacteria, Chloroplast, Waste material. The enzymes involved in production of hydrogen. Use of H_2 as a fuel and its role in sustainable development has also been discussed.

The authors Dr Deepak Vyas and Dr Rajan Kumar Gupta great fully acknowledge their respective institutions, *i.e.* Department of Botany Dr Hari Singh Gour University, Sagar, M.P. and Govt P.G. College Kotdwar, Uttarakhand. And family members and well wishers for their constant encouragement and cooperation.

Dr. Deepak Vyas
Dr. R.K. Gupta

Contents

Chapter 1

Introduction

Microorganisms present a great opportunity for energy science, and hence are a natural focus for the Department of Energy. Microorganisms are simpler than plants; they have smaller genomes and proteomes, and are easier to manipulate and culture. The enormous biodiversity of microorganisms presents a broad palette of starting points for engineering. Microorganisms already make many metabolic products, some of which are useful fuels.

Experimental measurements of efficiencies of fuel production must account for all system inputs and losses, including (but not limited to) pumping and sweeping out of products, stationary state relative to standard state, and the light intensity dependence of product yield. Current microorganisms are likely not optimized for energy production of useful fuels. For example, hydrogen production from algae is arguably operating at present, at 0.05 per cent efficiency.

Biofuels are advantageous because they inherently solve the storage problem posed by the diurnal fluctuation of sunlight. Additionally they make carbon-carbon bonds, which are high-value mobility fuels. Biofuels repreent a significant opportunity to address energy issues. For example, Brazil has a successful ethanol market at the present time, and ethanol is an important transportation fuel in Brazil. There is however a gap between what biofuels can currently do and what we need them to do to become a viable material component to global energy (energy friendly, carbon neutral, and economic.) On the other hand, the science underlying biofuels is still in an early stage of development and much likely remains to be discovered and understood. There is probably room for significant improvement.

The efficiency of biofuel production is ultimately limited by the efficiency of photosynthesis for converting sunlight into fuels. Photosynthesis has an upper bound on its efficiency of ~ 10 per cent, of the total medium energy in sunlight into stored chemical energy based on the conversion efficiency of the primary photosynthetic proteins. On the other hand, the time-averaged primary productivity for C4 plants in the field is approximately 0.25 per cent in the. Current annual energy usage in the world energy consumption is increasing rapidly especially with the modernization of developing world countries. On the other hand there is serious concern about carbon dioxide emissions. The current level of 380 ppm compares with the preindustrial values of 280 ppm. Annual increases are currently at the level of 1.8 ppm/year–so that by the end of the century, if current levels were maintained and action is not taken, the CO_2 level will be 550 ppm, whereas even higher levels will be produced if fossil fuel consumption increases.

The only way of addressing this issue is to find a domestic (global) energy supply which can cope with demand, while limiting CO_2 emissions. There are, unfortunately, very few possibilities. Perhaps the only viable option for producing transportation fuels (as opposed to stationary energy sources) is energy production from photosynthetically derived products. These are carbon neutral in that the CO_2 burned from any carbon-based fuel generated from photosynthetic products is exactly that taken up by the plants to grow in the first place. A principal advantage of photosynthetically derived biofuels relative to traditional (man-made) solar electricity systems is that biofuels inherently solve the storage problem posed by the diurnal fluctuation of sunlight: they make carbon-carbon bonds, which are themselves high-value mobility fuels.Before examining the specifics of what would be required for biofuels to make a significant impact on the global energy market, it is worth remarking upfront that although progress is required for real economic viability, biofuels represent a very real opportunity, not a fantasy.

For energy from biomass to be a viable CO_2 neutral energy supply, it is necessary that this energy source is able to replace a substantial fraction of the current U.S. energy usage. We can therefore ask how much land area is needed to produce 1 Terawatt of energy. This area is given by the equation 200 W m^2 × Efficiency × LandArea = 1 TW. Hence, the land area depends inversely on the efficiency for the conversion from the incident energy to fuel.

Competing Solar Technologies

Before turning to our analysis of photosynthesis, it is worth summarizing the attributes of the technologies that are currently the most efficient for converting sunlight to fuels.

Photovoltaic Solar Cells

The highest efficiency route to fuel production from sunlight currently involves a photovoltaic (PU) cell connected in series to an electrolysis unit. In this approach, the photovoltaic cell is optimized for capture and conversion of sunlight into electrical energy.

Electrolyzers for formation of H_2 and O_2 from H_2O can have energy conversion efficiencies of in excess of 80 per cent based on the electrical energy input divided into the energy of the fuels output from the electolyzer. Hence, the overall system efficiency of fuel formation produced by connecting a photo-voltaic cell with an electrolysis cell is, to a good approximation, 70 - 80 per cent of the efficiency of the photovoltaic cell itself. Thus, overall system energy conversion efficiencies for fuel formation of in excess of 20 per cent can be obtained by a PV array electrically connected to an electrolyzer unit.

This approach is the benchmark for energy conversion efficiency because the photovoltaic cell can have its band gap optimized to match the solar spectrum, either for the situation of a single band gap, for multiple band gaps with multi-junction systems, and/or under concentration. Under concentrated sunlight, integration of the electrolyzer unit with the PV cell array and heat produced by the concentrated sunlight at the focal region can produce still higher system efficiencies, because some of the solar heat can be transferred to raise the temperature of the electrolyzer, reducing the voltage needed to split water (for an entropically favored reaction like $H_2O = H_2 + 1/2O_2$, increasing the temperature favors formation of the products). Other systems that also have very high efficiency involve intimate integration of the electrolysis function with the photovoltaic cell, and do not involve a separate electrolyzer unit. Instead, the electrocatalysts for reduction of protons to hydrogen and for oxidation of water to oxygen are plated onto separate sides of a multijunction photovoltaic cell, with the cell designed to produce sufficient voltage at maximum power (typically 1.4-1.5 V) to sustain the electrolysis of water, instead of the production of electricity. Such systems have overall energy conversion efficiencies of 10 - 17 per cent, depending on the exact details of the system.

Solar Thermal Systems

Another route to fuel formation from sunlight involves concentrated solar thermal systems. Under high optical concentrations, very high temperatures can be produced in the focal region of the optical path. Such systems require dual axis tracking of the sun, but replace relatively expensive photovoltaic cells with less expensive optics as the main areal component of the system. The use of heat instead of light absorption in a photovoltaic cell at the focal point is conceptually advantageous in the formation of fuel. The reason for this is that photon conversion devices all suffer efficiency losses due to their inability to absorb photons below the lowest band gap region in the device, and photons of higher energy than the band gap in any region of the device generally thermalize rapidly and are only converted as if they had band gap excitation energy. These well-known (Shockley-Quiessar) restrictions set a limit on the ultimate conversion efficiency of a photon conversion system. In contrast, a solar thermal system can be used to generate high temperatures in a reactor, which then in principle is limited in efficiency by the second law ($T1$ - $T2$)/$T1$ term; with $T1$ of 2000 K and $T2$ of 300 K, efficiencies of> 80 per cent are possible in principle for direct conversion of sunlight into chemical fuels.

The keys to making such systems work in practice involve: a) engineering design to allow for good optical paths and photon capture into the reactor, while minimizing re-radiation of absorbed, thermalized photons back to the environment; b) developing a set of closed-cycle chemical process steps that can make the desired fuels, and c) developing materials for the solar thermal chamber construction that are compatible with the high temperatures and reactants used in the thermochemical cycle. For water splitting, a two step process that looks especially promising involves $Zn + H_2O = ZnO + H_2$, and then $ZnO + H_2O = Zn + O_2$. The engineering issue is that the two process steps much be performed in batch mode and isolated temporally from each other. Additionally, the chamber materials must be compatible with the mass flows of reactants and products at the very high temperatures at which such systems need to operate to produce a rapid rate and a good conversion to products at each step. In principle, such cycles could offer 60 - 70 per cent over-all energy conversion efficiency, if they can be assembled successfully into a practical concentrated solar power system.

Direct Photolysis

A third approach is to use direct photolysis, either with a semiconductor as the photocatalyst or with organic or inorganic dye molecules as the photo-catalyst for the water splitting process. Semiconductor photoelectochemistry functions conceptually as a photovoltaic cell hooked in intimate series with an electrolyzer unit, except that there is no metallurgical junction between two solid phases in the photovolatic unit to produce an electric field and effect charge separation. Instead, the junction is formed by virtue of the contact between the solid and the water-containing electrolyte. Semiconductors that can spontaneously split water with sunlight, and which are stable in sunlight, are relatively well-known, and generally involve metal oxides with high band gaps, such as SrTiO3, KTaO3, or SnO_2. The large band gaps of such systems preclude obtaining high efficiencies for solar fuel formation, but there is sufficient solar photon flux in regions where they absorb that overall solar energy conversion efficiencies of c.a. 1 per cent, *i.e.*, greater than that of most plants, has been demonstrated. Generally, oxides and other semiconductors with smaller band gaps, that are better suited to capture photons from the solar spectrum, are not robust in water and either oxidize (like Si to SiO_2) or corrode (such as GaAs to Ga3+ and As3). Recently developed materials have expanded the light absorption into a portion of the visible region of the solar spectrum, and efficiences for water splitting of approximately 4-5 per cent have been reported. Work on semiconductor photoelectrolysis being performed currently emphasizes development of new materials or materials combinations that can combine stability with high efficiency for fuel formation in water, following on the 4 - 5 per cent efficient materials that have been developed recently.

Each of the above approaches has its own tradeoff between cost and efficiency.

In recent past the feasibility of the indirect biophotolysis process has been called into question (Benemann and San Pietro, 2001), mainly based on the view that the photobiological stage would require several-fold the area envisioned in the earlier studies (Benemann, 1998a) and because of the limited gas exchange capacity of low-cost photobioreactors (this Report). This led to the conclusion that future research in biohydrogen should emphasize dark fermentations, both with microalgae and, in particular, fermentative bacteria (Benemann and San Pietro, 2001).

One such project was under taken at HNEI in early nintees,biohydrogen project involved the development of the genetic tools required to investigate and improve H_2 production by hydrogenases in cyanobacteria. *Spirulina* (now called *Arhrospira*) was chosen as the organism for this project, as it did not fix nitrogen (and, thus hydrogen evolution would not be dominated by this enzyme), but did produce hydrogen, at least under anaerobic conditions in the dark, that is by some type of fermentations (Bylina, 1994, and literature cited therein). As discussed later, cyanobacterial hydrogenases evolve H_2 primarily if not exclusively in dark fermentations. The regulation, reductant source and electron transport pathway, of such cyanobacterial hydrogenases have so far been little studied.

The specific strain selected was *Spirulina pacifica*, the tradename given to the *Spirulina* cultured commercially in Hawaii by Cyanotech, Inc. The immediate requirements in the development of a genetics system, allowing genetic manipulation and engineering of this organism, were five-fold:

1. Characterize DNA restriction-modification enzymes in *S. pacifica* (needed for gene transfers).

2. Determine inhibitory antibiotics and isolate resistant strains (to provide genetic markers),

3. Construct a library of chromosomal *Spirulina* DNA, to serve as raw material for gene transfers;

4. Design a DNA vector which will be stably maintained in *Spirulina*, to allow gene transfers, and

5. Develop physical or biological methods to introduce foreign DNA into Spirulina cells Significant progress was made by this project during its initial stage (Bylina, 1994):

   ☆ Three different restriction enzymes were purified and characterized of recognition sites. They were isoschizomers of known restriction enzymes (*e.g.* they cut the same DNA sequence). Identifying these enzymes allows construction of vectors that will not be destroyed in these cells.

   ☆ *Spirulina* was most sensitive to chloroamphenicol and erythromycin. Vectors engineered from strains selected for resistance to these antibiotics can be used as selectable markers in *Spirulina*.

   ☆ A *Spirulina* genomic DNA cosmid library was constructed, as a source of *Spirulina* genes. 4.,5. Exploratory research was carried out in vector development and DNA gene transfer.

This provided a basis for initiating gene transfer studies in this organism.

At this point, the focus of this research changed to hydrogenases in *Anabaena* sp. PCCC 7120, a heterocystous cyanobacterium. Like almost all such algae, *Anabaena* has an uptake hydrogenase involved in recycling the H_2 produced as a by-product of nitrogenase action. Like all such uptake hydrogenases, it is a Ni-Fe containing enzyme, and there are several genes (and proteins) involved in their assembly. One such gene is hubB, which was isolated from a genomic DNA library using the polymerase chain reaction (PCR) with so-called degenerate primers obtained from hupB genes from other bacteria. Although of scientific interest, neither this organism nor this uptake hydrogenase are of direct interest in the development of a practical biophotolysis process. In general, nitrogenase-based systems are not of interest in biohydrogen (Benemann, 1998b). With the appointment in 1995 of a new P.I. for this project, Dr. Oskar Zaborsky, the direction of the research changed, to the development and demonstration an indirect biophotolysis process.

The arguments for and against direct and indirect biophotolysis systems have been presented before (Benemann, 1996). Essentially the fundamental feasibility of a practical direct biophotolysis process (simultaneous O_2 and H_2 production without intermediate CO_2 fixation) requires a highly O_2 resistant hydrogenase reaction, including reductant supply, not just an O_2 stable enzyme. Such a metabolic process would need to produce H_2 and O_2 simultaneously at atmospheric pressures. Such an enzyme systems have not been described. O_2 absorbers, reversible or irreversible have been suggested as a way to overcome the limitations of such direct biophotolysis processes. However, irreversible O_2 absorption requires a consumable substrate, which most plausibly means glucose or a derivative. Simply stated, this would double the amount of photosynthetic area required: for each square meter of direct biophotolysis, at least as large (and in practice somewhat larger) area would be required to photosynthetically produce the oxygen absorber. Essentially reducing overall solar energy conversion by over half, compared to a truly oxygen resistant direct process. Reversible O_2 absorbers would be limited by the process energy inputs and capital costs required to efficiently recycle the absorber. A workshop by Benemann and San Pietro (2001) concluded that direct biophotolysis processes required basic and fundamental investigations, rather than applied and process oriented, R and D.

Another suggested biophotolysis process was to the use of heterocystous cyanobacteria, in which simultaneous O_2 and H_2 production is coupled through CO_2 fixation (Benemann and Weare, 1974). However, this approach can also be rejected, on various grounds. First based on the energy inefficiency of the nitrogenase reaction, and second, and more fundamentally problematic, the high energy required to maintain heterocyst metabolism, including respiration (Benemann, 1978b). Finally, another major objection in any direct (or heterocystous cyanobacterial) biophotolysis process, would be in the need to separate the two gases being created simultaneously. Although technically feasible, gas handling and separation represent a significant (though not yet well quantitated) costs, as well as a safety concern. And these arguments do not even address the most fundamental issue in such single-stage biophotolysis concepts: the costs of the closed photobioreactors required to cover the entire areas of the process.

These fundamental problems of direct biophotolysis, led to suggestions, already many years ago, for indirect processes, separating the O_2 and H_2 reactions into separate reactors, coupled through an oxygen stable reductant, such as the pyridine nucleotides (*e.g.* NADPH), which are also involved in CO_2 reduction (Benemann and Weissman, 1976). More logically, the intermediate would be a CO_2 fixation product, particularly one which could be relatively easily converted by intermediate metabolism to H_2, specifically a polyglucose such as starch (accumulated by green algae) or glycogen (found in cyanobacteria). Nitrogen limitation is known to result in the accumulation of large amounts of such storage carbohydrates in green algae and cyanobacteria. A key issue is how to convert the stored glucose into H_2. Some of it can be produced in the dark, the remainder would need to be evolved in a light-driven reaction. The loss of O_2 evolving capacity during nitrogen limitation is well known, and provides a mechanism for light-driven H_2 evolution without O_2 inhibition. This was observed in the alternating cycles of N_2 fixation and O_2 production in non-heterocystous cyanobacteria (Weare and Benemann, 1974). Such cyclic metabolism was a model for the indirect biophotolysis process proposed by Benemann (1994, 1996).

Thie indirect biophotolysis process was subjected to an initial conceptual design and cost analysis (Benemann, 1998a). The main components of such a process were

1. Open algal ponds for production of algal biomass under N-limitation but at high productivity;
2. An algal settling (harvesting) tank, which becomes anaerobic by endogenous respiration;
3. A fermentation tank for dark production of about one third of the H_2;
4. A photobioreactor, for completion of the H_2 production in a light-driven anaerobic reaction;
5. Gas separation (removal of CO_2), clean-up, storage and other gas handling subsystem; and
6. Support systems, such as inoculum ponds, waste treatment, water supply, *etc.*

The economic analysis was based on a number of very favorable assumptions: that the algal biomass could be produced at very high productivities (essentially near the theoretical limit of about 10 per cent solar conversion efficiencies), that essentially 100 per cent of the reductant could be recovered in the form of H_2 fuel (about one third in the dark, two-thirds in the photobioreactor stage), and that the light-driven stage would require only one photon per H_2 evolved. The costs of the ponds (and subsidiary systems) was based on prior studies of large-scale algal production, the cost of the gas handling also derived from prior work, and the photobioreactor costs were assumed to be some $100/m^2$ ($130/m^2$ including contingencies and engineering costs). Even though the photobioreactors would only cover one-tenth the area of the open ponds, they represented about half the total capital costs (the remainder roughly divided between the open ponds and gas handling).

For the HNEI project proposal, *Spirulina*, was again chosen as the immediate organism of choice, for similar reasons as before: it contains hydrogenase, but is not nitrogen fixing, and there is a great amount of information available on this species, from genetics (see above) to mass cultivation. Indeed, there was little need to carry out open-pond cultivation with *Spirulina*, as this is well-established technology. The key objective in *Spirulina* cultivation proposed under this project was to develop techniques for simultaneously maximizing both productivity and glycogen accumulation. The main objective in the proposed hydrogen production research was to demonstrate both a relatively high-yielding dark fermentations and, more importantly, a highly efficient light-driven reaction. As no such reaction had been reported in the literature, that was one of the specific goals of this research.

A major emphasis for the new HNEI biohydrogen project was the demonstration and development of a closed photobioreactor design suitable for biohydrogen production. Selection of such a reactor was based on the likely lowest cost system. Based on a review of the photobioreactor technologies (Benemann, 1998c), the design of Prof. Mario Tredici of the University of Florence was selected. This system has many potential advantages in photobiological hydrogen production. It provides the largest practicable photon capture per photobioreactor area. Its internal gas exchange avoids the need for an external gas exchange device. It can be of relatively large scale, with each unit being some 100 to 200 m² in size, and it is potentially of very low cost (Tredici *et al.*, 1998).

Do Biological Systems Stand a Chance?

Given these very high efficiencies, do biological systems stand a chance? The hope is that the balance of systems costs will be significantly lower, to compensate for the strong likelihood that the overall energy conversion efficiencies to fuel formation will be significantly lower. The hope is that the cost to install a biological system that can grow and reproduce will be low, and that the cost of harvesting the fuel and collecting it back to the central station will be low as well. Other than conventional biomass, however, we know of no systems-based cost analysis for solar H_2 generation from, for ex-ample, large algae ponds, that would allow one to more precisely evaluate the degree to which the systems costs can in fact be lowered compared to those.

Redesigining of Photosynthesis

It is important in considering the low photosynthetic efficiency to recognize that photosynthetic plants and microorganisms have not evolved to produce energy for us in the form that we want it—indeed it is the very real opportunity for bioengineering to substantially improve the situation that is the subject of the present report. The canonical wisdom is that plants and microorganisms are designed to maximize reproductive capacity: for plants this presumably requires optimizing seed production and dispersal, a very different requirement than biomass productivity. Damage control (against the production of excess free radicals) is also of paramount importance.

Later on in this book, we will discuss in some detail the physical and biological constraints on photosynthetic energy yield. We will see that the actual 0.25 per cent efficiency is well below the physical constraints. It is also well below the 10 per cent light processing constraint imposed by the primary proteins of photosynthesis (Photosystems I and II). On the other hand, we will argue that (i) for carbon based fuels CO_2 supply constraints are likely to keep the upper bound below 10 per cent; and (ii) we believe that it is premature to conclude that the only biological constraint is associated with these primary proteins.

To flesh this second point out, we first give an analogy, in the form of a parable. Imagine that Benjamin Franklin reappeared today (300 years from his birth), and came upon a modern day computer. Although he does not know how the computer works, he marvels at what it can do. At some point, he realizes that the computer would be more useful if it would only compute faster; and for this reason he decides to take the computer apart to discover what determines and limits the speed. The initial dissection leads him to the discovery of a remarkable oscillator (the clock) at the heart of the computer, which can switch very fast indeed. He experiments with different computers made by different companies at different times and finds that the clocks have different speeds. He notices that the number of operations that the computer carries out per unit time has a positive correlation with the speed of this oscillator. He therefore reasons that making the computer faster is just a matter of improving the clock–and designs a research program to improve the speed of this essential component.

On the other hand, in reality, we know that computing is not just about clock speed. The design of a computer requires understanding and coordinating a myriad of other issues including memory access time, bandwidth, software, hardware architecture, interconnects, etc. Indeed, increasing the clock speed too much can lead to instabilities in the processing of the computer. Improving the computer requires synergistic changes in the entire system–the modification of an individual component is in general not effective, or, in any case, has limited leverage.

The photosynthetic production of biomass or useable energy likewise requires coordination between the various active components–from the proteins converting light into electrons to the enzymes catalyzing the conversion of CO_2 to sugars. Tuning any individual component should therefore be expected to have limited leverage. Indeed, in a biological system, feedback loops are how evolution ensures robustness.

Despite the complexity of the problem, there are grounds for being optimistic that increased energy production can be achieved. A pertinent example is the case of food crops. Crop yields have increased (essentially linearly) over time to the present day, and have shown no sign of saturation. Improvements have come from combinations of nutrition, pesticides and breeding. One could imagine that a concerted effort at improving energy producing organisms (crops or microorganisms) would also lead to substantial improvements over current yields.

The Specific Case of Hydrogen Production

Having now given a detailed summary of photosynthesis and photosynthetic constraints, we turn to a discussion of the specific case of hydrogen production by microorganisms. What is particularly interesting about this example is that bacteria and algae certainly did not evolve to produce hydrogen. Indeed, the physiological function of a hydrogen metabolism is still unknown and debated. The most common idea is that the hydrogen metabolism assists survival under extreme conditions. In any case, it seems clear that in present day organisms with a hydrogen metabolism, the most prominent of which are the cyanobacteria and the green algae *Chlamydomonas*, hydrogen production is not a primary determinant of the organisms's fitness. Hence, trying to engineer these organisms to scale up hydrogen production provides an excellent specific case to discuss the challenges and bottlenecks that could come up when engineering a microorganism for fuel production.

The Technical Hurdles

The technical hurdles and the state of the art for biohydrogen production have been well summarized in a number of recent reviews, as well as in an excellent older review. Here we outline the main points and explain how they fit into our overall argument.

Oxygen Sensitivity of Hydrogenase

Hydrogen production in both cyanobacteria and algae relies on specialized enzymes to catalyze the reactions. As mentioned above, there are two possibilities. Cyanobacteria have nitrogenases, which catalyze reactions producing both NH3 and H_2. These reactions require an energy input of at least 2 ATP per electron. Additionally many cyanobacteria have so-called uptake hydrogenases that reconvert the electrons in H_2, degrading the efficiency further. Even if the uptake hydrogenases could be eliminated, the energy overhead of the nitrogenase reactions makes them an unlikely competitor for producing hydrogen.

Another option for catalyzing the reaction are hydrogenases. There exist hydrogenase enzymes which catalyze the reaction without requiring any ATP. These enzymes have high specific activities, but their main difficulty is that they are extremely sensitive to oxygen–small amounts of oxygen cause them to irreversibly shut off. (Indeed, nitrogenases suffer from this same difficulty.)

Figure 14 shows the specific activity of hydrogenase enzymes from several photosynthetic algae. The laboratory work-horse *Chlamydomonas reinhardtii* is poisoned upon exposure to air in less than a second. In contrast, *Clostridium* has a substantially larger half life, though still only several minutes.

The question is whether a hydrogenase can be either discovered or engineered with highly reduced oxygen sensitivity. Otherwise, the hydrogen production reaction can only take place in the absence of oxygen. Indeed, the early demonstration by Greenbaum of the hydrogenase reaction in *Chlamydomonas* was done by continously purging the oxygen out of the sample cell.

There are reasons for both optimism and pessimism in the quest for an oxygen insensitive hydrogenase. On the optimistic side, substantial progress has been made recently in understanding the structure and hence the mechanism of operation of the hydrogenases, and this has led to both careful thinking and modelling (*e.g.*, we were briefed in this regard by M. Ghiradi) for why the enzymes are oxygen sensitive. Such understanding will lead to a natural opportunity for the possible rational redesign of the enzymes.

Another reason for optimism is that the number of photosynthetic algae that have been carefully studied is relatively small; recent efforts of Venter and collaborators could lead to the discovery of new algal species and associated hydrogenases that are markedly less sensitive to oxygen. The fact that in Figure 14 *Clostridiium* has a 400-fold higher half life than *Chlamydomonas* well illustrates the opportunity. Are there organisms with even higher half lives?

The main reason for pessimism is that there is presently no known chemical system, living or nonliving, that can evolve hydrogen without reacting with oxygen. Although we know of no fundamental principle that says such a system cannot exist, we also do not know of any principle that says such a system can exist. It is perhaps worth recounting here the experience of evolution with rubisco, the central enzyme in the dark reactions of photo-synthesis. Rubisco also reacts with oxygen, leading to photorespiration. C_4 plants managed to limit photosrespiration by evolving a compartment where the oxygen concentrations are much lower than that of C3 plants; it did not, however find a way to re-engineer the enzyme.

Producing Hydrogenase

The hydrogenase enzyme is not produced under normal situations. Classically, the enzyme is produced in the dark, when the culture is deprived of oxygen. However, the genetic circuits underlying this switch are not known. If for example the promotor for the hydrogenase enzyme were discovered, it could be possible to turn on production of the enzyme under normal conditionsest cases. There is thus a fortyfold decrease between principle and practice. Photosynthetic efficiencies are far below that of man-made solar devices, and are likely to remain this way for a long time. However, systems and material costs for photosynthesis are much less expensive so biofuel can possibly form an economically attractive energy production option.

Plants are not necessarily optimized to be energy conversion machines. For example food crops have been genetically improved to increase food production, and the efficiency of food production has not yet plateaued.

However, the reengineering of plants to improve biomass energy yield is a multi-axis problem, and will likely require more than single (*e.g.*, genetic) modifications of single proteins. The photosynthetic machinery has evolved to optimize fitness in a complex environment: Biological systems are intrinsically complicated because of the multiple feedback and control loops that must be present to guarantee robust survival. As a result, modifying any one property will likely have a limited leverage. In attempting to improve the system, it is important to think about the whole system (organism, environment, product, process). Progress bridging these gaps requires a dedicated commitment to breeding and/or molecular and systems level analysis. These two approaches should be synergistic.

Strategies for Improving and Producing New Fuels

As discussed above, there is a substantial gap between the currently attainable bioenergy yield and the estimated upper bound of ~ 10 per cent based on the known properties of the components of the photosynthesis machinery. In this section, we examine various aspects of the known photosynthesis pathway in order to identify bottlenecks and explore strategies to improve the energy yield. We will illustrate the issues in the context of biohydrogen evolution, which is a simpler conceptual problem and is also somewhat more distinct (compared *e.g.*, to bioethanol production) from issues addressed in the traditional metabolic engineering context.

Hydrogen Evolution as an Application of Metabolic Engineering

It is desirable for hydrogen evolution to occur in aerobic conditions in order to minimize the investment of extra energy needed to stringently maintain anerobic conditions. This task is akin to those routinely encountered in metabolic engineering, where a specific chemical compound, *e.g.*, amino acids or carotenoids, is to be produced in large quantities. Such metabolic engineering tasks typically consist of several key challenges:

1. To increase the input flux (in this case, photosynthetic electrons)
2. To increase the efficiency of product synthesis (in this case, hydrogenase activity),
3. To re-route the input flux towards product synthesis.

Metabolic engineers use a variety of methods to meet these challenges, including directed evolution of key molecular components, combinatorial, and rational design of alternative metabolic pathways. Below, we will describe these methods in the context of the hydrogen evolution problem. Conceptually, hydrogen evolution is simpler than canonical metabolic engineering applications in that the terminal pathway is of only one step, which branches immediately from the main input pathway. On the other hand, the photo-synthesis pathway is highly regulated, as organisms generally transduce just enough energy to satisfy their metabolic needs. Moreover, compared to most biosynthetic pathways, regulation of the photosynthetic pathways is not understood as well and involves molecular components (*e.g.*, membrane proteins and co-factors) that are not as easy to manipulate by genetic means. The discussion below serves both to illustrate applications of possible metabolic engineering strategies and to expose problems specific to photosynthesis and hydrogen evolution.

Increasing the Yield of Photosynthetic Electrons

As discussed above, a major obstacle preventing the more efficient capture of photoenergy is the saturation of the photosynthesis rate for incident irradiation beyond 0.1–0.2 solar flux. This effect is attributed to the slow electron current out of PSII, which limits the ability of the reaction center to convert the photoenergy captured by the chlorophylls into additional electron current. As reengineering the Photosystems themselves is beyond the reach of current knowledge and abilities, strategies to overcome this obstacle have centered around ways to avoid light saturation. These include the physical method to dilute the incident irradiation, and the genetic approach to reduce the antenna size (the number of chlorophyll molecule per reaction center) as explored by Melis *et al.* [2000] and discussed above. By distributing the incident irradiation to a larger number of cells, these methods aim to improve the overall yield of photosynthesis by reducing the photoenergy waste (the photoenergy that chlorophyll absorbs but the reaction center cannot process is dissipated into heat). However, they do not address the specific yield (rate of photosynthetic electron production per cell), which determines the size of the culture that needs to be actively maintained. This may be the appropriate strategy for the time being, as the current bottleneck for hydrogen evolution is the conversion of photosynthetic electron current to hydrogen, rather than the electron current itself. However, as hydrogen evolution becomes more efficient, it will eventually be useful to find ways to improve the specific yield of photosynthetic electrons. This may be accomplished in principle by increasing the number of reaction centers which are fitted with smaller antenna sizes. In practice, this will not be an easy task as little is known about genes and regulatory mechanisms that control the number of reaction centers and the size of the antennas. Basic molecular biology research is needed here before engineering efforts can be attempted. Needed here are strong selectable markers that can be used to identify the number of reaction centers and the antenna sizes. For example, the Melis lab identified the tla1 mutant of *Chlamydomonas* based simply on differences in shades of greenness shown by the different mutants. By adopting more quantitative optical characterization in a high throughput capacity, it should be possible to screen a much larger number of mutants with desired reaction centers and antenna sizes.

Increasing the Efficiency of Hydrogenase Activity

A potential bottleneck of hydrogen evolution is the activity of the [Fe] hydrogenase: In principle, does the hydrogenase produce H_2 quickly enough to process photosynthetic electrons? The specific activity of the hydrogenase was measured to be ~ 100nmol H_2/(mg - min). Let us assume a cell mass of 10-9 g and ~ 106 reaction centers per cell; if the hydrogenase is say x per cent of the cell mass, this corresponds to ~ 1 H_2 molecules evolved/sec per reaction center, about 10-2x of the rate of photosynthetic electrons generated by the Photosystem at maximum capacity. Hence, hydrogenase with this level of activity can apparently only keep up with photosynthetic electron production if of order the entire cell mass is filled with hydrogenase. Moreover, a glaring shortcoming of the hydrogenase of *Chlamydomonas* is its extreme sensitivity to oxygen, with a half-life of < 1 sec when exposed to oxygen. Interestingly, a bacterial [Fe]-hydrogenase (*Clostridium pasteurianum*) was shown to be two orders of magnitude more stable in the presence of oxygen, although the half-life itself was still quite short (< 10 min). *in vitro* and *in vivo* coupling of the clostridial hydrogenase with the cyanobacterial photosynthetic system via cyanobacterial ferredoxin was demonstrated in the presence of light. Recently, H_2 evolution by the [NiFe]-hydrogenase of cyanobacteria has also been studied. A Synechocystis strain deficient in its native NAPDH-dehydrogenase complex was shown to evolve a significant amount of H_2 in light. Significantly, an O_2-tolerant [NiFe]-hydrogenase from R. gelatinosus was recently identified with a half-life of 21 hours *in vivo* and 6 hours in purified form *in vitro*. *In vitro* experiments demonstrated that this O_2-tolerant [NiFe]-hydrogenase could work with ferredoxin of red algae as the electron donor, although at a diminished activity level.

Currently, there is an effort at NREL to reengineer the hydrogenase of *Chlamydomonas* to make it more O_2-tolerant. The work is computa 3Here the activity is measured per unit purified protein-tional, based on homology modeling of the [Fe]-hydrogenase of *Clostridium pasteurianum* whose crystal structure is available.

Directed Evolution of Hydrogenase

Given the availability of hydrogenases with improved O_2-tolerance in various organisms, we suggest that directed evolution may be an effective way to find the desired hydrogenase with both O_2-tolerance and strong coupling to the ferredoxin of *Chlamydomonas* to yield high specific activity. Directed evolution is an iterative scheme of generating genotypical variations and selecting for those with a desired phenotype. This principle has been used successfully in breeding animals and plants throughout the history of mankind. It has also been used in metabolic engineering to obtain enzymes with various desired properties. In laboratory-scale protein evolution, genotypical variations are typically generated by random mutagenesis and/or recombination, followed by screening or selection. The applicability of the directed evolution approach is dependent largely on the existence of a powerful selection scheme or a high-throughput screening assay. It is especially useful if the selection/screening assay can identify small changes in phenotype so that the corresponding mutants may be amplified in subsequent rounds of evolution. Recently, Posewitz *et al.* (2005) showed the feasibility of high-throughput screening of H_2 production in C. reinhardtii, by using a library of 6000 colonies on agar plates with sensitive chemochromic H_2-sensor films. Such methods enable the application of directed evolution methodology to finding improved hydrogenases in *C. reinhardtii*.

The effectiveness of the directed evolution approach is dictated to a large extent by the size of the viable mutant pool. On the one hand, it is desired for the population to acquire as much mutation as possible, as evolution is driven by the variability of the population. On the other hand, too large a degree of mutation tends to make most mutants not viable. This conundrum is addressed to a large degree by the method of DNA shuffling, which randomly fragments a population of homologous DNA sequences and then reassembles them into full-length, chimeric sequences by PCR. The idea is to increase the diversity of the viable population by mixing a starting pool of proteins that have been proven to work through natural selection. Over the decade since it was introduced, DNA shuffling (combined together with random mutagenesis by error-prone PCR) has contributed to dramatic increases in the efficiency with which large phenotypic improvements are obtained. We believe the existence of a large number of hydrogenases from bacterial and algal species with varying degrees of performance (ranging from high specific activity but O_2-sensitive to low specific activity but O_2-tolerant) make this problem well suited for directed evolution by DNA shuffling. Two independent approaches may be adopted: One is to evolve the [Fe]-hydrogenases for improved O_2-tolerance, the other is to evolve the O_2-tolerant [NiFe]-hydrogenase for increased activity with the *Chlamydomonas* ferredoxin. The initial phase of either approach may proceed *in vitro*, which allows exploitation over a larger library (103 - -106 for high-throughput screening and > 1012 for display methods). Eventually, iteration of *in vitro* shuffling/mutagenesis and *in vivo* selection may be used to ascertain the effectiveness of the *in vivo* function while still imposing a high mutation rate. For the purpose of increasing the efficiency of hydrogenation, it may be useful to evolve both the hydrogenase and the ferredoxin. In this case, it is important to maintain the selection *in vivo* to ensure that the mutated ferredoxin functions properly with the rest of the electron transfer system of the photosynthetic pathway.

Re-routing the Photosynthetic Electron Current

The end product of the photosynthetic electron transfer system is the reduced ferredoxin (Fd*). It is normally used to charge the canonical electron carrier NAPDH, which together with ATP fuels the Calvin cycle. Until a time when the first two goals above can be accomplished, *i.e.*, an increased number of reaction centers with truncated antenna sizes is installed and an oxygen-tolerant hydrogenase with high specific activity for hydrogen evolution can be expressed in large quantities, it will be necessary to limit the electron flux to the Calvin cycle in order for there to be any significant flux for hydrogen evolution. Below we will address the Calvin cycle flow first, while noting that according to the analysis in Section 6.2, CO_2 flux will be a natural limiting factor if the photosynthetic electron current can be increased substantially, through more efficient light harvesting. Afterwards, we will address the regulatory issues to insure that the reduced demand for electron flux by the Calvin cycle leads to an increased flux into hydrogen evolution rather than an overall repression of the photosynthetic electron current.

Limiting the Electron Flux to the Calvin Cycle

The demand for photosynthetic electron current arises primarily from the need to reduce CO_2 in the Calvin cycle. One straightforward way to reduce this demand is to reduce the CO_2 partial pressure in the environment. As this may be energetically costly for large-scale implementations, we discuss various genetic strategies.

Reduction of CO_2 Uptake

CO_2 loss via diffusion across the cell membrane is a potentially serious problem for unicellular organisms such as algae. A strategy the algae adopt to overcome this problem is to convert the CO_2 into HCO- 3 which does not diffuse easily across the membrane, due to its charge. HCO- 3 is actively sequestered by Na+-dependent and other transporter systems, and is converted back to CO_2, by the enzyme carbonic anhydrase, at the site where Rubisco is packed. This ATP-consuming process effectively increases the affinity of Rubisco for CO_2 sufficiently for CO_2 fixation. Given the knowledge of this pathway, one can in principle reduce the CO_2 uptake rate by reducing the expression of key enzymes in this path-way; see below. Of course, disabling the Calvin cycle itself is another way to turn down the demand for photosynthetic electrons. This may be done by reducing the expression of genes encoding key Calvin cycle enzymes, down-regulating the electron flux to NADPH. Merely reducing the Calvin cycle current may lead to an elevated level of NADPH, which may have other undesirable effects, *e.g.*, reduction of the photosynthetic current itself, through negative feedback. Since NADPH is the designated end product of the photosynthetic electron current, downregulating its conversion from Fd* is the most direct way of reducing the competing electron flow. This can be done by reducing the expression of FNR which catalyzes the transfer of electrons from Fd*. A diminished NADPH level resulting from a reduction in the rate of electron transfer from Fd* to NADPH also has the added benefit of naturally reducing the activity of some Calvin cycle enzymes that use NADPH as their allosteric activator.

Effect of the Redox-Dependent Feedback

The photosynthetic system has an intricate set of feedback controls which ensures that in the situation of low light and low CO_2, the Calvin cycle is turned off and the captured photoenergy is directed primarily into ATP synthesis, while in the situation of high light and high CO_2, the Calvin cycle is turned on and ATP synthesis is maintained at a level to support the demand by the Calvin cycle. To understand the effect of re-routing the photosynthetic electron current from the Calvin cycle to hydrogen evolution, it is crucial to understand the feedback system that gives rise to the robust two-mode behavior.

To illustrate this effect, we describe below two possible models of this feedback control; both are based on the known facts reported in the literature, and both result in very different conclusions for understanding the consequence on the re-routing of the electron current. Without understanding enough about the regulation to understand which model is correct it is impossible to proceed with confidence.

Model 1

In this model, the two-state nature of the photosynthetic system is realized in two distinct modes of electron flow. As described above, the linear flow corresponds to the electrons generated from PSII are directed via the electron transfer system and PSI to reduce ferredoxin and ultimately NADPH. This pathway prevails in the high light situation. The cyclic flow describes the alternative situation in which few electrons are generated by PSII; instead, the energetic electron (excited by PSI) is transferred from the reduced ferredoxin (Fd*) to plastoquinone (PQ), and then recycled through the electron transport system. In this case the energy of the electron absorbed from PSI is used to pump protons into the lumen and synthesize ATP.

For a given environment, the actual mode of electron flow is selected by the redox potential, indicated by the level of Fd*. As will be argued below, the feedback regulation is such that the system supports either a high or low level of Fd*. Fd* is a powerful reducing agent; a high level of Fd* activates regulators such as ferredoxin-thioredoxin reductase (FTR) and thioredoxin (Td), which in turn activates both the expression and the activity of a large number of Calvin cycle enzymes. Thus, the redox level controls the carbon flux through the Calvin cycle. When the redox level is low, the Calvin cycle enzymes are not activated. This shuts off the carbon flux, and also the linear electron flow. On the other hand, at low redox level a state transition is known to occur, where up to 85 per cent of the light harvesting complex LHCII switches association from PSII to PSI. This transition virtually shuts off PSII. The few electrons generated are fed into the enlarged pool of PSI to generate Fd*. Since the flow to the Calvin cycle is shut off at low Fd*, the electron current recycles from Fd* back to PQ, thereby completing the cyclic flow. Thus the transition between linear and cyclic electron flow is driven by the state transition in Chlamydomonas.

Crucial to this model of alternate electron flow is the coordination of the state transition and the transition in Calvin cycle activity, both controlled by the redox potential. The occurence of a state transition requires the phosphorylation of LHCII by LHCII kinase (LK). LK is activated by a surplus of the reduced form of PQ (PQ*) when Fd* level is low. When the redox level is high, however, LK is inactivated by the reduction of its disulfide bond; consequently, the state transition can no longer occur even if the PQ* level is high. A key feature of this regulatory circuit is an effective positive feedback of the level of Fd* on the linear electron flow, mediated through the double negative effect of Fd* on LK and LK on PSII. Together with the aforementioned positive dependence of the Calvin cycle flow on Fd*, this circuit with positive feedback has the potential of supporting two steady states, one with a low and one with a high value of Fd* (and correspondingly a low and high linear electron flow) depending on the input light intensity. (The states may also be selected by the availability of CO_2 and/or ATP, since they determine the magnitude of the Calvin cycle flux.) If this positive feedback effect is indeed the driving force of the observed high-light/lowlight behavior, then not much re-engineering of the upstream electron flux is needed for hydrogen evolution, since the latter amounts to replacing the native output module (*Fd~ ~ NAPDH ~ Calvincycleflow*) by *Fd~ ~ H_2* without affecting the core feedback loop. To mimic the control of redox potential on Calvin cycle enzymes, a regulatory control of the redox level on the expression of the hydrogenase may be added. In this way, the circuit will produce the desired behavior of expressing HydA and evolving hydrogen only under the condition of high light, without being affected by the CO_2 status.

Model 2

The situation can be very different if the feedback loop identified will not be the driving force of the two-state behavior. As an alternative scenario, it was note that the Michaelis-Menton kinetics of catalysis by FNR produces a nonlinear sigmoidal dependence of the NAPDH flux on Fd*, as it takes two Fd* molecules to reduce one NADP+ molecule to NADPH. The balance of the Fd*-depdenent Calvin cycle flux with the NAPDH flux then may in itself be sufficient to generate the two-state behavior. If this is the main cause of the bistable electron flow, then replacing this module by the one step reaction Fd~ ~ H_2 will likely remove the bistabilty feature. Given how much of the chloroplast function depends on correctly discriminating the high-light/low-light environment, it may not be possible to significantly reduce the Calvin cycle flux by hydrogen evolution. There are likely many other alternatives consistent with known facts; these must be sorted out for real progress in metabolic engineering to proceed.

Altering the ATP/pH-dependent Feedback on Photosynthesis

The Calvin cycle flux requires a balance of 2 NADPH and 3 ATP per CO_2 molecule. This is approximately the NADPH:ATP ratio provided by the linear electron flow. If the ATP flux available to the Calvin cycle is reduced due to demand by other cellular processes, the photosynthetic circuit can adjust the NADPH:ATP ratio to provide the additional ATP flux. This may be accomplished by the system by sending a portion of electrons from the linear flow to the cyclic flow, presumably within the linear flow state without entering the state transition. In the opposite situation, when there is a surplus of ATP flux, the system responds by reducing the activity of PSII; the NADPH:ATP ratio can be balanced by adjusting the conductance of ATPsynthase. The physiological function of this interesting response is believed to protect the photosynthetic reaction centers from the catastrophic consequences of photodamage, which may occur even upon brief exposure to intense radiation. This process is mediated by the increased proton level in the lumen, through the energy-dependent nonphotochemical quenching (qE) process, which harmlessly dissipates excessively absorbed light energy as heat. The innate photoprotection pathway may present a major challenge for hydrogen evolution, which if successfully implemented, would utilize only the electron flux and create a huge surplus of proton flux. Some of this proton flux can be converted to ATP flux demanded by cellular maintenance and biosynthesis. The remainder will need to be eliminated to reduce the proton level in the lumen and thereby to avoid the onset of photoprotection. Lee *et al.* (2002) from ORNL propose to insert uncoupler proteins, such as UCP-1 and UCP-2, into the thylakoid membrane to reduce the proton level in the lumen. They propose to add a thylakoid targeting sequence upstream of the coding sequence of the uncoupler protein and place the recombinant gene under the control of the promoter of the hydrogenase gene. This is certainly a worthy effort to try, although possible post-transcriptional control mechanisms may hamper the expression of the recombinant construct. We suggest the use of the directed evolution method (above) to screen for variants of the uncoupler proteins (along with the target peptide and the UTR sequences) for those with superior performances.

Another issue of general concern regarding the uncoupler strategy involves the reduced viability of cells having reduced ATP synthesis activity, even if the gene is placed under the control of a condition-specific promoter. It may be useful to put the expression of this gene under multiple control, so that it is, *e.g.*, rapidly degraded when not expressed. More specifically from the metabolic pathway perspective, it is desirable to control the activity of the uncoupler protein to the ATP usage, so that the protein is only activated when the proton level in the lumen is the high. Otherwise, strong uncoupling activity may deprive the chloroplast of important maintenance activities that require ATP (*e.g.*, replenishing the supply of the D1 protein in the reaction center), while weak uncoupling activity may not relieve the lumen of the undesirable proton accumulation. This will require detailed knowledge of the alga gene regulation mechanisms. Possibly, molecular evolution strategies may be employed to discover such regulatory pathways.

Chapter 2

Methods Used for Estimation of Hydrogen and/or Hydrogenase

A number of menthods have been developed to estimate H_2 evolution or uptake by whole cultures or isolated hydrogenases. Each method has its own advantages and disadvantages. However, amperometric method of H_2 estimation is widely used (Rao and Hall, 1988). The choice of methodology depends upon the types of estimation and duration of experimentation. A brief description of different methods is given below:

1. (a) Manometry

Peck and Gest (1956) had first of all used manometric method for the assay of bacterial hydrogenase. Details of this technique may be found in the manual of Umbreit *et al.* (1972). Gilson submarine volumoneters or conical Wardurg flasks with center wells and one or two side arms are used. Alkaline pyrogallol is put in the center well. A mixture of methylene blue and 5 per cent palladium on aluminium oxide in the side arm is used for identification of H_2 released. Whole filaments (intact) of algae are put in the flask and flushing by required gases is done before putting the rubber stopper. H_2 uptake or evolution is observed at different intervals in manometer.

(b) Modified (Syringe Method) Manometry

Basically this method is based on manometry but rather than Warburg flask, glass syringe is used. For this modified glass syringes (normal capacity 80 ml) serve as the growth vessel; the usual external end of the hollow plunger is cut off and the plunger is inverted. This provides a device in which gas produced by the culture (within the syringe body and plunger) accumulates in the plunger, displacing it upwards. The plunger is precalibrated by injection of known quantities of air. and the volume of accumulated gas is directly indicated. During the syringe assembly procedures, the medium and inoculum are continuously gassed with a stream of argon bubbles to maintain anaerobiosis. The syringe is completely filled with inoculated medium, residual gas is expelled through the syring needle, and the syringe needle is then stabbed into a solid rubber. Syringes prepared in this way are incubated in a glass-sided water bath maintained at required growth temperature and light intensity.

2. Gas Chromatography

This method is very sensitive and less time consuming. H_2 is detected in a gas chromatograph fitted with thermal conductivity detector and molecular sieve column (5A = 500 pm, 30-60 mesh, 2 m, 1/8 inches = 3.175 mm). A number of other columns *viz.* Spherocarb, Silica Gel etc. are also used. Argon as a carrier gas gives best results. in terms of H_2 separation and detection. Depending upon the column, the oven temperature, detector temperature, carrier gas flow rate and bridge current are maintained. Incubation of cultures is generally done in serum vials or Warburg flasks, fitted with ground glass joints with stopcocks and butyl rubber stoppers for the side arms. Desired gases are filled in the vials after evacuation and flushing. The vials or flasks are incubated at desired temperature and light intensity preferably in a shaker and the H_2 formed or consumed is estimated after injecting 0. 2 to 0.5 ml of gas phase by gas tight syringe into a GLC. The peak and concentration of H_2 are edermined after injecting ultrapure H_2 in the GLC. This method allows workers to estimate H_2 formed or consumed by any microorganisms for unknown periods but without disturbing the cultures.

3. Amperometric Method

(a) Clark-type Electrode

Amperometric method for the detection of H_2 was originally developed by Wang *et al.* (1977). In this system Clark-type of electrode is used for measurement of hydrogen in aqueous solution. This system has been found applicable to the determination of biological H_2 production. The electrode has a platinum anode and a silver-silver chloride cathode. The electrode surface is covered with a film of half saturated KCl and 25 u thick Teflon membrane before use. During operation, the polarizing potential across the electrode is set at + 0.06 V versus the Ag-AgCl cathode. The output current from the electrode is amplified by using suitable microvolt-ammeters and fed into a 1.0 mV (full span) recorder.

The electrode is centered in a OX 705 glass water- jacketed cuvette whole total volume is around 1.9 ml. However, cuvettes of different volumes are available. The reaction chamber is sealed during use with a ground glass ball mounted on a stem; the ball sealed firmly into the constricted neck of the reation chamber whose contents are mixed with a magnetic stirring bar. The electrode response is absolutely specific for H_2. No interfernce by N_2, O_2, or argon is observed. The electrode is calibrated by injection of ul portions of H_2- saturated distilled H_2O or buffer. The final concentration of H_2 is calculated from the value for H_2 at 25oC of O.0175. For the determination of H_2 uptake or evolution by live cultures, desired suspension of culture is directly placed in the cuvette. Light is provided by a projector. Sparging of culture by any gas is done in the cuvette itself.

This method is widely used because of its extreme sensitivity and in this system O_2 evolution or uptake may also be measured by the same culture by inserting O_2 electrode or otherwisw by reversing the polarity of electrode.

(b) Zirconium Oxide High Temperature Electrode

This technique was developed by Greenbaum (1977). Actually this method was employed to elucidate the molecular mechanism of photosynthetic O_2 evolution. In this method the main apparatus is the fabricated flow system. The flow apparatus is capable of absoute calibration and/or hydrogen evolved from photosynthetic organisms illuminated with single turnover, saturating flashed. The sensing element in flow apparatus consists of a calibration of the appratus is achieved by placing an electrolysis cell in tandem with the reaction cuvette containing the algae. This system aldo provides an opprotunity to measure O_2 and H_2 simul taneously with the saqme culture.

4. Enzyme Electric Cell

Enzymic electric cell system was developed originally to separate hydrogenase-catalyzed hydrogen- producing system (cathode reaction) from the electron donating system (anode reaction) in order to protect hydrogenase from the by products of electron-donating reactions (Yagi, 1976). The principle is based on the changes in gas volume (H_2) and in the short circuit current of the enzyic electric cell. This method of H_2 estimation works efficiently with immobilized cultures. In brief the H_2 estimation is made by the following steps. The reaction mixture in the cathode contains about 12 pieces of the glutaraldehyde treated hydrogenase gel and 0.5 ml of the methylviologen phosphate stock solution in 3.0 ml. The anode is a zine electrode inserted in ammonium chloride solution. N_2 serves as a gas phase. The reaction is started by closing the circuit and the change in gas volume and current are recorded. No gas-evolution is observed in the absence of hydrogenase. A good account of this methodology has been described earlier by Yagi (1977).

5. Mass Spectrometry

Since the enzyme nitrogenase can catalyze the formation of HD from deuterium, a number of workers have studied the hydrogen-deuterium exchange reaction more frequently in N_2-fixing bacteria. Amongst cyanobacteria tritium exchange method has been employed mostly during the assay of hydrogenase. During the assay, required amount of $3H_2$ (with high specific activity) is injected into reaction vessel containing algal culture but with helium in the gas phase. The samples are incubated in light or dark as per the experimental protocol. After appropriate time interval, a known volume of the liquid phase is withdrawn, diluted three fold with water, and mixed vigorously to remove any dissolved $3H_2$. $3H_2$ present in the sample is ocunted in a liquid scintillation counter, after the addition of suitable cookail.

6. Spectrophotometric Method

The reduction of dyes by H_2, catalyzed by hydrogenase, can be followed by measuring absorbance changes in a standard spectrophotometer. The electron acceptros usually used and their extinction coefficients are as follows; methyl viologen : 12,000 M^{-1} cm^{-1} at 600 nm: benzyl viologen, 8,100 M^{-1} cm^{-1} at 555 nm and methylene blue, 7,000 M^{-1} cm^{-1} at 601 nm. The reaction mixture (usually 3 ml) of 20 mm porassium phosphate buffer, pH 7.0, and the electron acceptor (1 mM) are added to the main compartment of an anaerobic cuvette and the hydrogenase or culture in the side arm. Suitable flushing and degassing are done before the start of reaction. As such reduction of methyl viologen by hydrogenase in the presence of H_2 is the classic method for detecting and locating hydrogenase activity in a mixture of proteins.

Table 2.1: Techniques Used for Hydrogen/Hydrogenase Assay

Methods	Time Scale	References
1. Manometry	Minutes-hours	Davis and Stevenson (1977); Peck and Gest (1956); Umbreit et al. (1972)
Modified Manometry (Syringe method)	Minutes-days	Hillmer and Gest (1977)
2. Gas Chromatography	Minutes-days	
3. Amperometric Method :		
(a) Clark-type Electrode	Sec-minutes	Wang et al. (1977)
(b) Zirconium oxide high temperature Electrode	Seconds	Greenbaum (1977)
4. Enzyme Electric Caell	Minutes	Yagi (1976, 1977)
5. Mass Spectrometry	Sec-minutes	Jouanneau et al. (1980), Stuart et al. (1972), Hartman and Krasna (1963)
Hydrogen Isotope (Exchange reaction)	Minutes	Krasna (1978)
6. Spectrophotometry	Minutes-hours	Krasna (1978); Yu and Wolin (1969)

Chapter 3

Hydrogen Production by Bacteria

1. Introduction

The physiological 'properties' of bacteria are expressions of metabolic pathways present in the respective species. As far as types of overall energy metabolism are concerned, the relations to oxygen and light of a given bacterium decides whether it performs photosynthesis, fermentation or respiration as the main energy-yielding.

The ability to utilize light as the sole source of energy separates all phototrophic bacteria (including the cyanobacteria but has been dealt separately) from all other prokaryotes, this ability is associated with the possession of the numerous properties that constitute a functioning of photosynthetic apparatus. Strict anaerobiosis is also linked to a unique set of properties. In addition, there is a large number of bacteria with the ability to grow in the absence of oxygen as well as in its presence; the Facultative anaerobic bacteria, that possess more than one energy yielding system.

Hydrogen evolution usually occurs in anaerobic microorganisms and serves to get rid of excess reductant when protons are the only available oxidant, whereas H_2 utilization can occur in aerobic and anaerobic bacteria, and is linked to ATP-producing electron transport systems. Anaerobic bacteria oxidize H_2 using sulphate, sufur, CO_2, nitrate or fumerate as the terminal electron acceptor, and the photosynthetic bacteria use H_2 and other compounds instead of H_2O as the reductant for CO_2 fixation. Aerobic hydrogen oxidizing bacteria can grow with H_2 and O_2 as the sole energy and carbon sources, respectively.

In the present section an attempt has been made to review the works conducted on bacteria in relation to hydrogen. For the sake of simplicity, the bacteria have been divided into two major groups on the basis of energy requirements: the photosynthetic bacteria and the chemotrophic bacteria. Those bacteria which from symbiotic associations with higher plants have been discussed in a separate chapter. No attempt has been made to look into oxygen relationships of a bacterium.

2. Photosynthetic Bacteria

Many photosynthetic bacteria possess the capacity to produce molecular hydrogen by a process directly linked to this ability to capture light energy. The possibility of transforming solar rediation into stable chemical energy, in the form of hydrogen gas has greatly stimulated the study of H_2-metabolism in photosynthetic organisms. Many excellent review articles on H_2 metabolism in photosynthetic bacteria have appeared (Clayton and Sistrom, 1978; Pfenning, 1967, 1977; Kondratieva, 1977;/Kumazawa and Mitsui, 1982; Mitsui *et al.*, 1977, 1980, 1985; Vignais *et al.*, 1985; Gest, 1972; Mortenson and Chen, 1974; Mortenson and Chen. 1974; Kondratieva and Gogotov, 1976; Kondratieva, 1977; Omerod and Gest, 1962; Gray and Gest, 1965; Yoch and Arnon, 1974; Zaijic *et al.*, 1978; Siebert *et al.*, 1980; Van Niel, 1962).

(1) The Organisms and their Systematic Position

The photosynthetic bacteria are found in a wide range of environments, including marine and freshwater systems. They are gram-negative organisms and represent an important component of aquatic ecosystems because of their ability to utilize solar energy to fix CO_2 and N2. The principal light harvesting pigments are bacteriochlorophylls (pfennig and Truper, 1974; Truper and pfennig, 1981). They possess only one photosystem (Figure 5; Chapter II) and perform anoxygenic photosynthesis (Van Niel, 1932; Clayton and Sistrom, 1978) in contrast to cyanobacteria which, like green plants and algae, contain two photosystems enabling them to split water and therefore carry out oxygenic photosynthesis. The phototrophic green and purple bacteria are also known as photosynrhetic or anoxygenic phototrophic bacteria (Truper and Pfennig, 1981) or anoxy-phototrphic (Gibbons and Murray, 1978). A broad outline of different groups of phototrphic bacteria based on characteristic features is presented in Figure 4.1 and their physiological properties are mentioned in Table 3.1.

(2) Thermodynamic Considerations of H_2 Production from Photosynthetic Bacteria

The operation of anaerobic Krebs cycle for hydrogen photoproduction was discovered by Gest *et al.* (1962) and Hillmer and Gest (1977b). Complete dissimilation of compounds similar to glucose ($C_6H_{12}O_6$) through the light dependent anaerobic Krebs cycle yields 12 moles of hydrogen. An oxidation of 12 moles of hydrogen produces 680 Kcal of energy.

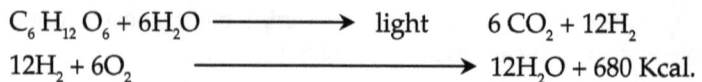

$$C_6 H_{12} O_6 + 6H_2O \longrightarrow light \quad 6 CO_2 + 12H_2$$
$$12H_2 + 6O_2 \longrightarrow 12H_2O + 680 \ Kcal.$$

Further Hillmer and Gest (1977b) have indicated a 72 per cent and 32 per cent energy conversion efficiency from lactate and glucose to hydrogen, respectively. On these considerations it appears that the light dependent process is more efficient than the dark fermen tative process of hydrogen production.

(3) Species Distribution of Hydrogenase Activity

Hydrogenase activity has been found in almost all families of photosynthetic bacteria. A list of these species has been presented in Table 3.2. Nitrogenase activity which is usually related to hydrogen production (Madigan and Gest, 1980) has been detected in many of the photosynthetic bacteria (Gordon, 1981). However, no reports on hydrogenase activity in members of Chloroflexaceae are available. In an intensive investigation.

Table 3.1: Physiological Characteristics of Photosynthetic Bacteria

Family	Physiological Characteristics
Rhodospirillaceae	Facultativa anaerobes, and require B-group vitamins for growth, CO_2 as the main source of carbon, unable to use inorganic electron eonors other than H_2 for growth. Few can oxidize sulphide, thiosulphate (but not elemental sulfur).
	Sulphide, sulphate, tetrathionate or elementals. (deposited in the medium outside the cell).
	The preferred phototrophic electron donors and carbon sources are simple organic compounds.
Chlorobiaceae	Strict anaerobic photoautotroph, some species require B12. All species oxidize hydrogen sulphide and sulphur, some can use thiosulphate or H_2 as electron donors. Some organic acids can be used but only in presence of carbon dioxide.
Chromatiaceae	Obligate anaerobic phototrophs: several species can grow in the darf in presence of oxygen (*Thiocapsa roseopersicina, Lamprobacter modestohalophilus, Ectothiorhodospira mobilis*). All species can utilize hydrogen sulfide and elemental sulfur as electron donors. Some can use thiosulphate, reduced sulphur compounds or H_2, only a few can use organic compounds as electron donors.
Chloroflexaceae	Filamentous forms with gliding motility, flexible cell walls, bacteriochlorophyll a, d or c. photosynthesis anoxygenic and species are thermophiles, sulphide as electron donor, and other reduced sulfur compounds (H_2S, Hs⁻, S_2^-).

Mitsui *et al.* (1980) isolated about 260 isolates from marine ecosystem and found hydrogenase activity in all of the members.

(4) Electron Donor Compounds for Hydrogen Production

A list of common electron donors used for hydrogen production by photosynthetic bacteria is mentioned in Table 3.3. The photosynthetic bacteria are capable of utilizing a wide spectrum of organic substances such as carbohydrates, lipids, fatty acids and some sulfur compounds, such as thiosulfate. The substrate specificity for hydrogen production varies greatly from species to species. A comparative and systematic study in the area of conversion efficiency for different electron donor compounds is very much needed.

Table 3.2: Occurrence of Hydrogenase Activity in Photosnthetic Bacteria (Kumezawa and Mitsui, 1980)

Species	References
Rhodospirillaceae (purple non-sulfur bacteria)	
Rhodospirillum rubrum	Omerod and Gest (1962), Bose and Gest (1962), Burns and Bullen (1966), Schick (1971), Gorell and Uffen (1978), Meyer *et al.* (1978a), Norlund and Erikson (1979), Zurreer and Bachofen (1979), Weaver *et al.* (1980), Miyake *et al.* (1982), Seger and Verstraete (1983), Vatsala (1987).
Rhodospirillum tenue	Pfennig (1969).
Rhodomicrobium vannielli	Hoare and Hoare (1969), sager and Verstraete (1983).
Rhodopseudomonas sp.	Gogotov and Kondratieve (1969).
Rhodopseudomonas sp. KCTC1437	Seol and Kho (1969).
Rhodopseudomonas capsulata	Klemme (1968), Weaver *et al.* (1975), Hillmer and Gest (1977a,b), Colbeau *et al.* (1980), Colbeau and Vignias (1980), Khanna *et al.* (1980), Song *et al.* (1980), Kelley and Nicholas (1981), Tsygankov *et al.* (1984), Willison *et al.* (1984), Zhu *et al.* (1987).
Rhodopseudomonas acidophila	Sietert and Pfennig (1974),
Rhodopseudomonas palustris	Klemme (1968), Gogotov *et al.* (1974), Mcler *et al.* (1979), Kim *et al.* (1980), Vincenzeni *et al.* (1985).
Rhodopseudomonas sphaeroides et al. (1981), Lee (1986).	Pfennig and Truper (1974), Serebryakova *et al.* (1980), Watanabe
Rhodopseudomonas sulfidophila	Hansen and Veldkamp (1933).
Rhodopseudomonas sphaeroides	Klemme(1968), Watanabe *et al.* (1981).
Rhodopseudomonas sulfidophilus	Jee *et al.* (1987), Vaseleva *et al.* (1988).
Rhodopseudomonas sulfidophilus	Stevens *et al.* (1986), Peng *et al.* (1987)
Rhodobacter capsulatus	Liessens and Verstraele (1986).
Chromatiaceae (purple sulfur bacteria)	
Chromatium sp.	Hendley (1955).
Chromatium sp.	Thiele (1968).
Chromatium sp. (marine)	Mitsui (1976).
Chromatium minutissimum	Nakamura (1939).
Chromatium vinosum	Serra *et al.* (1984).
Ectothiorhodospira mobilis	Truper (1968).
Ectothiorhodospira Shaposhnikovil	Pfennig and Truper (1974).
Thiocapsa floridana	Thiele (1968).
Thiocapsa roseopersicina	Gogotov *et al.* (1974, 1978), Zorin (1983), der *et al.* (1985), Zorin *et al.* (1984).
Chlorobiaceae (green sulfur bacteria)	
Chlorobium limicola	Larsen (1952), Lippert AND Pfennig (1969).
Chlorobium limicola f. *thiosulfatophilum*	Bernstein AND Olson (1981).
Chlorobium thiosulfatophilum	Larsen (1952), Lippert AND Pfennig (1969).
Chloropseudomonas ethylica	Kondrotieva and Gogotov (1969).

Table 3.3: Electron Donors Used for Hydrogen Production by Photosynthetic Bacteria

Electron Donor	References
RHODOSPIRILLACEAE (purple non- sulfur bacteria)	
Rhodospirillum rubrum	
Acetate	Gest *et al.* (1962)
lactate	Kohlmiller and Gest (1951), Zurrer and Bachofen (1979)
pyruvate	Gest and Kaman (1949), Gest *et al.* (1950), Garrell and Uffen (1977), Schon and Biederman (1973), Uffen *et al.* (1977), Voelskov and Schon (1978)
Succinate	Bose and Gest (1962), Gest and Kamen (1949), Gest *et al.* (19621,1950), Kohlmiller and Gest (1951)
fumarate	Gest and Kamen (1949), Gest *et al.* (1962, 1950), Kohlmiller and Gest (1951)
malate	Bose and Gest (1962), Bergoff and Kamen (1952), Gest and Kamen (1949), Gest *et al.* (1962,1950), Gogotov and Zorin (1972), Kohlmiller and Gest (1951), Omerod *et al.* (1961), Paschinger (1974), Schick (1971), Paschinger (1974), Schick (1971)
oxaloacetate	Gest and Kamen (1949), Gest *et al.* (1950)
Rhodopseudomonas acidophila lactate	Siefert and Pfennig (1978)
Rhodopseudomonas capsulata	
Propionate	Hillmer and Gest (1977)
lactate	Hillmer and Hest (1977), Kelley *et al.* (1977), Serebryakova *et al.* (1950)
Pyruvate	Hillmer and Gest (1977), Serebryakova *et al.* (1980)
succinate	Hillmer and Geat (1977)
fumarate	Hillmer and Geat (1977)
malate	Hillmer and Gest (1977), Kelley *et al.* (1977), Serebryakova *et al.* (1980)
butyrate	Hillmer and Gest (1977)
glucose	Hillmer and Gest (1977)
fructose	Hillmer and Gest (1977)
succinate	Hillmer and Gest (1977)
Rhodopseudomonas palustris	
formate	Quadri and Hoare (1968)
plyruvate	Gogotov *et al.* (1974)
-Ketoglutarate	Gogotov *et al.* (1974)
succinate	Gogotov *et al.* (1974)
malate	Gogotov *et al.* (1974)
oxaloacetate	Gogotov *et al.* (1974)
glucose	Gogotov *et al.* (1974)
thiosulfate	Gogotov *et al.* (1974)
Rhodopseudomonas sp.	
D, L-malate	Kim *et al.* (1981)
L- glutamate	Kim *et al.* (1981)

Contd...

Table 3.3–Contd...

Electron Donor	References
Rhodopseudomonas spaeroides	
glucose	Moler *et al.* (1979)
pyruvate	Serebryakova *et al.* (1980)
malate	Serebryakova *et al.* (1980)
lactate	Serebryakova *et al.* (1980)
	CHROMATIACEAE (Purple sulfur bacteria)
Chromatium D	
Thisoulfate	Arnon *et al.* (1961), Omerod *et al.*,
pyruvate	Bennet *et al.* (1964)
Chromatium sp.	
malate + CO_2	Newton and Wilson (1953)
Chromatium sp. (marine)	
thiosulfate	Mitsul (1976), Ohta, Frank and Mitsui (1980)
sulfide	Ohta, Frank and Mitsui (1980)
succinate	Ohta and Mitsui (1980)
acetate	Ohta and Mitsui (1980)
fumarate	Ohta and Mitsui (1980)
malate	Ohta and Mitsui (1980)
Thiocapsa roseopersicina	
thiosulfate	Gogotov *et al.* (1974)
acetate	Gogotov *et al.* (1974)
pyruvate	Gogotov *et al.* (1974), Gogotov (1978)
oxaloacetate	Gogotov *et al.* (1974)
	CHLOROBIACEAE
Chloropseudomonas ethylica	
lactate	Kondratieva and Gogotov (1969)
pyruvate	Kondratieva and Gogotov (1969)
citrate	Kondratieva and Gogotov (1969)
- Ketoglutarate	Kondratieva and Gogotov (1969)
xylose	Kondratieva and Gogotov (1969)
mannitol	Kondratieva and Gogotov (1969)
glucose	Kondratieva and Gogotov (1969)
Chloropseudomonas sp.	
formate	Kondratieva and Gogotov (1969)

B. Hydrogen Production

The first observation on photochemical production of molecular hydrogen by a photosynthetic bacterium *R. rubrum* was made by Gest and Kamen (1949,b). They demonstrated that photoevolution of hydrogen proceeded readily under an atmosphere of 100 per cent hydrogen, or when noble gases were used to create anaerobiosis, but did not occur in an atmosphere of N_2. Although growth was stimulated in medium containing ammonium chloride of high concentrations of yeast extract, the H_2 production was inhibited (Gest and Kamen, 1949a) Subsequent works using 15N indicated clearly that R. rubrum contains a N_2-Fixing system (Gest and Kamen, 1949b; Gest *et al.*, 1950). Subsequently, N_2-fixation was demonstrated in *Chromatium, Chlorobium* (Lindstrom *et al.*, 1951) and in several species of purple non-sulfur bacteria, induction of *R. palustris, R. spheroides, R. capsulata* and *R. gelationsa* (Lindstrom *et al.*, 1951). Since no H_2-production was observed in N_2-fixation repressed cells, a close relationship between these two processes was realized.

While studying nitrogen fixation by cell free preparations of R. rubrum, Bulen *et al.* (1956) demonstrated that the ATP and low potential reductant required for N-2 fixation were also necessary for the production of H_2 by these preparations. This ATP dependent hydrogenase activity differed from that of the 'conventional' hydrogenase activity as the activity was independent of the partial pressures of H_2, was inversible, and was insensitive to carbon monooxide. From their studies on aerobic and photosynthetic microorganisms, Bulen *et al.* (1965) concluded that the nitrogenase catalyzed both N_2 reduction as well as ATP dependent H_2 evolution.

C. Features of Hydrogen Evolution

Light dependent hydrogen evolution is now accepted as a genaral property of photosynthetic bacteria (Gest, 1972). The species that have been shown to evolve H_2 in light are summarized in Table 3.4. A detailed study regarding effects of various factors on the photoproduction of H_2 by *R. capsulata* was conducted by Hillmer and Gest (1977a,b). In growing cultures, the highest rates of H_2 production (130 ul hr-1 m culture -1) were obtained with DL-lactate or pyruvate as carbon source, and either glutamate, serine of alanine as growth limiting nitrogen source (Hillmer and Gest, 1977a). when the ratio of concentrations of glutamate and lactate was less than 1.0, hydrogen production was observed. On the other hand at higher ratios net production of NH^+_4 from glutamate occurred reslting in the inhibition of nitrogenase. The highest yield of H_2, 72 per cent of the theoretical maximum for the complete dissimilation of carbon substrate to H_2 and CO_2, was obtained with D,L-lactate and succinate. The sugars gave much lower yields than organic acids. In a standard 'lactate -glutamate' system, it was observed that H_2 production began prior to exhaustion of glutamate from the medium, but a maximum rats was attained only when the nitrogen source was completely exhausted. Hydrogen production continued untill lactate had been depleded from the medium. with the increasing light intensity, the rate of H_2 production increased, reaching a plateau at a similar intensity (600 foot candles or 6480 lux) to that which was saturating for growth. The rate of H_2 production by *R. sphaeroies* was found to be proportional to light intensity up to 12,000 lux (Macler *et al.*, 1979).

Under appropriate experimental conditions, resting cells of *R. rubrum* (from glutamate cultures) photometabolize organic substrates with the formation of large quantities of both hydrogen and carbon dioxide. The following (approximate) stoichiometries have been demonstrated (Gest, 1972).

Table 3.4: Hydrogen Phoroproduction by Photosynthetic Bacteria

Family / Species	Conditions	Electron Donor	Strain	Activity ml H₂/h/g Dry Wt	Activity µM H₂/mg Protein h	References
Rhodospirillacea						
Rhodospirillumrubrum		Lactate	S-1	109		Omerod et al. (1961)
		Lactate	S-1	146		Weaver et al. (1980)
	10,800 lux	Lactate		20		Zurrer and Bachofen (1979)
	135 W/m²	Malate			1.7	Stiffer and Gest (1954)
	46 W/m²	Malate			1.68	Schick et al. (1971a)
		Malate			0.96	Gogotov and Zorin (19)
Rhodopseudomonas capsulata	Nitrogen limited growth	Lactate	B-10	115		Meyer et al. (1978a)
		Lactate	SCJ	168		Weaver et al. (1980)
		Lactate	LB2	176		Weaver et al. (1980)
		Malate	B 100	69		Takakuwa et al. (1983)
		Malate	ST410 (B100 Hup-)	100		Takakuwa et al. (1983)
		Malate	IR 4 (B10 Hup)	127		Willison et al. (1984)
		Malate	B 10	95		Willison et al. (1984)
		Malate	N-3	119		Song et al. (1980).
Rhodopseudomonas sphaeroides	10,800 lux	Lactate			1.2	Hillmer and Gest (1977a,b)
		Pyruvate				Watanamb et al. (198)
		Malate	S	90	0.54	
Rhodopseudomonas acidophila	8,000 lux	Glucose	Mutant DMS 137	49		Mcler et al. (1979)
		Lactate				Siefert and Pfennig
Chromatiaceae Chromatium sp. chromatium	50,000 lux	Thiosufate	D	134	1.2	Arnon et al. (1961a,b)
		Succinate & thiosulfate	Miami PBS1071	20		Ohta and Mitsui (1981)
		Pyruvate	BBS			Ggotov (1978)
Thiocapsa roseopersicina Ectothiorhodospira shaposhnikovii		Sulfide	E.Ba1011	8		Mtheron and Baulaigue (1983)

Acetate : $C_2H_4O_2 + 2H_2O \longrightarrow 2CO_2 + 4H_2$

Succinate : $C_4H_6O_4 + 4H_2O \longrightarrow 4CO_2 + 7H_2$

Fumarate : $C_4H_4O_4 + 4H_2O \longrightarrow 4CO_2 + 6H_2$

L-malate : $C_4H_6O_5 + 3H_2O \longrightarrow 4CO_2 + 6H_2$

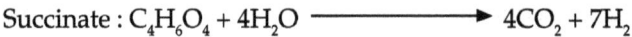

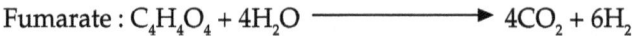

The assumed role of H_2-photoproduction is to dissipate reducing equivalents and ATP when these are produced in excess; *i.e.* possibly to regulate the intracellular redox balance (Gest, 1972). However, the observation that H_2 production occurs even when light intensity is growth limiting suggested that H_2 production, at least under some conditions, constitutes an energetic burden to the cells (Hillmer and Gest, 1977a).

The mechanisms involved in the regulation of H_2 production are least understood. The effect of carbon and nitrogen source on the rate of H_2 production probably reflects both differing rates of utilization of carbon substrates, and differing nitrogenase activities under various growth conditions. The enhanced H_2 production with increased light intensity could be due to enhanced synthesis of ATP leading to an increased activity and/or synthesis of nitrogenase. In *R. capsulata* strain B10, the increase in nitrogenase activity following illumination was inhibited by chloramphenicol. This suggested that nitrogenase synthesis is induced by light (Hillmer and Gest, 1977b, Meyer *et al.*, 1978b).

There is only partial dissimulation of most organic substrates to H_2 and CO_2. However, the fate of residual substrate carbon to different pathways of carbon substrate utilization has received very little attention. A mutant of *R. sphaeroides* was isolated by Macler *et al.* (1979) which was able to produce H_2 from glucose with an mutant did not accumulate gluconate during growth on glucose. It was suggested that the mutant had acquired an increased capacity to convert gluconate into C_3 intermediates, via the enzymes of the Entner-Doudoroff pathway (Mcler *et al.*, 1979). Mutants of *R. capsulate* unable to grow photoautotrophically Aut⁻ phenotype) have been isolated which also show an increased stoichimetry of H_2 production from various organic substrates such as DL-malate (Willison *et al.*, 1984). Analysis of the carbon balance in H_2-producing cultures of these strains indicated that, in the mutants, a lower proportion of the organic carbon source provided was excreted into the medium in the form of unidentified end products. The excretion of acidic end products by H_2 production cultures has been confirmed by high performance liquid chromatography (Takakuwa *et al.*, 1983).

A linkage between carbon metabolism and H_2 production might be provided by changes in the redox state of nicotinamide nucleotides. This has also been suggested for the regulation of H_2-production photoreduction of CO_2 (Schick, 1971a,b; Hillmer and Gest, 1977b), although direct evidence has not yet been provided. It is clearly important to identify both the end products excreted by H_2 producing cultures and the metabolic pattern involved in their formation, in order to fully understand the regulation of the H_2 production process.

In the early stiuiies, where organic reductants were used and CO_2 evolved simultaneously with H_2, CO_2 was considered necessary for the photoevolution of

H_2. Omerod *et al.* (1961) showed that an inorganic electron donor, thiosulphate, could provide the reduction equivalents for the production in light. It appeared, therefore, that the features of H_2-photoevolution from thiosulphate (response to light and dark and to the nitrogen source) were similar in *Chromatium* to those observed in *R. rubrum* with carbon compounds as electron donor, suggesting that the same system was involved in both cases (Losada *et al.*, 1961).

The relationship between nitorgenase mediated H_2 evolution and carbon metabolism in the Chromatiaceae is not yet fully understood. Ohta and Mitsui (1981) showed that the marine *Chromatium* sp. Miami PB 5 1071 produced hydrogen gas at two to three times greater rates when two donor substrates, succinate and thiosulfate (or succinate and sulphide) were used together than when the substrates were used individually. Matheron and Baulaigue (1983) reported that the photoevolution of H_2 from sulphide by purple bacteria, *Chromatium*, *Thiocapsa*, *Thiocystis* and *Ectothiorhodospira* was tightly linked to CO_2 assimilation. They also observed that the utilization of endogenous substrates was markedly stimulated by sulphide in *Ectothiorhidospira*.

D. Production of Molecular Hydrogen in Dark

Purple bacteria have the capacity to ferment their endogenous substrates in the dark with production of organic acids, CO_2 and H_2 (Van Niel, 1944). Formative H_2 production was stimulated by the addition of pyruvate (Kohlmiller and Gest, 1951).

E. Hydrogen Production Rates

There is considerable variation in the rates of H_2 evolution reported by various workers. This conclusion is exemplified following inspection of Table 3.4. These rates do not necessarily represent those which might be achieved under optimum conditions. The high rates observed in same species provide enough impetus for further studies of utilization.

F. Genetics of Hydrogen Production

The genetic studies of hydrogen metabolism in the photosynthetic bacteria are restricted to certain members of the Rhodospirillaceae and in particular to Rhodopseudomonas capsulata. It is apparent from the results of various investigators that mutants affected in enzyme systems involved (*i.e.* nitrogenase and hydrogenase) are readily obtainable and they have provided informations on the regulation of H_2 metabolism in the photosynthetic bacteria. Several methods are now available for the genetic analysis of these mutants, including transduction like systems (gene transfer agent) and conjugation.

The first gene transfer system to be described in a photosynthetic bacterium was the gene transfer agent (GTA) in *R. capsulated* (Marrs, 1974). This GTA is a bacteriophage-like particle containing about 3×10^6 dalton of double stranded DNA. It seems that it is entirely derived from the bacterial chromosome (Solioz *et al.*, 1975). The amount of genetic material carried by GTA is equivalent to five or six genes. It, therefore, can be used for fine structure mapping of closely linked genes such as photosynthesis genes of *R. capsulata*.

Transfer of chromosomal genes by conjugation, using broad-host-range drug resistance plasmids, has been demonstrated in *R. capsulata* and *Rhodopseudomonas sphaeroides* (Marrs, 1981; Pemberton and Bowen, 1981). Further, evidences for the transformation of *R. spheroides* by isolated plasmid or bacteriophage DNA have been produced (Farnori and Kaplan, 1982).

The method of genetic engineering and transform mutagenesis have been applied to the photosynthetic bacteria. These techniques have helped in the physical mapping of the structural genes for the nitrogenase complex in *R. capsulata* (Avtges *et al.* (1983) and have revealed the presence of multiple 'pseudocopies' of these genes in *R. capsulata* (Sconlnik and Haselkorn, 1984).

Hydrogenase deficient mutants in photosynthetic bacteria have been isolated from *R. capsulata* strain B 100 (Takakawa *et al.*, 1983) or strain B 10 (Willison *et al.*, 1984).

G. Use of Photosynthetic Bacteria as Biological Solar Energy Converters

The photosynthetic bacteria are a tool of great potential in various fields of biotechnology. Their metabolic versatility and rapid growth enable them to survive and proliferate in a wide variety of environments. They can synthesize large amounts of biomass from organic materials. Consequently, the cultivation of photosynthetic bacteria for single-cell protein production for animal feed may have potential use.

By oxidizing low molecular weight carbon and sulfur compounds (including mercaptans) to gaseous or innocuous products, the photosynthetic bacteria may contribute significantly to the purification of polluted water supplies. The treatment of industrial waste waters by processes involving photosynthetic bacteria has been advocated for several year (Kobayashi, 1975, 1977; Kobahashi and Tohan, 1973; Kobayashi *et al.*, 1976).

The capacity of photosynthetic microorganisms to produce molecular hydrogen has attracted attention and has stimulated researches into the possibility of using these organisms as solar energy converters. The subjects has been reviewed by Mitsui (1978), Weaver *et al.* (1980).

For the purpose of solar energy conversion into hydrogen, the utilization of blue-green algae (cyanobacteria) was initially considered since these organisms use water as reductant. Photosynthetic bacteria, which cannot split water but require organic compounds as electron donor, were thought to be economically less viable. Bennet and Weetall (1976) evaluated the influence of substrate cost on H_2 produced from glucose by immobilized *Rhodospirillum rubrum* cells and concluded that H_2 could not be produced economically from a substrate having a cost of more than US $ 0.10 per pound.

In spite of the above fact, the photosynthetic bacteria present several advantages over the cyanobacteria as H_2-producing organisms. Firstly, they generally show much higher rates of H_2-photoproduction. Secondly, as they conduct anoxygenic photosynthesis, the gas produced is free from contaminating oxygen. Thirdly, they are much more amenable to genetic manipulations. Lastly the possibility of producing H_2 as a byproduct of processes such as biomass production or purification of water also should be considered.

(a) Strain Screening and Selection

Significant differences in the ability to photoproduce H_2 have been observed in different bacterial isolates from nature. This capacity is linked not only to the nitrogenase content and activity of the cells but also to the ability of these cells to degrade organic substrates. Additional characteristics (*e.g.* capacity for growth in salt water or tolerance to high temperature) may be valuable in developing a competitive and viable process.

It has been general observation that faster growth of a bacterium is accompanied by aging or the ultimate decline in hydrogen production rates. A general practices has been to supply fresh nutrients intermittently or continuously (continuos culture techniques) so that bacterium is always in the active phase of hydrogen production.

To circumvent the problem it is of paramount interest to identify/isolate/ cultivate a bacteria that could exhibit sustained hydrogen production with respect to time. Considerable progress has bee made in this direction by Miyake *et al.* (1982) who isolated a strain of *Rhodospirillum rubrum* showing inherent capacity to evolve hydrogen over a large span of time. Further efforts by Miyake and Kawamure (1987) led to the isolation of *Rhodobacter spheroides* that could evolve hydrogen over extended periods with light to hydrogen converstion efficiency of about 8 per cent.

1. Marine Strains

A survey of tropical marine environments for species of hydrogen producing capacity was undertaken by Mistul and coworkers, at the University of Miami's Marine Schools. The advantages of using marine species are that large part of the earth's surface is covered by oceans, and seas water contains many nutrients (*e.g.* magnesium, sulfate, potassium and other elements) which are assential for growth. The collected hundreds of strains.

Table 3.5: Hydrogen Production by Marine Photosynthetic Bacteria (from Mitsui *et al.*, 1983).

Family	No. of Strains Capatile of H_2 Production	Rates of H_2 Production (mmol/mg dry wt/h)
Rhodospirillaceae	40	0.10-4.0
Chromatiaceae	40	0.05-3.0
Chlorobiaceae	3	0.02-0.26

Of marine bacteria (both purple and green photosynthetic bacteria) (see Table 3.5). Among these, they selected a *Chromatium* sp.Miami PSB 1071, whose charateristics make it a good candidate for H_2 photorpdoction. This is one of the fastest growing strains (doubling time 1.75 hours) and can sustain a high rate of H_2 photoproduction capability (134 ml H_2 h^{-1} g biomass^{-1}) (Mitsui, 1976c; Ohta and Mitsui, 1981). This strain can use various organic and sulfur compounds such as malate, acetate, succinate, thiosulfate, and sulfide as electron donor (Mitsul *et al.*,

1980). It has also the faculty of taking up sulfide and depositing elemental sulfur inside the cells and therefore can be used for environmental treatment of sulfide rich salt waters (Ohta *et al.*, 1980). Thus the concept of a bacterial from could be very easily realized using marine bacteria.

2. Thermophilic or the Thermostable Strains

The optimum temperature for growth and H_2 production is around 30°C in most strains of photosynthetic bacteria. In climates, this temperature may be exceeded. In such as situation if would be desirable to grow thermophilic, or at least thermostable, strains as solar energy bioconveters. Watanabe *et al.* (1981) have succeeded in isolation of *Rhodopseudomones sphaeroides* strains from Thiland which are as active in H_2 production at 40°C as at 30°C. A similar effort by Buranakarl *et al.* (1987) resulted in successful isolatation of four strains of non-sulphur photosynthetic bacteria like TR-22 T-A, TR-22 RA-B, TR-22 R-C and TR-22 R-D (all being species of Rhodopseudomenas) that could grow well and produce hydrogen at temperatures between 40-45°C.

3. Strains with High Nitrogenase Content and Activity

It is well known that nitrogenase has high energy requirement and low torunover number as compared to hydgorenase, suggestions that the latter enzyme might be more suitable for H_2 production. The adenosine triphosphate (ATP) requirement for functioning of nitrogenous is provided by light, which is a primary energy source. It is therefore, anticipated that it should be possible to increase the rate of H_2 production by increasing the nitrogenous content of bacterial cells. This could be achieved in several alternative ways.

A. Since nitrogenase is an inducible enzyme, it is possible to adjust the growth conditions to have maximal expression of the enzyme. To met such a condition, Ohta and Mitsui (1981) grew *Choromatium* sp. Miami PBS 1071 in batch cultures on molecular nitrogen and observed hydrogen production rates as high as 6 umol H_2 hr^{-1} as mg protein^{-1} in cells taken from the cultures in middle of the logarithmic phase. Continuos cultures of *R. rubrum* exhibited increased H_2 photoproduction in presence of molecular nitrogen, provided that N_2 was in limited amounts. Zurrer and Bachofen (1982) pointed out that this makes possible the production of hydrogen in growth media lacking combined nitrogen.

The other physiological means to maintain the nitrogenase fully active is to allow its regeneration during the course of the experiments. For example, Miyake *et al.* (1982) were able to maintain the nitrogenase activity of R. rubrum cells for more that 2 weeks in an experiment with as simulates day and night rhythms of 12 hours light and 12 hours dark, by supplying the system with small amounts of ammonium suplhate or nitrogen gas at the beginning of each dark period.

B. The nitrogenase synthesis cab be stimulated by an increase in light intensity. At saturating light intensities, the nitrogenase content of cells of *R. capsulata* grown in continuous cultures represents upto 25 per cent of the total soluble proteins and the cultures maintained a steady state rate of 45 ml H_2 h^{-1} litre^{-1} for at least 10 days (Vignais *et al.*, 1985).

The above observation have implication in outdoor cultivation vessels as light may be as limiting factor for nitrogenase activity. This has been shown with outdoor batch cultures of *R. sphaeroides*. Two cultures (33 litres) containing 53 mM DL-lactate and 5 mM glutamate were placed on the ground, one vertically and the other inclined at 30°C to receive more light. After 25 days, the total H_2 production was 155 litres in the first case and 177 litres in the second. Their productions corresponded to conversion efficiency of lactate to H_2 of 69 and 78 per cent, respectively (Kim *et al.*, 1982).

C. An improved understanding of the regulation of nitrogenase activity and synthesis will make possible the construction of de-repressed cells. Weare (1978) has already obtained a mutant of *R. rubrum* which is not repressed by NH_4^+ for the photoproduction of ammonia and hydrogen. Glutamine autotrophs of *R. capsulata* also synthesize nitrogenase and produce hydrogen in presence of ammonia.

4. Hydrogenase

(a) Deficient (Hup⁻) Strains

Since H_2-uptake hydrogenase is active in the recycling of enhanced production of hydrogen. Hydrogenase negative mutant of *R. capsulata* was obtained by Takakuwa *et al.* (1983) that produced 4.6-6.2 mol H_2 for each mole of sugar compared with 1.2-4.3 mol for the wild type strain. Willison *et al.* (1984), using Hup⁻ mutant, concluded that the increased H_2 evolution by the Hup⁻ mutant oxidizing DL-malate resulted from an enhanced ability of the mutant to consume D-maslate. It was observed by these authors that some non-autotrophic mutants with apparently normal levels of hydrogenase activity also exhibited increased H_2-production on DL-malate and were unable to demonstrate, with the wild-type strain in presence of excess of organic substrate, a recycling of hydrogen as observed by Takakawa *et al.* (1983).

The other way to minimize the role of the uptake hydrogenase would be to place the cells under conditions where the enzyme is not functional, namely in absence of O_2 and in the opresence of excess of organic compounds (so that CO_2 photoreduction does not occurs) or at pH values below 7.5 where the uptake activity of the enzyme is negligible (Vignais *et al.*, 1985).

(b) Economical Substrates

As indicated earlier, photosynthetic bacteria have diverse metabolism that enables them to produce hydrogen from a great variety of carbon compounds. From an economic point of view, only very cheap substrates cab be employed. Adequated electron donors may be found in organic wastes from paper mils, fruit or milk processing factories, and sugar refineries among others.

Since lactate is a good electron donor for hydrogen production, *R. rubrum* and *R. capsulata* have been tested for their ability to produce H_2 gas continuously form lactic acid containing wastes. *R. rubrum* strain S-1, in batch cultures, produces hydrogen gas continuously at an average rate of 6 ml hr^{-1} (g dry wt cells)$^{-1}$ over a period of upto 80 days when supplied periodically with whey or yoghurt wastes (Zurrer and Bachofen, 1979). Nine strains of *R. capsulata* were tested for their capacity to produces H$_2$ anaerobically in the light from either DL-lactate, acetate or butyrate. Out of these substrates, lactate proved to be the best and a maximum rate of 1630 ml of H$_2$ day^{-1} litre reactor^{-1} was achieve. The strains differed in their efficiency of conversation from the carbon substrate into H$_2$ and CO$_2$, but produced H$_2$ at about the same rate. These authors suggested that the choice of a strain for H$_2$ production from wastes waters can be made only on the basis of the availability of hydrogen donors.

Mitsul *et al.* (1983), in course of their search for new species, isolated *Rhodopseudomonas* strains able to grow on carbohydrate polymers such as cellulose, soluble starch or pectin. In this contexts, Buranakarl *et al.* (1988) isolated strains of non-sulfur purple bacteria which could utilize raw strarch from corn, potato and cassava as a source of electron donor for hydrogen-production. These authors showed that amylases were secreted by these strains. Similarly, Billinger *et al.* (1985) isolated several photosythetic non-sulfur bacteria from waste water pond of a sugar refinery. These strains could utilize untreated waste substrate where higher rates of H$_2$ production were obtained. The hydrogen evolution from ethanol, n-butanol, malate and mixed substrates was investigated by Fujii *et al.* (1987). The yield and rate of evolution at 40°C was 70-80 l H$_2$/mg cell/hr.

Odom and Wall (1983) have described a different strategy for cellulose degradation, they grew mixed culture of *Cellulomonas*, which degrades cellulose to cellobiose by fermentation, and *R. capsulata* which can photoevolve H$_2$ from cellbiose. Working on a similar line, Miyake *et al.* (1985) used *Clostridium butyricum*, which converts glucose to butyrate, and *Rhodopseudomoas* sp., which produced hydrogen from butyrate. On the other hand Miyamoto *et al.* (1987) obtained about 4 fold H$_2$ evolution and 5 fold H$_2$ molar yield (mol H$_2$/mol glucose) in a mixed culture of *Chlamydomonas reinhardtii* and *R. rubrum*. The increased rate of H$_2$ production was due to the consumption of fromate by the alga.

In some cases pretreatment of the substrate has been found to increase the hydrogen production. Salhi (1989) found that *Rhodospirillium rubrum* produced hydrogen when grown on cheese whey pretreated with lactose fermenting bacterium *E. coli*. A correlation between amount of lactic acid formed and quantity of hydrogen evolved was observed. The total amount of H$_2$ formed was 3.6 times higher in pretreated whey than untreated one.

(c) Cell Stabilization by Immobilization

Entrapment of microbial cells in polymeric matrics (immobilization) is a mean to increase the stability of biological material with a view to use in biotechnological applications.Suitable techniques have now been developed to prepare entrapped cells which are fully viable and biosynthetically active so that they can catalyze multi step reactions. Besides the use of immobilized cells is compatible with a continuos process and eliminates the problems of product separation and harvesting of biomass. The systems using immobilized microbial cells to produces H$_2$ have been documented. Results of some of the observations on such systems are presented in Table 3.6.

Rhodopseudomonas sp. Miami 2271 cells, immobilized with agar on an agarose coated polyster films, produced H_2 at rates as high protected from inhibition by O_2 and N_2 (Matsunaga and Mitsul, 1982). Immobilized cells exhibited wider salt tolerenace than aquenous cell suspensions and photoproduced H_2 at the same rate over 10 d period.

Cell immobilization technique has been successfully utilized for achieving maximum and sustained H_2 production by bacteria in co-culture experiments (two different bacteria) by Miyake *et al.* (1984). It is with this sort of entrapment of bacterial cells that light to hydrogen conversion efficiency has gone upto 7.9 per cent (Miyake and Kawamura, 1987). Similarly Joe and Hoseon (1985) observed enhanced hydrogen yield and conversion efficiency in alginate-immobilized cells of *Rhodopseudomonas capsulata* 10006 and *Rhodospirillum rubrum* KS 301 over their free cells courterparts.

Feltin *et al.* (1985) examined various hydrohpilic gels such as agarose, calcium and barium alginate, calcium pectin, K-carrageenan and gelatine for the immobilization of *R. rubrum* in order to optimize hydrogen evolution from lactate. Agar proved to be the best immobilization medium, calcium alginate and calcium pectinate to be the worst, because they release the cells to the medium, *Rhodospirillum* colonized the agar beads homogeneously. Hydrogen evolution ran for 3000 h (4 months). The mean rate for the first 60-70 h was 5.9-9.4 mmol kg^{-1} (dm) s^{-1}.

(d) Outdoor Hydrogen Production

The next step in achieving the goal of a viable hydrogen production tgechnology will be to tests and implement actual systems for hydrogen production in the natural environment. Mitsui *et al.* (1983) tested *Rhodopseudomonas* sp. Miami PBE 2271 for outdoor hydrogen production. This photosynthetic bacterium was immobilized in a series of thin agar plates in a reaction. Waste water from an orange processing factory, diluted with sea water to give a final concentration of 430 ppm of total organic carbon (TOC), was used as a substrate for hydrogen production. Hydrogen was continuously produced for seven hours from 9 AM to 4 PM. In such a system the overall input was solar energy and organic substances. Under these conditions 3.9 mmol of H_2 was produced. Outdoor experiments were also performed with Chromatium sp. Miami PBS 1071. The overall hydrogen produced in 4 days was aproximately 2000 ml (Mitsui *et al.*, 1984).

(e) Engineering Aspects of Hydrogen Production from Photosynthetic Bacteria

The subject has been reviewed by Herlwich and Karpuk (1982). They considered the design requirements for the solar bacterial reactor and suggested that the critical elements of such a system should meet the criteria of low cost, temperature control, transparency of the reactor, H_2 collection and hydrogen impermeability. They proposed a system to produce 28,000 m^3/day (1×10^6 ft^3/day) at a conceptual level and including hydrogen clean up, substrate storage and waste disposal. The most critical component in the design is the solar bacterial reactor. Several designs

Table 3.6: Hydrogen Production by Immobilized Photosynthetic Bacteria

Species/Strain	Condition	Electron Donor	Rate of H_2 Evolution	References
Rhodospirilliym rubrum	Planer agar matrix bounded by microprous membrane filter	Synthetic waste with malate and glutamate light 15 Klux.	565 mm³ H_2 per h cm³ agar	Planchard *et al.* (1989)
R. capasulated B-10	Agar/carrageenan beads.	Lactate	54 ml/h/mg dry wt.	Francov and Vignais (1984)
Rhodoseudomonas Miamin PBE 2271	Agarose coated polyester film with agar (2 per cent).		3.96 umol/protein/h	Matsunaga and Mitsui (1982)
Chromatium Miami PBS sp. 1071/Mismi	Agar gel matrix nitrogen or ammonia as nitrogen source	Sulfide	2.5 ml/mg dry wt after 300 h	Ikemoto and Mitsui
Rhodopseudoas/Polytics L Palustris 420	Agar entrapped cells Agar entrapped cells Agar entrapped cells	Lactate Sugar refinery waste Straw paper mill effluent	42 ml H_2/h/g dry wt 43 ml H_2/h/g dry wt 50 ml H_2/h/g dry wt	Vincenzeni *et al.* (1982)
Rhodospirillium Molischianum	Agar entrapped cells	Straw paper mill effluent	139 ml H_2/h/g dry wt	Vincenzeni *et al.* (1982)

were developed and analyzed, such as tubular reactor. Among such reactors a large covered pond concept appears to be most attractive cost estimates for the design showed favourable economics. The authors believed that 5 per cent light-to-hydrogen conversion efficiency was achievable in 5-10 years while a 10 per cent efficiency could be managed in 10-15 years.

3. Hydrogen Metabolism and Production by Heterotrophic Bacteria

A. Introduction

Gray and Gest (1965) considered that bacteria which evolve molecular hydrogen fall into three groups: (i) obligate anaerobes which do not synthesize cytochromes (*e.g.* clostridia), (ii) heterotrophic facultative anaerobes and (iii) the genus *Desulfovibrio*. The subject has been reviewed by Cole (1976).

A list of heterotrophic microorganisms which possess the enzyme hydrogenase and nitrogenase (ATP-dependent H_2 evolving microbes). For the sake of convenience the organisms have been grouped on the basis of their physiological properties mentioned below:

1. The Hydrogen Oxidizing Bacteria

The group aerobic hydrogen oxidizing bacteria is physiologically defined and comprises bacteria from different taxonomic units. This group is defined by the ability to utilize gaseous hydrogen as electron donor with oxygen as electron acceptor and to fix CO_2 *i.e.* to grow chemolithautotrophically. They are different from other bacteria (*e.g. Acetobacter, Azotobacter,* Enterobacteriaceae and others) that also oxidize hydrogen under aerobic conditions, but without autotrophic CO_2 fixation. They are also different from the bacteria that utilize hydrogen under anaerobic conditions, with sulfate or carbon dioxide as hydrogen acceptors (*e.g. Desulfovibrio, Clostridium aceticum, Acetobacter woodii* and *Methanobacterium thermoautotrophicum.* It is the combination of these two abilities-to accept hydrogen as electron donor and to use CO_2 as the sole carbon source that is biologically unique. The overall equation for gas consumption is

$$6H_2 + 2O_2 + CO_2 \longrightarrow (CH_2O) + 5H_2O.$$

Many of the microorganisms use molecular hydrogen preferentially as an electron and energy donor and compete for the hydrogen available (Schlegel, 1974). Under anaerobic conditions, the major part of hydrogen gives rise to the production of methane, hydrogen sulfide and gaseous nitrogen.

2. The Methanogenic Bacteria

These microorganisms are diverse morphologically consisting of such forms as short of long rods, spirilla, cocci and various other arrangements. In general, the methanogenes share the properties of strict anaerobiosis and the ability to reduce CO_2 with molecular hydrogen to produce methane. Some have the property of forming methane from formate, methanol, methylamine or acetate (Mah *et al.*, 1978; Wolfe, 1971; Zhilin, 1976).

Table 3.7: Chemotrophic Microorganisms in which Hydrogenase/Nitrogenase Activity has been Demonstrated

Physiological Group/ Family	Genus	Species
Hydrogen oxdizing bacteria:	Alkaligenes	eutrophus, latus, paradoxux, ruhlandii
	Aquaspirillum	autotrphicum
	Pseudomonas	carboxydovorans,
		facilis, flava
		pseudoflava, hydrogenovora, hydrogenother-movora, pallronii, saccharophila, thermophila.
	Selebaria	carboxydrogenase
	Flavobacterium	autothermophilum
	Paracoccus	denitrificans
	Xanthobacter	autotrophicus
	Arthrobacter sp.	
	Mycobacterium	gordonae
	Nocardia	autotrophica, opaca
Methanogenic bacteria:	Methanobacterium	formicum, bryantii, thermoautotrophicum.
	Methanobacter	ruminantium, smithii, arboriphilus.
	Methanomicrobium	mobile
	Methanogenium	cariaci, marisnigri
	Methanospirillum	hungagtel
	Methanosarcina	bsarkeri
	Methanococcus	vannielii, voltae
Chemotrophic bacteria : Spirillaceae	Spirillum (Azospirillum)	lipoferum, brasiliensis
	Aquaspirillum	faciculus
Azotobacteriaceae	Azotobacter	chroococcum, beijerinkii, vinelandii, paspali
	Azomonas	agilis, insignis, macrocytogenes
	Beijrinckia	indica, moblis, fluminensis, derxia
	Derxia	gummosa
Rhizobiaceae*	Rhizobium	leguminosaru, phaseoli, trifolii, meliloti, japonicum, lupini and cowpea rhizobia
Methylomonadaceae	Methylosinus	trichosporium
	Methylomonas	methanitrificans
	Methylobacterium	organophilum
	Methylococcus	capsulatus

Contd...

Table 3.7–Contd...

Physiological Group/ Family	Genus	Species
Enterobacteriaceae	Escherichia	coli, intermedia
	Citrobacter	freundii, intermediums
	Klebsiella	pneumoniae, aerogenes
	Enterobacter	aerogenes, cloacae
	Erwinia	herbicola
Bacillaceae	Bacillus	polymyxa, macerans
	Desulfotomaculum	ruminis, orientis
	Clostridium	butyricum, butylicum, beijerinckii, pasteurianum, madsonii, kluyveri, lactoacetophilum, felsineum, pectinovorum
The coryneforms	Corynebacterium	
Frankia sp.	autotrophicum	
Other N_2-fixing bacteria	Desulfovibrio	desulfuricans, vulgaris, gigas
	Pseudomonas	azotogensis
	Thiobacillus	ferrooxidans

A. The Nitrogen Fixing Bacteria

Metabolic variations among the N_2-fixing organisms are reflected in the number of processes by which energy for N_2-fixing may be generated : fermentation, aerobic or anaerobic respiration. They may exists primarily as independent organisms or in loose associations or intimate symbioses with micro-or macro-organisms. The only property shared by these organisms is their prokaryotic cell type and possession of the enzsymes nitrogenase and hydrogenase.

B. The Hydrogenase

There are at least two types of hydrogenases involved and their distribution is or taxonomic importance. The two types of hydrogenase can be differentiated with respect to localization in the cell and the hydrogen acceptor. A cytoplasmic NAD-specific hydrogenase known as hydrogen dehydrogenase (hydrogen NAD* oxidoreductase EC 1. 12. 1.2) is present in *Pseudomonas saccharphila* (Bone, 1960), *Alkaligenese eutrophus* (Eberhardt, 1966; Schneider and Schlegel, 1977; Wittenberger and Repaska, 1958), and *A. rublandii* (Vishniac and Trudinger, 1962). These species also contain a record hydrogenase which is membrane bound and does not reduce pyridine nucleotides. The majority of hydrogen bacteria contain only a membrane-bound hydrogenases; this is their sole hydrogen activating enzyme (Schneider and Schlegel, 1977). A NAD-specific soluble hydrogenase not accompanied by a membrane bound hydrogenase has been encountered only in *Nocardia opaca* (Aggus and Schlegel, 1974).

C. Thermodynamic Considerations

Hydrogen is an endproduct in the fermentation of many chemotrophic bacteria. Hydrogen formation via fermentation has therefore, been discussed as a possible process for energy production. Thauer (1977) pointed out that at most 33 per cent of the combustible energy of the organic compounds can be conserved in H_2 via fermentation. In case of glucose fermentation Thauer (1977) predicted a maximum yield of 4 mol of H_2 via the following reaction.

$$\text{Glucose} + 2H_2O \longrightarrow 2 \text{ acetate}^- + 2H+ + CO_2 + 4H_2$$

Infact 4 mol of H_2 is the highest amount ever reported to have been obtained from hexoses via fermentation (Hunguts, 1974).

Fermentation of hexose to acetate, CO_2 and H_2 has been reported for many bacteria under the conditions only when the H_2 partial pressure is maintained very low (10^{-3} atmosphere) by hydrogen utilizing bacteria (Iannotti *et al.*, 1973). At a pressure of one atmosphere, the amount of H_2 formed decreased to approximately 2.6 mol per mol of hexose fermented (Iannotti *et al.*, 1973; Jungermann *et al.*, 1973). Thus the practically available hydrogen in organic matter is considerably smaller than the amount that can maximally be formed via fermentation.

Based on the above consideration, Thauer (1977) was of the opinion that H_2 formation from organic matter by chemotrophic should not be considered as an efficient process for energy production.

D. Hydrogen Production

1. Hydrogen Production by Nitrogen Fixing Bacteria

Hydrogen formation is essentially an anoxic process, even in an obligate aerobe such as Azotobacter. Partiridge *et al.* (1980) studied the relationship between hydrogenase and nitrogenase in *A. chroococcum*. They showed that hydrogenase activity was significantly higher in N_2-fixing than NH_4^+ or NO_3^- dependent cultures under carbon limitation or in early stationary phase batch growth. In a subsequent study, Walker *et al.* (1981) showed the H_2 evolution occurred under air and was augmented when argon replaced N_2-pretreatment of each culture with 40 per cent acetylene in air. That the H_2 evolution occurred as a result of ATP hydrolysis *A. vinelandii* was demonstrated by Hageman *et al.* (1980).

Chain *et al.* (1980) studied production of hydrogen by *Azosprillum brasiliwensis* under N_2-fixing condition. These workers observed net H_2 production only when the gas phase contained CO. nitrogenase activity and H_2 evolution (in the presence of 5 per cent CO) showed a similar response to O_2 and were highest at 0.75 per cent dissolved O_2.

2. Hydrogen Formation by Obligate Anaerobes

Obligate anaerobes such as clostridia are essentially formative bacteria which accumulate hydrogen as one of many fermentation products. It has been indicated that hydrogenase is essentially a constitutive enzyme in clostridia which functions predominantly in the direction of hydrogen evolution (Gray and Gest, 1965).

Joyner *et al.* (1977) studied hydrogen production in the following rumen anaerobes : *Bacteroides clostridiformis, Butyrivibro fibrisolvens, Eubacterium limsum, Furobacterium necrophorum, Megasphaera eledonii, Ruminococcus albus, R. flavefaciens, Clostridium pasteurianum.* They showed that H_2 production in these rumen microorganisms was apparently similar to that of saccharolytic clostridia. It has been observed that presence of reducing agents such as sodium sulficde or cysteine favoured the formation of butyrate and hydrogen by *Glostridium acetobutylicum* (Rao and Mutharasan, 1988).

Search for newer organisms which can produce hydrogen has been conducted by various workers. For example, H_2 production by *Selenomonas ruminantium* was not reported untill Scheifinger *et al.* (1975) showed that the organisms ferments carbohydrates to lactate, propionate, acetate CO_2 and trace amounts of H_2. They also showed that the H_2 production could be increased to about 100 fold if the organisms are conclutured with methanogenic bacteria.

3. Hydrogen Formation by Desulfovibrio

Although hydrogen is not formed during normal growth of the sulfate reducing bacteria, they contain higher activities of hydrogenase and will decompose formate to hydrogen and carbon dioxide (Gray and Gest, 1965). Traore *et al.* (1981) investigated *Desulfovibrio vulgaris* grown in medium containing lactate or pyruvate plus a high concentration of sulfate. This strain formed 0.5 mol of H_2/mol lactate and 0.1 mol H_2/mol of pyruvate. Influence of growth conditions on hydrogen production and hydrogenase activity of *Desulfovibrio desulfuricans* strain 2198 was investigated by Zolotukhina *et al.* (1987) using calcium lactate and CO as electron donors. These authors showed that the amount of hydrogen produced and hydrogenase activity of the cell depends both on the nature of substrate used and presence of sulfate in the medium. The maximal amount of hydrogen was produced in a sulfate free medium containing lactate.

Hydrogenase from *D. gigas* or *D. desulfurcans* can be released into the medium by osmotic shock treatment, and therefore, it is located in the periplasmic space. The hydrogenase from the latter organism has been reported to catalyse the production of more than 9000 µmol H_2/min/mg protein (Glick *et al.*, 1980).

E. Genetic Aspects of Hydrogenase

In *E. coli*, hydrogenase participates in the metabolism of hydrogen in two distinct pathways. When grown anaerobically in the absence of exogenous electron acceptors, one pathway is the formate-hydrogen lyase pathway which converts formate to CO_2 and H_2 (Gray and Gest, 1965). It also participates in the energy conserving oxidation of hydrogen, which allows the organisms to grow anaerobically in non-fermentable carbon sources such as fumarate or malate under hydrogen-containing atmosphere (Macy *et al.*, 1976; Bernard and Goltaschalk, 1978). Ballentine and Boxer (1985) demonstrate that there are two hydrogenases present in *E. coli* when grown anaerobically in the absence of exogenous electron acceptors. However, both enzymes are not present under all growth conditions. Analysis of cells growht aerobically revealed that predomonantly only hydrogenase 2 was present. Since hydrogenase 1 was present only in absence of electron acceptors, conditions under which hydrogen-lyase pathway was operative, it was suggested that this hydrogenase functioned in this pathway with an H_2 evolving role. This enzyme was a single polypeptide (membrane bound) of Mr about 64,000.

Mutant strains of *E. coli* deficient in hydrogenase activity have been isolated and the lesions mapped close to minute 59 on the chromosome (Pascal *et al.*, 1975; Graham *et al.*, 1980). Hyd⁻ mutants in *Salmonella* have been also isolated and shown to map in the corresponding region of the chromosome (Chippaux *et al.*, 1972). Evidence also showed the presence of a single hydrogenase in *E. coli* (Adams and Hall, 1979; Graham *et al.*, 1980) and its was suggested that structural gene for hydrogenase was located at 59 min. Waugh *et al.* (1985) reexamined the mutants to determine whether one or both the hydrogenase isozymes are deficient. Analysis of various mutants supported the existence of two hydrogenase isoenzymes in *E. coli*.

Hisanori and Isao (1987) have succeeded in transferring hydrogenase gene from the bacterium *Citrobacter freundii* to the colon bacterium *Escherichia coli* using recombinant plasmid like pc BH4 (6.2 kb) and pc BH6 (5.7 kb). The hydrogen producing activity in the plasmid bearing *E. coli* was almost two times over the wild type *E. coli* (ca. 600). These investigations have looked into the involvement of the repressor gene suppressing the synthesis of hydrogenase gene in the bacterium.

The information concerning the genetics of hydrogenase in the hydrogen bacteria is currently in a state of development. From the available biochemical and genetic data, several conclusions can be made regarding the hydrogenase in *A. eutrophus*. A large plasmid contains information necessary for the expression of both hydrogenases (Anderson *et al.*, 1981), and probably codes for the structural genes themselves. Although each hydrogenase may function independently of the order, certain aspects of their function are coordinately regulated. Whether this involves structural or electron transport proteins that are required by the two hydrogenase has not yet been determined. It is also apparent that regulatory relationship exist between genes involved in the fixation of CO_2 and the hydrogenase.

Schlegel and Maria (1985) isolated hydrogenase regulatory mutants of hydrogen-oxidizing bacteria by colony-screening method. Such mutants were depressible for hydrogenases (Hoxd) in comparison to wild type *Alkaligenes hydrogenophilus* which is inducible for hydrogenases (Hox i). When mutants of *A. hydrogenophilus* and *A. eutrophus* H16 lacking the Hox-encoding plasmids pH G21-a and pH G1, respectively were used as recipients and Hox OZHo-encoding plasmids pH G21-a and pH G 1, respectively were used as recipents and Hox d mutant M 201 A. hydrogenophilus as donor, transconjugants appeared which had received Hox d character and the magaplasmid pH G21-a.

F. Immobilization of Hydrogenases

Any practical application of hydrogenase is only feasible if the enzyme is sufficiently stable, *i.e.* retains its catalytic activity over a prolonged period of time. The minimal requirement for such a stability is that those hydrogenase that are O_2-labile be stabilized against oxygen inactivation. Accordingly several approaches to stabilization of hydrogenases have been elaborated, especially in relation to

Table 3.8: Immobilization of Hydrogenases (Cammack et al., 1985)

Source	Immobilization Agent and Conditions	Result	References
Clostridium pasteurianum	Succinycarbodiimide galss	6 per cent specific activity, improved O_2 stability.	Lappi et al. (1976)
	Ferredoxin bound to Sepharose 4B	Hydrogenase stabilized, No activity for ferredoxin.	Rao et al. (1978)
	DEAE cellulose, Tris/Cl" buffer	Half life for resistances to O_2 inactivation increase 25 fold.	Klibnov et al. (1978)
	PEI cellulose, phosphate buffer	Half life for resistance to O_2 increased 400 fold.	
	Glutaral dehyde (GA) – activated amino spherocil	Half likfe increased 20 fold, H_2 evoloved from chlorophasts ferredoxin system.	Plasterk et al. (1981)
	Glutaral dehyde-activated aminopropyl glass.		
Desulfovibrio gigas	Activiated aliphatic and aromatic amino spherocil with or without cytochrome C_3.	O_2 stability improved. Immobilization yield and specific activity higher with coimmobilized. C_3	Hatchikian and Monsan (1980)
Desul fovibrio desulfuricans Norway	Activated amino spherocil and aminoropyl glass with and without cytochrome C_3.	Storage stability in O_2 improved, No improvement in specific activity with C_3 coimmobilized.	Plasterk et el. (1981)
Desul fovibrio desul furicans NRC 49001	GA- treated cells in calcium alginate.	GA prevents leakage of enzyme Ziamek et al., From perifplasm, storage stability improved.	Zinmek et al. (1982)
Alkaligenes uthrophus soluble enzyme	GA-treated aminopropyl glass	90 per cent activity immobilized, Lagpfase for NAD reduction abolished, Storage at 0°C improved.	Simon et la. (1978)
Alkaliges euthrophus cells	Calcium alginate	20 per cent specific activity in calcium alginate. 60 per cent activity after 1 week at 25°C. Used for regeneration of enzyme.	Klibnov and Puglist (1980)

various factors such as oxygen, pH and proteolysis by immobilization in solid matrices (Klibnov, 1983). Immobilization studies have been performed on many hydrogenases mainly to improve their oxygen stability. In some cases the enzyme was coimmobilized with choloroplasts and a suitable electron carrier and then used in hydrogen production reactions. Immobilization of hydrogenases usually resulsts in a decreases in specific activity of the enzyme. Once immobilized, the enzyme activity such more stable to oxygen than that of free, soluble enzyme employed for immobilization of hydrogenases.

Chapter 4
Hydrogen Production by Algae

1. Introduction

The pioneering discovery of hydrogen production by photosynthesis was made by Gaffron and Rubin (1942) using the fresh water green alga *Scenedesmus*. They showed that *Scenedesmus* could produce hydrogen in two ways: (i) in the dark via a fermentation reaction and (ii) via a photochemical reaction. The photoreaction produced hydrogen was ten times faster than the dark reaction. Subsequent research by Gaffron and his coworkers (Bishop sand Gaffron, 1963; Kaltwasser *et al.*, 1969; Stuart and Kaltwasser, 1970; Stuart and Gaffron, 1971, 1972a, b) demonstrated that dark hydrogen evolution is ATP dependent, while the light driven reaction is not, uncouplers of photophosphorylation stimulated the light reaction.

Hydrogen evolution by a photosystem containing organisms suggests that the source of reductant could water. However, Gaffron and Rubin showed that glucose could stimulate hydrogen evolution, and they concluded that the source of hydrogen was either the dehydrogenation of an organic substrate of possibly the photolysis of water. The substrate for photosynthetic hydrogen evolution became a controversial subject. Spruit (1958) in an experimental 'tour de force' demonstrated simultaneous hydrogen and oxygen transients in adapted *Chlorella*. Improving on Spruits technique Bishop *et al.* (1977) used polarography to measure hydrogen and oxygen production by anaerobically adopted algae. Due to the build up o photosynthetically produced oxygen and subsequent inactivation of hydrogenase, hydrogen production ceased after a few minutes. In order to overcome inactivation of hydrogenase Greenbaum (1980, 1981) constructed a flow system that continuously purged oxygen from the reaction cell with an inert carrier gas (such as helium or nitrogen). This technique can be used to simultaneous photoproduction of hydrogen and oxygen for many hours in anaerobically adopted algae. It is currently believed that water serves as the ultimate source of reductant for hydrogen photoproduction as well as organic reductant. However, as discussed above the specific mechanisms for hydrogen production may vary considerably depending on the species of algae and experimental conditions.

2. Photosynthetic Hydrogen Production in Cell-Free System

The production of hydrogen by a cell free system of photosynthetic organisms represents a potential source of energy that does not exploit traditional energy resources but utilizes available solar radiation. Although an intact cell approach to hydrogen production is the most logical first step to attacking this area of solar energy bioconversion. It is not necessarily the only avenue to success. It is possible that hydrogen gas production could be achieved through 'a cell-free' approach (Mitsui, 1975). More specifically, this method of solar energy bioconversion would entail extracting the essential elements of the hydrogen production pathway from living cells and recombining them in an *in vitro* environment. The principal advantage of this approach is that, it frees the processes of H_2 production from the demand and limitations put upon it by the organisms metabolism. Removing these "demand and limitation showed resulted in a dramatic increase in the potential solar conversion efficiency, successful cell free hydrogen production has been achieved by Mitsui (1974), by applying these theoretical considerations to a practical hydrogen production system. Therefore, the cell free system is a more long range approach than the intact cells system. Abeles (1964) obtained active cell free hydrogenase preparation from *Chlamydomonas equametos* by means of sonic oscillation. It was observed that when cell free preparation of *Chlamydomonas* were illuminated, light-dependent hydrogen was evolved. The hydrogen production could be stimulated by anion when methylviologen reduced by dithionate in cell free extract of *C. reinhardtii*. The most stimulatory anion tested 1, gives a 6-fold increase in activity at a concentration of 0.5 N. However, when reduced ferredoxin is used as the electron donor to hydrogenase, there is strong inhibition of H_2 production by salts, the inhibitoriest salt tested KI, which decreased hydrogenase activity by 93 per cent at a concentration of 0.2 N.

3. Regulation of Hydrogen Production by Different Ions

The rate of H_2 production is affected in the presence of ions which not only accelerate the rat of H_2 production but also inhibit metabolism of the organisms. Roessler and Lien (1982, 1984) showed that when cationic methylviologen was used as an electron mediator, anions stimulated the maximum velocity of H_2 production (*e.g.* 320 per cent increase in the presence of 1 M NaCl) but had little effect on the Km for methylviologen. Conversely, when hydrogenase activity is mediated by anions (*e.g.* 70-77 per cent of inhibition by 0.2 M NaCl). This inhibition is primarily due to reduced affinity of hydrogenase for these mediators as evidenced by a large increase in Km values rather than a change in the maximum velocity of the reaction. Anions have little effect on the kinetics of hydrogenase activity mediated by Zwitter ionic sulfenatopropyl viologen, a redox agent with a nearly neutral net charge. These results suggest the presence of cationic region near the active site of hydrogenase.

4. Light Regulation of Hydrogen Production

The growth of aquatic photosynthetic organisms is limited to a narrow zone of surface illumination known as the euohotic zone. The exact depth of this region depends on incident light intensity and a myriad of factors related to the transparency of water. In the tropics, afternoon surface irradiance, normally exceeds 100 Klx throughout the year and the euophotic zone (*i.e.* depth at which 1 per cent of incident light penetrate) can extend down to depths greater than 100 m. On the other hand surface illumination in polar water seldom exceeds 20 Klx during the summer, while winter intensities are very low and daylength are short. Photosynthetic organisms have been found to successfully inhabit this entire range of light environment. It was found that kinetics of fluorescence quenching in adopted *Chlorella vulgaris* cells under nonaxenic conditions correlates with the kinetics of H_2 photoproduction. In the presence of hydrogenase inhibitor Co, kinetics of fluorescence quenching are strongly delayed (Arkhipov *et al.*, 1980). Greenbaum (1977) showed that the absolute photoproduction of hydrogen by *Chlorella vulgaris* with single turnover flashes of light indicates that the Emerson and Arnold photosynthetic unit has the value chlorophyll : oxygen 1700:1, the hydrogen analogue of this unit has the value chlorophyll : hydrogen 1400:1.

During the illumination of *Chlorella pyrenoidosa* evolution of H_2 and O_2 has been observed. Reversible transition of O_2 was also observed. The action spectrum of photoproduction of molecular H_2 by *C. pyrenoidosa* was studied in argon atmosphere by gas chromatography and a correspondence to the absorption spectrum of chlorophyll a and has been observed. Comparison of the effect of specific inhibitors, electron donors, and change of temperature, intensity and quality of light as well as experiments with mutants showed that the function of O_2 and H_2 evolution are competitive (Oshchepkov and Krasnovoskii, 1972, 1974, 1976).

5. Hydrogenase Catalyzed Hydrogen Metabolism

This can be summarized by the following reactions :

In the Light

Hydrogen Photoproduction

$$H_2O \longrightarrow H_2 + 1/2O_2$$
$$RH_2 \longrightarrow H_2 + R$$

Photoreduction

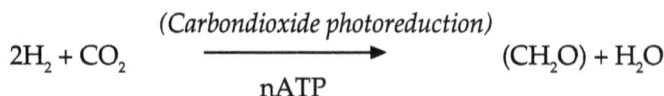

$$2H_2 + CO_2 \xrightarrow[nATP]{(Carbondioxide\ photoreduction)} (CH_2O) + H_2O$$

In the Dark

Hydrogen Production

$$RH_2 \longrightarrow H_2 + R$$

Carbon-dioxide Reduction

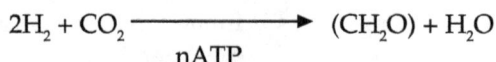

$$2H_2 + CO_2 \xrightarrow[nATP]{} (CH_2O) + H_2O$$

Oxyhydrogen Reaction

$$H_2 + Y_2O_2 \xrightarrow{\hspace{2cm}} H_2O$$

Hydrogen Uptake

$$R + H_2 \xrightarrow{\hspace{2cm}} RH_2$$

In addition to the above hydrogenase reactions, nitrogenase, the enzyme responsible for dinitrogen fixation, participates in hydrogen production (Gest and Kamen, 1942). Hydrogenase is a ferredoxin-type nonheme iron-sulphur protein and nitrogenase is a complex of two proteins component I (Fe protein) and component II (Fe-Mo protein). These enzymes are basically nonheme iron-sulphur proteins (Orme-Johnson, 1973; Hardy and Burns, 1973).

Most hydrogenase can catalyze both forward and reverse reactions. Some are unidirectional *e.g.* catalyze only the forward reaction. Absolute catalytic activities of hydrogenase vary widely. Most hydrogenases are catalytically active and stable upto about 50°C but inactive at higher temperature (Schlegel and Sohneider, 1978).

Occurrence of hydrogenase in algae has been reviewed (Table 4.1) (Kessler, 1974; Spruit, 1962). Kessler classified algae with and without hydrogenase activity and observed that 50 per cent of the algae so far tested exhibited activity. however, the presence of hydrogenase in cells may not correspond to the capability of an organism to carry out hydrogen photoevolution (Ben-Amotz *et al.*, 1975; Ward, 1970; Krasna and Rittenberg, 1959; Hartman and Krasna, 1963).

Table 4.1: Occurrence of Hydrogenase in Algae

Group	Reference
Euglenophyceae	
Euglena gracilis	Hartman and Krasna (1963)
Euglena sp	Krasna and Rittenberg (1954)
Chlorophyceae	
Ankistrodesmus sp	Gaffron (1940); Kessler and Czygan (1967)
Ankistrodesmus braunii	Kessler (1956)
Ankistrodesmus falacatus	Bishop *et al.* (1977)
Chlamydomonas debaryana	Healey (1970)
Chlamydomonas dysmos	Healey (1970)
Chlamydomonas eugametos	Ables (1964)
Chlamydomonas intermedia	Kessler (1967)
Chlamydomonas moewusii	Frenkel (1949)
Chlamydomonas reinhardtii	Hartman and Krasna (1963)

Contd...

Table 4.1–Contd...

Group	Reference
Chlorella fusca	Ward (1970)
Chlorella homosphaera	Kessler and Zweier (1971)
Chlorella kessleri	Kessler (1967)
Chlorella prothecoides	Bishop *et al.* (1977)
Chlorella vacuolata	Bishop *et al.* (1977)
Chlorella vulgaris of tertia	Kessler (1967)
Chlorella vacuolatum	Kessler (1974)
Coelastrum sp	Kessler *et al.* (1969)
Coelastrum proboscideum	Bishop *et al.* (1977)
Crucigenia apiculata	Kessler (1974)
Kirchneriella lunaris	Bishop *et al.* (1977)
Rhophidium sp	Gaffron (1940); Kessler and Czygan (1967)
Scenedesmus sp	Gaffron (1940); Kessler and Czygan (1967)
Scenedesmus obliquus	Gaffron (1940)
Scenedesmus quadricauda	Healey (1970)
Scenetrum gracile	Kessler and Maiforth (1980)
Scenetrum sp	Bishop *et al.* (1977)
Ulva faciata	Ward (1970)
Ulva latuca	Frenkel and Rieger (1951)
Phaeophyceae	
Ascophyllum nodosum	Frenkel and rieger (1951)
Rhodophyceae	
Ceramium rubrum	Ben-Amotz *et al.* (1975)
Chondrus crispus	Ben-Amotz *et al.* (1975)
Corallina officinalis	Ben-Amotz *et al.* (1975)
Porphyra umbilicalis	Frenkel and Rieger (1951)
Porphyridium aeruginum	Ben-Amotz *et al.* (1975)
Porphyridium cruentum	Frenkel and Rieger (1951)

6. Measurement and Hydrogenase Activity

The methods used for measuring hydrogenase activity, manometry and gas chromatography are the most widely used. Amperometric methods mass spectroscopy can be used for monitoring change in activity over second to minute intervals. Redox reaction of viologen dyes also is used for measuring hydrogenase activity spectrophotometrically. Short term measurements can be used as an index of potential capability for hydrogen. Photoevolution while long-term measurements can be used as an index of capacity for hydrogen photoevolution for application (Tables 4.2 and 4.3).

Table 4.2: Methods Used for Measuring Hydrogenase Activity

Method	Time Scale for Measurement	References
Manometry	Minutes-hrs	Umbereit, *et al.* (1972)
Modified Manometry		Davis and Stevenson (1977)
Syringe Method	Minutes-day	Hillmer and Gest (1977)
Gas Chromatography	Minutes-day	
Amperometric method	Second minutes	Wang *et al.* (1971)
Clark type electrode		
Ziroconium oxide high temperature electrode	Seconds	Greenbaum (1977)
Enzyme electric cell	Minutes	Yagi (1976)
Mass spectroscopy	Second-Minutes	Stuart *et al.* (1972)
Hydrogen Isotope	Minutes	Hartman and Krasna (1963)
Exchange reaction between hydrogen and deuterated or friliated water		Krasna (1978)
Spectroscopic method	Minutes-hrs	Krasna (1978) Yu and Wallin (1969)

7. Role of Photosystem I and II in Hydrogen Photoproduction

Boichenko (1980) showed that the kinetics of H_2 and O_2 photoproduction in the course of algal adaptation to nonaxenic suggest that hydrogenase activation considerably enhances the dark reduction of the acceptor pool and the inactivation of photosystem II, but it activate the reoxidation in the light of the pool by photosystem I.

Stuart (1971) showed that photosystem I of *Scenedesmus* has contribution in hydrogen production. Photosynthesis, photoreduction, the p-benzoquinone Hill reaction and glucose uptake by whole cells, as well as cyclic photophosphorylation with PMS (phenozine methosulfate) by chloroplast particles were strongly inhibited by 10^{-2} M salicylaldoxime or by hearting whole cells for 1-2 min at 55°C. In contrast, H_2 production by whole cells of mutant No. 8 and wild type of *Scenedesmus*, PS I (photosystem I) mediated MR (methyl red) reduction by chloroplast particles was either stimulated or not significantly inhibited by these agents. H_2 production by mutant No. 8 was slightly depressed by salicxylaldoxime. DCMU inhibited H_2 photoproduction with 10^{-2} M salicylaldoxime by approximately 20 per cent, indicating some contribution of electrons by endogenous organic compounds to photosystem II between the O_2 evolving mechanism and the DCMU-sensitive site. It is concluded that photohydrogen production by PS I of *Scenedesmus* does not require cyclic photophosphorylation but is due to noncyclic electron flow from organic substrate(s) through PS I to hydrogenase where molecular H_2 is released. The contribution of PS II to H_2 photoproduction by several unicellular green algae was measured both when O_2 evolution and photophosphorylation were unimpaired and also when these processes had been eliminated by Cl-CCP. The effect of DCMU on

Table 4.3: Hydrogen Production Rate in Algae

Algae	H2 Produced/hr (µmol)	Method*	Light Intensity	Reference
Ankistrodesmus braunii	0.49/mg Chl	M	16.7 w/m²	Stuart and Gaffron (1972)
	11.1/mg Chl	A		Bishop et al. (1977)
Chlamydomonas reinhardtii	0.25/mg dry wt/17/mg Chl**	M	300 lux	Healey (1970)
	3.2/mg Chl	A	16.7 w/m²	Stuart and Gaffron (1972)
Chlorella fusca	0.16/mg dry wt/0.49/mg Chl	M	16.7 w/m²	Stuart and Graffon (1972)
Chlorella protothecoides	16.4/mg Chl	A		Bishop et al. (1977)
Chlorella sp.	0.16/mg dry wt/11/mg Chl**	M	400 lux	Healey (1970)
Coelastrum sp.	24.6/mg Chl	A		Bishop et al. (1977)
Selenastrum sp.	11.8/mg Chl	M		Bishop et al. (1977)
Scenedesmus obliquus	0.22/mg dry wt/15mg Chl	M	500 lux	Healey (1970)
	0.4/mg Chl	M	16.7 w/m²	Stuart and Graffon (1972)
	20.2/mg Chl	A		Bishop et al. (1977)
Kirchneriella lunaris	32.1/mg Chl	A		Bishop et al. (1977)

*Abbreviation used **M** – Manometry, **A** – Amperometry.

**: Converted chlorophyll basis with following factors for dry wt to chlorophyll.

PS II contribution was found under both sets of experimental conditions for several strains of *Chlorella, Ankistrodesmus* and *Scenedesmus* (Stuart and Gaffron, 1972).

8. Simultaneous Photoproduction of Hydrogen and Oxygen

The simultaneous photoproduction of hydrogen and oxygen represents a physiological stress for algae. However, green algae are fairly rugged microorganisms with respect to the biophotolysis problem. They can survive hundreds of hour of anaerobiosis and irradiation a procedure that selects for strains of algae with enhanced properties for fuel production (Reeves and Greenbaum, 1985; Ward *et al.*, 1985) (Table 4.4).

Table 4.4: Simultaneous Photoproduction of H₂ and O₂ in Selected Marine Algae (Greenbaum *et al.*, 1983)

Species	Strain	Hydrogen (µmol/mg Chl/min)	Oxygen (µmol/mg Chl/min)	Isolation Data
Chlamydomonas sp.	F-9⁺	0.029	0.018	Falmouth Great Pond, M.A. VI/6/56 R. Guillard
Chlamydomonas sp.	0.5	0.13	0.007	Oyster Pond Marthas Vineyard M.A. Vi/6/56, R. Guillard.
Chlamydomonas sp.	11/35	0.018	0.005	Milford, C.T. (1950) R.A. Lewin (Cambridge Collection 11/35).
Chlamydomonas sp.	D	0.17	0.29	Milford, C.T. (1957) R. Guillard.
Halochlorocococcum	Fla-9	0.058	0.063	Pinellas point, Tampa Bay F.L. 1963. S.J. Erickson.
Chlamydomonas sp.	CPChl	0-	0.011	Charleston Salt Pond. R.I. (1961) T.J. Smayda.
Chlorella autotrophica		-	0.004	Milford C.T. (1950), R.A. Lewin Lu. Texas Coll. No. 580.
Chlorella sp.		-	0.005	Oyster Pond, Martha's Vineyard M.A. Vi/6/56 R. Guillard.

The production of O_2 and H_2 by anaerobically incubated *Chlorella* has been studied with rapid and sensitive electrochemical as well as volumetric method (Spruit, 1958). Continuous (35) illumination was studied in *C. vulgaris* with an apparatus for simultaneous fluorescence and H_2 or O_2 transients monitoring, O_2 evolution rate strongly decreased from the initial burst. These complementary changes indicate that electron flow blocks on the reducing side of photosystem II and I are formed. After prolonged (1-2 h) anaerobiosis, fluorescence shows pronounced delay without the transient. These changes are caused by increased H_2 photoevolution rate in the course of hydrogenase activation (Boichenko *et al.*, 1983), Kessler (1973) also reported simultaneous photoproduction of hydrogen and oxygen.

9. Photosynthetic Unit of Hydrogen Production

Photosynthetic is linear at less light intensities and saturates at higher intensities. A similar pattern is observed in the simultaneous photoevolution of hydrogen and oxygen by anaerobically adapted green algae. The photosynthesis saturates because, with increasing light intensity, the rate of quantum excitation of the reaction centers can exceed the kinetics rate of thermally activated electron transport of the non light dependent biochemical reaction that serially link the light reaction. The turnover time of photosynthesis was first measured by Emerson and Arnold (1932a). Who demonstrated that it was 40 ms at 2.5°C. A measurement of turnover time of photosynthetic hydrogen production and its comparison with the turnover time of normal photosynthesis is an important consideration in the development of a practical engineering system.

According to Diner and Mauzerall (1973), the turnover time of photosynthesis can be measured by two methods. The first used by Emerson and Arnold and the second is the repetitive double flash method in which the frequency of flash pair is varied. The turnover time of photosynthetic hydrogen evolution has been determined with each technique.

The experimental data indicate that the turnover times of photosynthetic hydrogen evolution are comparable to those of normal photosynthesis. These results suggested that the low value of light saturated hydrogen evolution by anaerobically adapted algae are not due to inherently slow turnover kinetics of hydrogen evolving biochemistry. Instead, the limitation is primarily associated with the number of apparent functional photosynthetic units (Table 4.5).

Table 4.5: Summary of Turnover Time for Light Driven Photosynthetic Hydrogen Production at 20°C Based on the Repetitive Double Flash Technique

Sl.No.	Alga	Turnover Time (m 8)*
1.	*Chlamydomonas reinhardtii*	0.3
2.	*Chlorella vulgaris*	0.1
3.	*Scenedesmus obliquis*	0.3
4.	*Chlamydomonas reinhardtii* (TAP)	1.0

*: The turnover time is defined as the time between two flashes for which there is a 50 per cent increase between the initial steady state flash pair yield and the final steady state flash pair yield form Greenbaum (1970).

According to Emerson and Arnold (1932b), there are in the green alga *Chlorella vulgaris* approximately 2,400 chlorophyll molecules per molecule of oxygen evolved per single turnover saturating flash of light. This number is a useful absolute figure which can be used to compare a variety of photosynthesis system and their light driven products. Two approaches are used to determine the size of the photosynthetic unit for hydrogen evolution. The first consists of measuring individually resolved bursts of hydrogen following single turnover saturating light flashes at a low flash repetition rate (0.1 H_2) (Greenbaum, 1977). The second, which is analogous to the Emerson and Arnold method, involves measuring the continuous steady state rate of hydrogen production under faster and varying repetitive flash illumination.

Theoretically, the value of the photosynthetic unit size, for hydrogen evolution (the ratio of chlorophyll to hydrogen per single turnover flash) should be predictable from a knowledge of the Emerson and Arnold, photosynthetic unit size for oxygen evolution and Z scheme model of photosynthesis. If all the reducing equivalents that are expressed as molecular hydrogen are derived from the mainstream of the electron transport chain of photosynthesis, the unit size value for hydrogen evolution should be half that for oxygen evolution. For *Chlamydomonas reinhardtii* and some other algae, using the individual flash yield technique, this if fairly close to what is actually observed (Greenbau, 1977; Greenbaum and Reeves, 1985). At higher flash rates under steady state rates of hydrogen evolution, the yield of hydrogen per chlorophyll molecule per flash falls by a factor of 0 100. The low light saturated steady-state rates of hydrogen and oxygen evolution can to a large extent, be explained by a loss in the number of apparent functional photosynthetic units at higher equivalent light intensities. A fundamental understanding of this phenomenon would be an important first step in the development of a practical gaseous fuel producing system based on algal water splitting.

10. Maximization of H_2 Production

Yanyushin (1981) studied the effect of Cl-CCP, diuron and dibromothymoquinone on light induced H_2 evolution by *C. reinhardtii* cells in synchronous culture. It was found that, when the cells were treated with Cl-CCP, H_2 production enhanced with increasing light intensity upto 22 w/m² whereas diuron lowered H_2 evolution in light, down to the level, that slightly exceeded the dark rate dibromothymoquinone also lowered H_2-production. Yanyushin (1982) also studied H_2 evolution and activation of hydrogenase enzyme in synchronous cultures of *Chlamydomonas reinhardtii*.

11. Application of Mutants as a Tool

An anaerobically adapted with type cultures of *C. reinhardtii* photoevolved hydrogen at an initial rate of 15 to 10 mmol H/mg algae/h. The photoevolution of H_2 ceased when the atmospheric O_2 concentration approached 1 per cent. DCMU, at 10 mM inhibited the photoevolution of H_2 in the wild type while the mutant remained uninterrupted and photoevolved H_2 at a rate equal to 10 per cent of that of the wild type (Mcbride *et al.*, 1976).

A comparative study was made by Oshchepkov *et al.* (1970) on H_2 and O_2 photoproduction by mutant of green algae. The rates of O_2 and H_2 evolution upon illumination of different strains of *Chlorella pyrenoidosa*, *Scenedesmus obliquus* and *Chlamydomonas reinhardtii* were measured by the gas chromatography technique in argon atmosphere. Photoproduction of O_2 and H_2 were obligate and alternative function for the cells containing chlorophyll regardless of its spectral forms. H_2 evolution was a primary function in the greening experiments. Those strains which synthesized chlorophyll in the absence of light developed the O_2 and H_2 evolving enzyme system in the dark. The maximum rate of H_2 photoproduction correlated with the activity of RUDP carboxylase. Photoproduction of molecular hydrogen by green algae apparently is a type of photorespiration.

The capacity of H_2 photoproduction and photoreduction was determined in the greening pigment C-2_A of *Scenedesmus obliquus* In the dark grown culture the capacity for H_2 photoproduction is low and for photoreduction was not detectable Both the processes increase with chloroplast development in the light photoreductuon increases in parallel to the photosynthetic capacity using H_2 instead of water as electron donor. H_2 photoproduction reaches optimal values during the stage of highest quantum efficiency of photosynthetic oxygen evolution but does not increase further. Hydrogenase of adapted cells is most active in dark grown cultures and declines during greening This indicates that hydrogenase activity is not the limiting factor in the hydrogen metabolism studied in the investigation (Randt *et al.*, 1985b).

According to Boichenko *et al.* (1986), the mutants of green algae *Chlamydomonas* which had lost pigment, protein complexes and photochemical activity of photosystem I are capably of photosynthetic H_2 formation by photosystem II complexes. The wild strain of *C. reinhardtii* and the mutant ASS 238 (contains photosystem II complex and light harvesting chlorophyll a protein complex) and ASS-1H (contains only photosystem I chlorophyll a protein complex). O_2 and H_2 photoproduction by algal cell was measured using polarography. Randt and Senger (1985), observe the participation of the two photosystem in light dependent hydrogen evolution in wild type and mutant cell of *Scenedesmus obliquus*.

12. Marine Algae as Hydrogen Producers

Studies have been made by different workers on marine unicellular green algae. In 1987, Kumazawa *et al.*, studied the relation between dark anaerobic hydrogen (H_2) evolution of marine unicellular green algae and the energetic state of cells. One of the 4 investigated strains produced H_2 continuously for 48 h and maintained its adenylate energy charge (EC) at about 0.6, whereas the H_2 evolution of the other strain stopped within 24 h and their EC decreased at low values.

Dark hydrogen production by marine green alga *Chlamydomonas* MGA 161 was studied by Miura *et al.* (1987). This alga was a halotolerant not a halophylic, grew well in both natural and artificial sea water media. With the experiment cycloheximide, it was found that the hydrogenase reaction in this large was not a rate limiting step of dark hydrogen evolution. But accumulation of starch increased at a low NH_4Cl concentration (0.5 mM), at a low temperature (20°C) or at a high NaCl concentration (7 per cent). Hydrogenm evolution was correlated with starch degradation rather than starch accumulation, and the molar yield of hydrogen from starch or glucose was very high, at about 2 mol H_2/mol glucose.

Greenbaum *et al.* (1983) reported first measurements of the simultaneous photoproduction of hydrogen and oxygen in marine green algae. Eight species in the genera *Chlamydomonas*, *Chlorella* and *Halochlorococcum* were tested in CO_2-free sea water, four of the five species of *Chalmydomonas* were able to produce hydrogen in the light after a period of 3-4 h of dark anaerobic adaptation only one of the two Chlorella species test was able to photoproduce hydrogen, intrace amounts, *Halochlorococcum* fla-9 gave positive results and *Chlamydomonas* species (clone f-9) has a steady-state rate of hydrogen and oxygen production during irradiation with a stoichlometric ration near 2:1. The intergrated yields of hydrogen and oxygen produced by this species corresponds to about 450 turnover of the photochemical reaction centers. This number exceeds (by about a factor of 20) the electron carrying capacity of the electron transport chain linking photosystem I and II. These data suggest that Chlamydomonas f-9 makes sea water a potential substrate for solar hydrogen and oxygen production.

13. Anaerobiosis Vs. H_2 Production

The role of anaerobiosis in H_2 production has been worked by many workers. The anaerobically grown or adapted cells or cell free extract of green algae shows their effect on hydrogen production or in hydrogen metabolism. Abeles (1964) obtain cell free hydrogenase prepared from anaerobically adapted cells of *Chlamydomonas equimetos* by means of sonic oscillation. He studied the effect of different compounds like methylviologen, benzsylviologen, triphenyl tetrazolium chloride, nicotinamide adenine dinucleotide phosphate and observed evolution of hydrogen when hydrogenase was incubated with reduced methylviologen.

Boichenko *et al.* (1983) made simultaneous measurements of fluorescence induction and hydrogen photoproduction in *Chlorella* vulgaris under anaerobic conditions. Kesseler (1973) detected the effect of anaerobic ($N_2 + CO_2$) preincubation in the dark on photosynthetic reaction (O_2 evolution, measured manometrically, and with the oxygraph fluorescence and photoproduction of H_2 ; measured with mass spectrometer), in algae with hydrogenase (stains of *Chlorella fusca, C. kesseleri, C. vulgarios, F. tortia* and *Ankistrodesmus braunii*) and in algae without hydrogenase (strain of *C. vulgaris* and *S. minutissima*). The inhibition by anaerobic incubation of photosynthetic O_2-evolution is much stronger in algae without hydrogenase than it is in algae with hydrogenase. The effect of anaerobiosis is most pronounced at rather low light intensity (about 1000 lux) in acid medium (pH 4), and after prolonged anaerobic incubation in dark (about 20 h). These results indicate that the presence of hydrogenase might be ecologically advantageous for algae under certain conditions. Chlorophyll fluorescence showed the fastest response to anaerobic incubation, and the most pronounced difference between algae with and without hydrogenase. After incubation of 30 min under $N_2 + CO_2$, fluorescence in algae with hydrogenase starts with a peak and decreases within 10-20 s to rather low steady-state level which is only slightly higher than that found under aerobic conditions. In algae without hydrogenase, fluorescence is rather low during the 1st 1-2 s and then rises to a higher steady state level which is much higher than that of the aerobic control. This indicates an inhibition due to anaerobiosis of photosystem II in algae without hydrogenase. Algae with hydrogenase can react in different ways during the 1st minute of illumination. In some cases where there is an immediate photoproduction of hydrogen which is followed after a few minutes by photosynthetic O_2 evolution; in other algae, there is a simultaneous production of H_2 and O_2 from the very beginning; in a few experiments there was no photosynthetic O_2 evolution either. Thus photoproduction of H_2 seems to be the process which normally enables algae with hydrogenase to oxidize and there activate their photosynthetic electron transport system after anaerobic incubation.

The relation between dark anaerobic hydrogen H_2 evolution of marine unicellular green algae and the energetic state of the cells, as reveled by H investigated strains produced H_2 continuously for 48 h and maintained its adenylatic energy charge (EC) at about 0.6, whereas the H_2 evolution of the other strains stopped within 24 h and. Their EC decreased at low values (Kumazawa *et al.*, 1987). An anaerobic incubation period of varying duration is required to induce hydrogenase activity on *C. reinhardtii*. Inclusion of sodium acetate, metabolizable carbonoeous substrate in the medium during anaerobic incubation, accelerate the activation process, thus in the presence of sodium acetate, H_2 photoproduction is detected within 7.15 min after the onset of anaerobiosis. The presence of uncouplers like Cl-CCP or sodium arsenate the rate of H_2 photoproduction decreases (Lien and Pietro, 1981). The concentration of ATP, NADH and NADPH were measured during a 5 hour period of anaerobiosis in the dark, and upon subsequent illumination with high light intensities (770 w/m^2). Conditions which favour optimal photoproduction of H_2 can be very well assessed in cells of *Chlorella* with active hydrogenase (Mahro *et al.*, 1986).

The significance of anaerobic pre-incubation periods and high light intensities for hydrogen photoproductivity of *Chlorella* fusca. Using sodium dithionite an oxygen scavenger the influence of different light intensities and periods of anaerobic preincubation in the dark on H_2 photoproductivity is studied. By measuring hydrogen production in the light using manometric and gas chromatographic methods the effectiveness of sodium dithionate in stabilizing photoproduction was established for high rates of H_2 photoproduction high light intensities upto 30,000 lux 580 w/m^2 were necessary. The results shows that the initial burst kinetics, the light saturation, and the obligate period of anaerobic adaptation H_2 photoproduction by *Chlorella* is apparently an anaerobic photosynthetic process which occurs in the absence of CO_2 and can be experimentally stabilized by exogenous O_2 scavengers (Mahro and Grimme, 1982).

14. The Economics of Hydrogen Production

From the standpoint of utilization it is essential that the hydrogen production system meet certain economical demands, *i.e.*, low cost and world-wide applicability. In additions, it must pass certain pollution standards aimed at protecting the ecosystem.

The last prerequisite is easily met since hydrogen gas is the cleanest burning fuel known, resulting in the production of water. However, attaining the other two goals is a considerably more complex problem.

15. The Choice of the Cheaper Hydrogen Donor

There are two essential elements needed as fuel for the hydrogen production system, sunlight and a hydrogen-electron donor. If hydrogen is to be produced on a massive scale and worldwide, the donor must be worldwide in distribution and readily available in large quantities. Fortunately, nature has partially solved our problem in this respect since may marine algae can utilize salt water as an electron and hydrogen donor. The benefits accrued from this fact are many : (1) salt water is the most ubiquitous substance on earth, (2) it is readily available to most countries of the world, (3) it could be inexpensively collected and (4) it contains the most nutrients necessary for the algae which will be used as hydrogen producers. Hence from both a logistics and economics point of view salt water is presently the most attractive source.

Chapter 5

Hydrogen Production by Cyanobacteria

1. Introduction

First of all Jackson and Ellms (1896) reported that *Anabaena* species from a Massachusetts reservoir immediately produced H_2 when placed in a sealed bottle. Whether H_2 production was by algae or by contaminating bacteria is still unclear. There was a considerable gap in this area of research since the first report appeared. It was in 1942, that Gaffron and Rubin demonstrated H_2 production by anaerobically incubated *Scenedesmus obliquus*, a green alga, and since then this alga has become a major focus of research for H_2 metabolism. Blue-green algae again came into focus when the occurrence of a reversible hydrogenase was demonstrated in a unicellular blue-green alga *Synechococcus elongates* obviously after several hours of anaerobic preincubation (Frenkel *et al.*, 1950). Studies on H_2 metabolism by blue-green algae restarted only after the unequivocal report of N_2 fixation by this group of algae. The real H_2 production mediated by the nitrogenase enzyme complex was first of all reported in *Anabaena* cylindrical under *in vitro* condition by Haystead *et al.* (1970). Subsequently H_2 production by intact filaments of *A. cylindrical* in the light under an atmosphere of argon and CO_2 was reported by Benemann and Weare (1974a,b). In the same year H_2 production by other cyanobacteria was independently reported by Russian workers (Oshchepkov *et al.*, 1974). Since then extensive work has been done on various aspects of H_2 metabolism employing a variety of blue-green algal strains.

2. Enzymes Involved in H_2 Metabolism

(a) Nitrogenase

Nitrogenase, a typical enzyme present in all the N_2-fixing microorganisms, forms H_2 both under *in vivo* and *in vitro* conditions. With normal growth at the expense of dinitrogen, nitrogenase reduces N_2 to ammonia under consumption of electrons, protons and ATP. Some of the reduced protons are evolved as H_2. H_2 formation by a blue-green algal nitrogenase was first of all demonstrated *in vitro* by Haystead and coworkers (1970). Nitrogenase has been isolated andpurified by a number of workers (See Stewart, 1980; Lambert and Smith, 1981a; Fay and Van Baalen, 1987).

(b) Hyrogenases

(i) Nitrogenase/Hydrogenase

When nitrogenase specifically produces H_2 in an ATP-dependent reaction, such nitrogenase is preferably called as hydrogenase. This hydrogenase is basically a nitrogenase enzyme and shares all the properties of typical nitrogenase.

(ii) Reversible Hydrogenase/Soluble Hydrogenase/Bidireactional Hydrognease

This enzyme may catalyze either uptake or evolution of H_2 and is called reversible hydrogenase. Reversible hydrogenase couples to a low potential electron carrier (with Em near 0.4 V). This is easily solubilized, evolves hydrogen from reduced methyl viologen, and catalyzes the reduction of phenazine methosulfate, methylene blue, and dichlorophenolindophenol with PMS being the most effective electron acceptor.

(iii) Uptake Hydrogenase/Unidirectional Hydrogenase

This enzyme is membrane bound, saturated by low H_2 tension, and is capable of reducing methylene blue but not low-potential acceptors. In the heterocystous, nitrogen-fixing cyanobacteria, possession of a unidirectional hydrogenase enables them to catalyze the H_2 consumption reactions typical of anaerobically adapted algae, and at the same time, probably increases the efficiency of N_2 fixation in these organisms. This enzyme also takes part in the oxy-hydrogen reaction in a number of blue-green algae (Lambert and Smith, 1981b). Evidence has been presented that this hydrogenase is capable of recycling some of the enrgy lost as hydrogen by supplying nitrogenase with some form of reductant (Benemann and Weare, 1974a; Lambert and Smith, 1981b). This uptake "hydrogenase" enzyme is particularly active in heterocysts and is weakly or not at all expressed in NH_4^+–grown and heterocyst-free filaments of cyanobacteria (Lambert and Smith, 1981b).

3. The Capacity to Metabolize H_2: Species Distributions

Table 5.1 lists the names of blue-green algal species capable of metabolizing H_2. Earlier there was a belief that only heterocystous cyanobacteria fix N_2 under aerobic and/or anaerobic conditions, the number of N_2–fixer species has increased.

Drastically, so far all the N_2-fixing species tested have shown the capacity of either H_2 formation or the uptake of H_2. Thus hydrogenase has now been reported from all sections of blue-green algae, *viz.*, and unicellular, non-heterocystous filamentous and heterocystous-filamentous species. Nevertheless, the number of hydrogenases present in different species shows considerable differences. Thus

Table 5.1: Occurrence of Hydrogenase in Blue-green Algae (Cyanobacteria)

Species	Reference
A. Unicellular	
Anacystis nidulans	Eisbrenner *et al.* (1981) Peschek (1979a)
Aphanothece halophytica	Belkin and Padan (1978)
Cyanothece sp. (7822)	Van der Oost *et al.* (1987)
Gloeocapsa species	Gallon *et al.* (1974)
Gloeothece sps. 6909	Van der Oost *et al.* (1987)
Cyanophora paradoxa	Eisbrenner *et al.* (1981)
Microcystis aeruginosa	Asada and Kawamura (1986)
Synechococcus elongates	Frenkel *et al.* (1950)
Synechococcus sp. Strain Miami Bg043511	Arai and Mitsui (1987)
Synechococcus cedorum	Gerasimenko an Zavarzin (1981)
Synechococcus sp. 7425	Van der Oost *et al.* (1987)
Synechocystis sp.	Frenkel and Rieger (1951)
B. Non-heterocystous filamentous	
Lyngbya sp. 108	Kuwada and Ohta (1987)
Myxosarcina chroococcoides	Lambert and Smith (1980)
Oscillatoria limnetica	Belkin and Padan (1978)
Oscillatoria brevis	Lambert and Smith (1980)
Oscillatoria thiebautii	Scranton *et al.* (1987)
Oscillatoria sp. Miami BG-7	Kumazaa and Mitsui (1981)
Plectonema boryanum	Padan and Schneider (1978)
Phormidium angustissimum	Gerasimenko and Zavarzin (1981)
Phormidium laminosum	Smith *et al.* (1982)
Spirulina sp.	Liama *et al.* (1979)
Schizothriz calcicola	Lambert and Smith
C. Heterocystous-filamentous forms	
Anabaena cylindrical	Hattori (1963)
Anabaena sp. N-7363	Asada *et al.* (1985)
Anabaena variabilis	Eisbrenner and Bothe (1979), Almon and Boger (1984)
Anabaena azollae	Peters *et al.* (1977) Chen Van *et al.* (1983)
Anabaena sp.	Jones *et al.* (1976)
Anabaena cycadeae	Perraju *et al.* (1986)
Anabaena sp. Strain CA	Xiankong *et al.* (1984)
Anabaena sp. Strain 1F	Xiankong *et al.* (1984)
Anabaena oscillaroides	Paerl (1980)

Contd...

Table 5.1—Contd...

Species	Reference
Anabaena sp.	Srivastava *et al.* (1989)
Ambaena CH₂	Chen *et al.* (1989)
Aphanizomenon species	Paerl (1982)
Cylindrospermum licheniforme	Hirosawa and Wolk (1979)
Cylindrospermum sp.	Srivastaa *et al.* (1989)
Fischerella muscicola	Lambert and Smith (1980)
Fischerella sp.	Srivastava *et al.* (1989)
Macrozamia communis	Daday and Smith (1987)
Mastigocladus laminosus	Ernst *et al.* (1979) Singh and Kashyap (1988)
Nostoc sp.	Srivastava *et al.* (1989)
Scytonema hofmanii	Srivastava *et al.* (1989)
Anabaena, Nostoc sp.	Vyas and Kumar (1995)

it is possible for an organism to possess methylviologen-dependent hydrogen formation only under the specified growth conditions, nitrogenase and uptake hydrogenase only, nitrogenase only, uptake hydrogenase only or none of these activities under the conditions tested. Species like *Anabaena cylindrical* possess all three activities *viz.*, nitrogenase, uptake hydrogenase and reversible hydrogenase under N_2-fixing conditions, but only the methylviologen dependent H_2 formation when grown in ammonium salts. Figure 5.1 outlines the general scheme showing the inter-relationship between vegetative cell and heterocyst.

4. Isolation, Purification and Characterization of Different Hydrogenases

Exceptive nitrogenase, very little work has been done on the biochemical characterization of hydrogenases from cyano bacteria (Lambert and Smith, 1981b; Houchins, 1984; Rao and Hall, 1988), This has been mainly due to oxygen sensitivity of hydrogenases which ultimately creates problem during isolation (Houchins, 1984). However, now methods have been developed which yield good preparation of hydrogenases from a selected cyano bacteria (Rao and Hall, 1988). Due to limited investigations on hydrogenases, especially reversible hydrogenase, there still exists confusion about the exact number and localization of hydrogenases among cyanobacteria (Houchins, 1984; Rao and Hall 1988). Extensive work is needed to obtain clearer picture about the hydrogenases in cyanobacteria. In this section a brief account is given for all types of hydrogenase including nitrogenase separately.

a. Nitrogenase (ATP-dependent hydrogen formation)

Nitrogenases from all organisms are very similar in their physical and chemical properties (Adams *et al.*, 1980; Bothe, 1982; Fay and Van Baalen, 1987; Mortenson, 1978; Stewart, 1980). The nitrogenase complex is comprised of two protein components. The larger component dinitrogenase (also called MoFe protein or component I) is responsible for reduction of substrate molecules. The MoFe component of *A.cylindrical* is n acidic tetramer (216000 mol wt); its isolectric point lies between pH 4.72-4.99, and the Mo:Fe:s per mole ratio is 2.2:20.4:20 per cent. In amino acid composition relatedness it resembles MoFe components from other N_2 fixers, particularly *Rhodospirillum rubrum* (Stewart, 1980; Fay and Van Baalen, 1987). The second component, dinitrogenase reductase, accepts electrons from donors such as ferredoxin flavodoxin or dithionite and transfers these electrons to dinitrogenase with the concomitant hydrolysis of two molecules of ATP per electron transferred. The Fe protein (component II) is a highly labile dimmer (60000 mol wt; two identical subunits). The MoFe and Fe proteins of *A. cylindrical* and *P. boryanum* cross react with 65-90 per cent of the efficiency of homologous crosses (Stewart, 1980).

The six electron reduction of N_2 to $2NH_3$ therefore requires a minimum of 12 ATP molecules, making N_2 fixation an energetically expensive process. In addition to reducing N_2 to NH_3, dinitrogenase can reduce a number of other substrates including H^+ and acetylene.

Both components of nitrogenase are rapidly and irreversibly inactivated by O_2 (Mortenson, 1978; Mortenson and Thorneley, 1979). Organisms that fix N_2 in aerobic environments, therefore, must provide some means of protecting nitrogenase from oxygen. Cyanobacteria are the only organisms capable of simulataneous O_2 evolution and N_2 fixation, demonstrating that these processes can be compatible under suitable conditions.

Like the reduction of N_2, reduction of protons by nitrogenase requires ATP and a strong reductant. Reductant can be provided as reduced ferrodoxin, flavodoxin or, as commonly employed in laboratory experiments, dithionite (Asada and Kawamura, 1985, 1986; Yu and Wolin, 1969). The evolution of H_2 by nitrogease always occurs during N_2 fixation and a minimum of 25 per cent of total electron flux through nitrogenase is diverted to the reduction of H^+ even at very high partial pressures of N_2 and optimal levels of ATP and reductant (Houchins, 1984). If the rate of dinitrogenase turnover is slowed owing to limitation in the supply of ATP or reductant, the relative proportion of electrons diverted to H^+ reduction increases (Bothe, 1982). The evolution of H_2 by nitrogenase, therefore, consumes a substantial amount of the reductant and ATP utilized by nitrogenase.

b. Reversible Hydrogenase

Unlike "uptake" hydrognease, reversible hydrogenase has been demonstrated in representatives of nearly every major cyanobacterial group (Houchins, 1984, Reactions catalyzed by reversible hydrogenase have been demonstrated in the unicellular organisms, *Synechococcus, Synechocystis* and *Aphanocapsa* (Asada and Kawamura, 1985; Asada *et al.*, 1987; Belkin and Padan, 1978; Mitsui *et al.*, 1979, 1985, 1987; Peschek, 1979a), in filamentous non-heterocystous strains such as *Spirulina, Lyngbya* and *Oscillatoria* (Belkin and Padan, 1978; Kumazawa and Mitsui, 1982; Kuwada and Ohta, 1987, 1988, 1989; Llama *et al.*, 1979; Mitsui, 1978, 1979, 1981; Phlips and Mitsui, 1983a,b; Ramchandran and Mitsui, 1984; Scranton *et al.*, 1987; Smith *et al.*, 1982); and in heterocystous organisms including several strains of *Anabaena, Nostoc* and *Mastigocladus* (Almon and Boger, 1984; Asada *et al.*, 1979; Benemann and Weare,

1974; Benemann *et al.*, 1982; Chen *et al.*, 1989; Eisbrenner and Bothe, 1979; Ernst *et al.*, 1979; Ewart and Smith, 1989; Houchins and Burris, 1981a, b, d; Laczko, 1980; Laczko and Barabas, 1981; Miura *et al.*, 1982; Miyamoto *et al.*, 1979, 1984; Spiller *et al.*, 1983; Tel-or *et al.*, 1977, 1978; Weare *et al.*, 1980). In most cases, reversible hydrogenase was demonstrated by measuring emthyl-viologen dependent H_2 evolution. The various physiological functions of reversible hydrogenase are represented in Figure 5.2.

Reversible hydrogenase has been isolated and partially purified from some cyanobacterial sources (Belkin *et al.*, 1981; Ewart *et al.*, 1989a, b; Houchins and Burris, 1981; Hallenbeck and Benemann, 1978; Llama *et al.*, 1979; Klentemich *et al.*, 1989; Rao and Hall, 1988). Various molecular weights have been reported for this hydrogenase. Llama *et al.* (1979) found a molecular weight of 56000 based on elution from a Sephacryl S-200 column for the enzyme from *Spirulina maxima*. The enzyme from *Anabaena* 7120 eluted in two peaks of molecular weight 113000 and 165000 (Houchins and Burris, 1981a, b). If this enzyme was first heated at 70°C for 1 h, an additional peak of 55000 appeared suggesting that the larger forms might be multimers of a single polypeptide. Hallenbeck and Benemann (1978) have reported a molecular weight of 230000 for the enzyme from *A. Cylindrical*. Multiple forms of reversible hydrogenase which may or may not correspond to different states of aggregation, have been reported. Multiple bands staining for hydrogenase activity are invariably observed on polyacrylamide gels (Ward, 1970; Ewart and Smith, 1989; Kentemick *et al.*, 1989). Some of the characteristics of reversible hydrogenase are represented in Table 5.6.

Partially purified preparations of reversible hydrogenase are usually quite resistant to irreversible inactivation by O_2. Samples of the enzyme from *Oscillatoria limnetica* (Belkin *et al.*, 1981) exposed to air for 50 days at 20 or -196°C retained about 80 per cent of the activity of an anaerobic control sample. Contrary to this result, the enzyme from *Mastigocladus* was considerably more O_2-sensitive and lost about 75 per cent of its activity relative to a control after 8 days under air (Rieder and Hall, 1981).

Recently two soluble hydrogenase activities were obtainable from cell extracts of the cyanobacterium *A. cylindrical*, one detectable by the tritium exchange assay, the other having a relatively low tritium exchange activity but catalyzing methyl viologen dependent hydrogen formation. Their molecular weights, by gel filtration chromatography, were 42000 and 100000 respectively (Ewart and Smith, 1989). Ewart and Smith (1989c) have further raised antibodies in mice against the 42 Kd subunit of the soluble hydrogenase purified from *A. cylindrical*. The antibody did not cross-react with the 50 Kd protein, which appears to be necessary to confer methyl viologen-dependet reductive hydrogenase activity. Using western blotting procedures they found no evidence for the 42 Kd protei in heterocyst cells (Ewart and Smith, 1989a.

Furthermore, although the protein was found only in the soluble fraction of vegetative cells from cultures grown in air/CO_2, antibody reactivity was also obtained with the particulate fraction when cultures were grown in nitrogen/4 per cent H_2/0.3 per cent CO_2. They have also developed an activity stain for the detection of cyanobacterial hydrogenase in polyacrylamide gels (Ewart and Smith, 1989b). In another study on *Anacystis nidulans*, Kentemich *et al.* (1989) have reported isolation and purification of reversible hydrogenase (250-fold) by classical methods. Activity staining on gels obtained by native PAGE allowed to identify two bands. Antibodies were raised against the electrophoretically homogeneous protein. Crude extracts from the unicellular *Anacystis* and from heterocysts and vegetative cells of *Anabaena variabilis* showed precipitation bands of 56 and 17 Kd. From their study it appears that *Anacystis* has two different hydrogenases; the reversible or bidirectional hydrogenase which is located exclusively at the cytoplasmic membrane of the cells and a thylakoid-bound enzyme which catalyzes only the uptake of H_2. The activity as per different purification steps is represented in Tables 5.4 and 5.6–5.8.

Inspite of many new reports appearing on reversible hydrogenase, the ambiguity still persists about the number and localization of this bi-directional enzyme. A systematic investigation is needed to reveal the exact nature and types of reversible hydrogenase in cyanobacteria.

c. Uptake Hydrogenase

Uptake hydrogenase has been demonstrated in all heterocystous cyanobacteria so far examined (Lambert and Smith, 1981b; Houchins, 1984; Rao and Hall, 1988). There exists great variations in activity with growth conditions and in some strains of *A.cylindrica* activity is nearly udetectable under certain grown conditions (Bothe *et al.*, 1978; Jones and Bishop, 1976; Lambert and Smith 1980). So far uptake hydrogenase has not been detected in any non-heterocystous N_2-fixers. The only non-heterocystous organism so far shown to posses uptake hydrogenase is the non-N_2-fixing unicell, *Anacystis nidulans* (Peschek, 1979 a,b,c).

Table 5.2: Profiles of Hydrogenase Activities in Different Cyanobacteria

Organism	Growth Conditions	Nitroge- nase Activity	H_2- Cons- mption	MV- Dependent H_2 Formation	Reference
Anacystisnidulan	Various	–	+	+[b]	Peschek (1979a,b,c)
Myxosarcina chrococcoides	NH_4^+-grown	–	–	–	Lambert & Smith (1980)
Oscillatoria brevis	NH_4^+-grown	–	–	+	Lambert & Smith (1980)
Plectonema boryanum	Air-grown	–	–	+	Weare & Benemann (1974)
Schizothrix calcicola	NH_4^+-grown	–	–	–	Lambert & Smith (1980)
Anabaena cylindrica	N_2-grown	+	+	+	Bothe *et al.* (1977)
	NH_4^+-grown	–	–	+	Lambert & Smith (1980)
Fischerella muscicola	N_2-grown	+	+	–	Lambert & Smith (1980)

[a] modified from Lambert and Smith (1981b); [b] Activity was demonstrated to be membrane bound

The properties of cyanobacterial uptake hydrogenase *in vitro* are similar to those from other bacteria. Houchins and Burris (1981a,b) have shown that when isolated heterocysts were broken with a French press, about 70 per cent of the recovered activity was membrane bound and could be solubilized by detergent treatments or by prolonged sonication. The remaining 30 per cent of the activity that was released into the soluble fraction during the French press treatment resembled the membrane bound uptake hydrogenase in physical and catalytic properties and is probably the same enzyme. The solubilized enzyme from *Anabaena* 7120 eluted from Sephadex G-100 with a molecular weight of 56000. Co-inhibition was competitive versus H_2 with Ki of 0.039 atm. Exposure to O_2 results in both reversible inhibition and irreversible inactivation, the sensitivity to O_2 increasing with increasing disruption of the system (Houchins an Burris, 1981a, b; Lambert and Smith, 1981b). Irreversible inactivation of solubilized uptake hydrogenase displays biphasic kinetics with half-times of 2 and 14 min. Optimal O_2 levels for oxyhydrogen activity were 20 per cent for whole filaments, 10 per cent for isolated heterocysts and 2.5 per cent for heterocyst membranes (Houchins and Burris, 1981d). In *Anacystis nidulans,* the oxyhydrogen reaction in isolated membranes was completely resistant to an O_2 concentration of 30 per cent.

Table 5.3: Purification of Hydrogenase from *Spirulina maxima**

Step	Total Protein (mg)	Total Activity (units)[a]	Specific Activity (units/mg protein)	Purification (-fold)	Yield (per cent)
Lysate	12800	12860	1.0	1.0	100.0
40000 g supernatant	1207	6265	5.2	5.2	48.7
100000 g supernatant	800	6027	7.5	7.5	46.8
DE-52 eluate	107	3750	35.0	35.0	29.0
Sephacryl S-200	6.6	1453	112.4	112.0	11.3

* Modified from Llama *et al.* (1979); [a] Activity expressed as µmol H_2 evolved/hr.

Table 5.4: Partial Purification of *Oscillatoria limnetica* Hydrogenases*

Step	Total Protein (mg)	Total Activity (units)[a]	Specific Activity (units/mg protein)	Purification (-fold)	Recovery (per cent)
Crude sonicate	5190	4418	0.85	1.0	100.0
35000 g supernatant	1875	3225	1.72	2.0	75.0
77000 g supernatant	1540	3004	1.95	2.3	68.0
75 per cent $(NH_4)_2SO_4$	311	928	2.98	3.5	21.0
First DEAE-Sephacel Form I	41	353	8.7	10.2	8.0
Form II	43	530	10.5	12.4	12.0
Second DEAE-Sephacel Form I	15	191	13.0	15.3	4.3
Form II	19	295	15.6	18.4	6.7
Sephacryl S-200 Form I	6	106	17.6	20.7	2.3
Form II	6	158	25.8	30.3	3.6

* Modified from Belkin *et al.* (1981); [a] Activity expressed as µmol H_2 evolved/hr.

Table 5.5: Properties of Reversible and Uptake Hydrogenases from Anabaena sp. Strain 7120*

	Uptake Hydrogenase	Reversible Hydrogenase
Cellular localization	Heterocysts	Both cell types
Subcellular localization	Membrane bound	Cytoplasmic
Effect of NH_3 during growth	Repressed	No effect
Response to anaerobic conditions during growth	Increased activity	Large increase in activity
Effect of H_2	Increased activity	No effect
Stability to atmospheric O_2 levels	Irreversible inactivation	Stable
Acceptor specificity	Positive Em_7 only	Both positive and negative Em_7
Km for H_2	0.9 μM	2.3μM
Molecular weight	56000	55000-230000
Ki for CO	0.039 atm	0.0095 atm
Stability to 70°C heat treatment	$t_1/2$ = 12 min	Stable
pH optimum for H_2 uptake	6.0, 8.5	9.0

* After Houchins (1984).

Table 5.6: Localization of the Reversible Hydrogenase in Aerobically and Microaerobically Grown *Anacystis nidulans* by the Immuno-gold labeling Technique

	Cells Grown Aerobically	Microaerobically
Specific H_2-evaluation activity [a]	269	408
Gold label associated with the cytoplasmic membrane	65 +11	108 + 9
Gold label associated with the cytoplasm	26 + 11	28 + 8

* After Kentemich *et al.* (1969); a The specific activity is given in nmol H_2 evolved/mg protein/h.

5. Nitrogenase-Mediated H_2 Formation

Extensive studies have been conducted on H_2 production employing a number of N_2-fixing cyanobacteria including both heterocystous and non-heterocystous forms. A number of factors have been reported to affect H_2 formation (Lambert and Smith, 1981; Houchins, 1984). Excepting a few species, when cultures are grown in air, net H_2 formation is almost zero (Lambert and Smith, 1981; Xiankong *et al.*, 1983, 1984). This is mainly8 due to recycling of H_2 formed via "uptake" hydrogenase (Adams *et al.*, 1980; Bothe, 1982; Bothe *et al.*, 1980; Peterson and Burris, 1978a, b; Spiller *et al.*, 1978; Xiankong *et al.*, 1974). The basic strategy to obtain net H_2 formation is either to remove N_2 from as phase or to prevent N_2 incorporation, so that the assimilating power available to nitrogenase is all channeled into H_2 formation. A number of methods have been employed to optimize H_2 formation, and these allow the proper conditions for optimal gas production to be deduced.

Table 5.7: Purification of the Reversible Hydrogenase from *Anabaena nidulans**

Purification Step	Specific Activity[a]	Enrichment	Total Activity (X10³ U)[b]	Yield per cent
Crude extract	855	1.0	329	100
Supernatant after 20% $(NH_4)_2SO_4$ addition	929	1.1	254	77
Linear KCl gradient on DEAE-cellulose fraction No.				
10	2631	3.0	69	21
11	3993	4.6	110	33
12	2170	2.5	24	7
Octyl-sepharose chromatography Fraction no.				
4	16136	18.9	53	16
5	214408	250.7	85	26

* After Kentemich *et al.* (1989); [a] The specific activity is given in nmol H_2/mg protein/h in the $Na_2S_2O_4$ and methyl viologen dependent evolution assay.

a. Selection of Species and Strains

Studies of H_2 formation have focused largely on the filamentous heterocystous cyanobacteria (Lambert and Smith, 1981b). *Anabaena cylindrica* has been consistently used as model organism for H_2 formation by a number of workers (Hallenbeck and Benemann, 1979; Lambert *et al.*, 1979). The rate of H_2 formation particularly in *A.cylindrical* obtained by various workers shows close similarity and reported differences if any, may be due to differences in culture and/or incubation conditions (Bothe *et al.*, 1978; Lambert and Smith, 1981b; Tel-Or *et al.*, 1978). In general, the capacity to photoproduce H_2 varies widely depending on the levels of nitrogenase and hydrogenase activities of the cyanobacterium used. Although, any strain showing highest nitrogenase activity may or may not show the highest rate of H_2 formation. *Anabaena CA*, the fastest growin cyanobacterium, is reported to have the highest nitrogenase activity amongst all the cyanobacteria so far tested but it does not show the highest rate of H_2 formation (Xiankong *et al.*, 1983, 1984). On the other hand a few marine non-heterocystous cyanobacteria though having moderate level of nitrogenase activity show a very high rate of H_2 formation (Kumazawa and Misui, 1981, 1982, 1985; Mitsui *et al.*, 1977, 1979, 1985). Thus it is evident that rates of H_2 formation can vary widely with different species (see Table 5.9). Therefore, screening of many cyanobacteria isolated from diverse ecosystems may provide more suitable H_2 producers or strains for investigation. Non-heterocystous strains may be useful for biochemical studies on H_2 formation as well as H_2 formation may be enhanced as all the cells may contain nitrogenase unlike in heterocystous cyanobacteria.

b. Factors Affecting Hydrogen Formation

As stated earlier a number of factors have been reported to affect nitrogenase-mediated H_2 formation. In this section a brief account is given for all those factors affecting the rate and duration of H_2 formation.

Table 5.8: Factors Affecting Hydrogen Formation

Species	Moles H_2 Produced per hour		Method	Reference
	Mg dry wt	Mg chl a		
Anabaena cylindrical	1.4	320	GLC[b]	Weissman and Benemann (1977)
		0.7	GLC	Bothe *et al.* (1977)
	0.17	11.0	GLC	Daday *et al.* (1977)
	0.58	–		Jeffries *et al.* (1978)
	0.10	6.7	GLC	Lambert and Smith (1977
Anabaena strain CA	40 micro l			
Anabaena ambigua	–	8.5	GLC	Srivastava (1990)
Anabaena cycadeae	–	1.56	GLC	Srivastava *et al.* (1989)
Anabaena (N-7363)	–	72	GLC	Asada *et al.* (1979)
Anabaena flos-aquae	0.025/ micro l pcv[d]		AMP[c]	Jones and Bishop (1976)
Anabaena doliolum	–	1.90	GLC	Srivastava *et al.* (1989)
Calothrix membr3anaceae	0.108		GLC	Lambert and Smith (1977)
Calothrix scopulorum	0.128		GLC	Lambert and Smith (1977)
Cylindrospermum majus	–	3.25	GLC	Srivastava (1990)
Fischerella muscicola	–	9.10	GLC	Srivastava (1990)
Mastigocladus laminosus	8.5 ul	70 mol	GLC	Benemann *et al.* (1982)
Nostoc muscorum	20 mg	Chl a/h	GLC	Weeisshaar and Boger (1983)
Oscillatoria brevis	0.168	–	GLC	Lambert and Smith (1977)
Oscillatoria sp.	0.26	260	GLC	Mitsui and Kumazawa (1977)
Oscillatoria BG-7	0.183	–	GLC	Kumazawa and Mitsui (1985)
Scytonema hofmanni	–	9.36	GLC	Srivastava *et al.* (1989)

[a] Incubation conditions are not similar in all the cases; [b] GLC–Gas Liquid chromatography;
[c] AMP – Amperometric; [d] pcv–Packed cell volume

c. (A) Growth Conditions Affecting H_2 Production

(a) Gas Atmosphere for Growth

The most commonly used gas atmosphere for the growth of N_2-fixing cultures is air supplemented with CO_2 to permit optimal growth rates. In the absence of added CO_2, growth is drastically slowed (Fay and Van Baalen, 1987; Stewart, 1980) and the characteristic blue-green coloration due to phycocyanin is reduced (Lambert *et al.*, 1979b). It has been demonstrated that cultures grown under limiting CO_2 conditions have H_2 photoproduction rates proportional to their growth rates (Jones and Bishop, 1976). The concentration of CO_2 also affects the rate of H_2 formation. A concentration below 5 per cent CO_2 is most suitable for growth as well as H_2 production (Benemann and Weare, 1974a; Lambert *et al.*, 1979b; Tel-Or *et al.*, 1978).

With *Nostoc muscorum* the effect of growing cultures with N_2/CO_2 rather than air/ CO_2 has been tested (Ernst *et al.*, 1979). It has been observed that both acetylene reduction and H_2 formation increased slightly. *Azolla* fronds when incubated in air alone (Banerjee *et al.*, 1989; Peters *et al.*, 1976). Tel-Or *et al.* (1977) have also reported the induction of hydrogenases by incubating cultures in $N_2/H_2/CO_2$. It was found that growth in $N_2/H_2/CO_2$ (75:20:5) enhanced nitrogenase and hydrogenase activities between days 4 and 8 of growth of *A. cylindrical* and *Nostoc muscorum* (Tel-Or *et al.*, 1977). The maximum enhancement of nitrogenase activity was 3- to 5-fold and of H_2 consumption and methyl viologen-dependent H_2 formation, 5 to 20-fold. This induction of hydrogenase has been observed in a number of other cyanobacteria (Eisbrenner *et al.*, 1979; 1981; Howarth and Codd, 1985). On the contrary increase in H_2 production has also been observed following CO_2 deprivation in *A.cylindrical* (Jeffries *et al.*, 1978). Furthermore, obligate requirement of CO_2 at least in aged cultures for prolonged H_2 production has been demonstrate in a number of cyanobacteria (See Lambert and Smith, 1981b). In summary it seems that CO_2 is essential for continued H_2 production and this effect seems to be mediated via net CO_2 fixation.

(b) Culture Medium

Culture medium used for growth of various cyanobacteria plays an important role in regulating the evolution of H_2 (Lambert and Smith, 1981b; Stewart, 1980). The most commonly employed medium has been that of Allen and Arnon (1955). However, this medium does not constitute the minimal nutritional requirement for cyanobacteria and some of the trace elements are possibly unnecessary. However, *Azolla* fronds when incubated with different synthetic media showed highest rate of H_2 evolution with Allen and Arnon's medium (Banerjee *et al.*, 1989; Kumar *et al.*, 1990).

A number of other media have been used for various cyanobacteria; however BG-11 seems to be the most suitable medium for growth and H_2 formation for a variety of cyanobacterial strains (Lambert and Smith, 1981b).

Many other constituents and added chemicals have been demonstrated to affect subsequent H_2 formation. Almost all the combined nitrogen sources *viz.*, NH_4^+, NO_3^-, NO_2^- and a few amino acids like glutamine, asparagines, arginine and glutamic acid inhibit H_2 formation. The inhibition of H_2 formation is solely due to loss of heterocyst and/or nitrogenase activity (Banerjee *et al.*, 1989; Bothe, 1982; Ernst *et al.*, 1979; Houchins, 1984; Lambert *et al.*, 1979c; rivastava, 1990; Stewart, 1980). Addition of exogenous $NaHCO_3$ to culture medium has been reported to stimulate H_2 formation in a number of cyanobacteria (Srivastava *et al.*, 1989). Similarly supplementation of a few metal ions and their concentration has been demonstrated to affect H_2 evolution. Jeffries *et al.* (1978) showed that cultures grown with 5.0 mg of Fe^{3+}/litre produced H_2 at a rate about twice the rate of cultures with 0.5 mg of Fe^{3+}/litre. Nickel addition significantly affects H_2 evolution although the effect is indirect most probably via uptake hydrogenase (Almon and Boger, 1984; Daday *et al.*, 1985; Xiankong *et al.*, 1984). Sodium being an essential element for nitrogenase activity indirectly affects H_2

evolution (Srivastava 1990; Stewart, 1980). Tungsten has been reported to inactivate nitrogenase and thus causing total inhibition of nitrogenase catalyzed H_2 formation (Lambert and Smith, 1979b; Srivastava, 1990). Magnesium, phosphorus and calcium have also been reported to influence H_2 formation although the effects are mediated though nitrogenase (Lambert and Smith, 1981b; Stewart, 1980).

(c) Light

Light quality and intensity both affect H_2 formation in a number of cyanobacteria. This effect is indirect and mainly through photosynthesis and nitrogenase activity (Stewart, 1980). A number of studies have demonstrated that pigment content of cyanobacteria can vary according to the wavelength characteristics of the light source used (Fay and Van Baalen, 1987; Frenkel *et al.*, 1950; Greenbaum, 1977; Kerfin and Boger, 1982; Lambert and Smith 1980; Philips and Mitsui, 1983b; Stewart, 1980). Chan Van *et al.* (1983) have demonstrated a close correlation between wavelength of light and photosynthesis together with H_2 production. Gogotov *et al.* (1976) demonstrated that light of low intensity (2.5, 10^3 erg/cm^2/s) stimulates H_2 production by cell suspensions of *A variabilis* in the presence of glucose, pyruvate and formate. The maximum rate of H_2 production in the presence of these substrates was observed at light intensities of 650, 1400 and 2250 erg cm^{-2}s^{-1} respectively. In another study it has been demonstrated that *A. cylindrical* sparged with argon gas produced H_2 continuously for 20 days under limited light conditions (6.0 Wm^{-2}) in the absence of exogenous nitrogen. Kumar and Kumar (1988) have observed that low light intensity favours H_2 production in *Anabaena CA*. In *Oscillatoria* sp. Strain Miami BG-7, the rate of H_2 production saturates at low light intensity (*i.e.* 15-30 uEm^{-2}s^{-1}). Although so far no systematic studies on the effect of light intensity during growth on subsequent H_2 formation have been performed but studies so far conducted suggest that lower light intensity favours H_2 production.

Under dark, rate of H_2 production only about 10 per cent of the light incubated cultures. As yet, not a single species has been found to show sustained H_2 production in the dark. H_2 production shows almost similar trend in terms of duration as observed with nitrogenase activity.

(d) Temperature

Excepting the thermophilic cyanobacterium *Mastigocladus laminosus*, in all the mesophilic cyanobacteria H_2 formation follows the pattern of growth temperature (Lambert and Smith, 1981b). In marine *Oscillatoria sp.* Strain Miami BG-7, the upper temperature limit for H_2 production has been found at 46°C.

H_2 evolution by the thermophilic cyanobacterium *Mastigocladus laminosus* has been studied by a few worker (Benemann *et al.*, 1982; Miura *et al.*, 1980, 1981, 1982; Miyamoto *et al.*, 1979, 1984; Smith *et al.*, 1982). The optimal temperature for H_2 production was 44-49°C (Miura *et al.*, 1980).

(e) Age of Culture

It has been realized lately that the age of cultures used for H_2 measurements is an important factor (Lambert and Smith, 1981). First of all Benemann and Weare

(1974a) showed that H_2 formation in *A. cylindrical* is optimal when harvested at about 0.3 mg dry wt/ml of cyanobacterial concentration, after which a sharp decline is observed. This decline in H_2 formation is more pronounced than the observed decline in nitrogenase activity suggesting that hydrogen recycling capacity remains relatively constant. Similar finding was later on reported by Daday *et al.* (1977). In *Nostoc muscorum* a similar optimum H_2 production in logarithmic growth was observed which correlated with changes in heterocyst frequency (Ernst *et al.*, 1979). The decline of H_2 -evolution in aged cultures seems mainly due to gradual loss of initrogenase activity and also the pigments of PS II (Lambert and Smith, 1981; Stewart, 1980).

(f) Concentration of Culture

Concentration of culture has been implicated in regulating the net H_2 formation. H_2 formation has been shown to increase linearly with algal concentration to about 1.5 mg dry wt/ml suspension for *A cylindrical* (Daday *et al.*, 1977). Subsequent decrease in H_2 formation is probably due to self-shading of the culture.

c. (B) Factors Employed to Enhance Nitrogenase Catalyzed H_2 Production

1. Choice of Gas Atmosphere during Assay

(a) Argon/CO_2

As most of the N_2-fixing cyanobacteria lack the capacity to produce H_2 under aerobic conditions, cultures are frequently incubated in argon in the absence of CO_2 to avoid generation of O_2 through photosynthesis during the experimental incubations (Lambert and Smith, 1981b). However, these conditions are not optimal for H_2 formation and generally result in lower rate of H_2 formation. It has been also demonstrated that the longer the incubation, the greater the requirement for CO_2 for prolonged H_2 formation (Chen *et al.*, 1989; Daday *et al.*, 1977; Ernst *et al.*, 1979; Kuwada and Ohta, 1989; Laczko, 1980; Lambert *et al.*, 1979a; Lambert and Smith, 1980, 1981a, b). The reason for lower rate of H_2 formation under argon alone is primarily due to depletion of stored reductant (Lambert and Smith, 1981b). With argon alone there is no CO_2 and hence stored reductant is gradually consumed in all other metabolic reactions including nitrogenase (Ernst *et al.*, 1979; Lambert *et al.*, 1979a).

To improve the efficiency of H_2, use of 99 per cent argon + 1 per cent CO_2 has been demonstrated to be the best alternative (Asada *et al.*, 1985; Banerjee *et al.*, 1989; Bothe *et al.*, 1978; Kerfin and Boger, 1982; Kosyak *et al.*, 1978; Kumazawa and Mitsui 1981; Scherer *et al.*, 1980; Tel-Or *et al.*, 1977, 1978). Preincubation under 99 per cent argon + 1 per cent CO_2 leads to multifold stimulation of H_2 production rate although depending upon the strains used (Lambert and Smith, 1981b). However, prolonged incubation under 99 per cent argon + 1 per cent CO_2 (a non-growing condition) subsequently causes nitrogen starvation lading to the death and decay of cultures. Thus, this treatment has been found suitable for short term H_2 evolution study (Lambert and Smith, 1981b). However, investigators have shown that continued production of H_2 may be attained in the same culture if a very low concentration of NH_4^+ is added to the cultures so as to keep them under growing condition (Lambert *et al.*, 1979b; Lambert and Smith, 1980; Mitsui, 1987; Mitsui *et al.*, 1985; Weaver *et al.*, 1980).

(b) Dinitrogen (N₂)

To attain anaerobic condition, a number of workers have used N_2 gas in place of argon (see Lambert and Smith, 1981b). Furthermore the effect of N_2 along with argon has also been tested. In either condition, N_2 inhibits H_2 formation due to competition of N_2 reductional with proton reduction (Houchins, 1984). N_2 at 4 per cent of the gas phase gas been found to inhibit H_2 formation in argon by about 50 per cent, with complete inhibition occurring at 25 per cent N_2 for *A. cylindrical* (Jones and Bishop, 1979).

(c) Oxygen

The effect of O_2 in argon or N_2 atmosphere has also been tested (Ernst *et al.*, 1979; Spiller *et al.*, 1978). Added oxygen inhibits H_2 formation in all the cyanobacteria (Asada and Kumazawa, 1986; Banerjee *et al.*, 1989; Daday *et al.*, 1977; Kumar and Kumar, 1990b, c; Lambert *et al.*, 1979c; Smith *et al.*, 1983, 1984). In an *Anabaena species*, it was found that 5 *per cent* O_2 in argon stimulated H_2 production but this very concentration was inhibitory under N_2 (Asada *et al.*, 1979). The inhibition of H_2 formation by O_2 seems to be due to O_2 senstivity8 of nitrogenase or by the facilitation of an oxyhydrogen reaction in heterocyst which diminishes net H_2 production (Ernst *et al.*, 1979; Spiller *et al.*, 1978; Xiankong *et al.*, 1983, 1984).

Carbon Monoxide/Acetylene

Carbon monoxide has been extensively employ8ed in H_2 formation studies because it specifically inhibits N_2 fixation and C_2H_2 reduction but does not inhibit the nitrogenase mediate H_2 formation reaction (Brad beer and Wilson, 1963). Similarly, C_2H_2, an inhibitor or hydrogenase in many bacteria, has been frequently used (Smith *et al.*, 1976). Stimulation of H_2 formation by the addition of CO and C_2H_2 in an argon gas phase was first of all shown in *A.cylindrical* by Bothe *et al.* (1977a). Subsequently, H_2 formation was obtained even in air with the use of CO + C_2H_2 for *A.cylindrical* (Daday *et al.*, 1977) and also in marine cyanobacteria (Lambert and Smith, 1981b). At optimal concentration of co and C_2H_2, H_2 formation occurred at the same rate in air after a lag phase of 2-3has in argon gas phase (Lambert *et al.*, 1979). A number of other workers have shown the same effect of CO + C_2H_2 on H_2 formation in various cyanobactiria (Miura *et al.*, 1981,1982; Miyamoto *et al.*, 1984) Srivastava, 1990; xiankong *et al.*, 1984; see Figures 5.3 and 5.4) Increase of H_2 formation by co + C_2H_2 seem to be medicated by a number of ways. Bothe (1982) provided evidence that C_2H_2 inhibits H_2 recycling and that this inhibits is increased some what by co. there is also a report that Pre-incubation with C_2H_2 alters the conformation of the nitrogenase complex. Still other the workers have suggest that the effect of C_2H_2 is to alter the H_2 tension at the nitrogenase complex and that it is the H_2 level rather than C_2H_2 itself that causes a conformational activation of the enzyme (Scherer *et al.*, 1980).

2. Metabolic Inhibitors

(a) DCMU (3-(3-,4-dichloropheny1)-1,1-dimethtylurea

Out of various metabolic inhibitors, DCMU, an inhibitors of PS II, has been frequently used to optimize H_2 production (Lambert and Smith, 1981b). The possible effect of DCMU may be: (a) a reduction in H_2 formation by decreasing reductant supply to PS I from PSII; or (b) an increase in H_2 formation due to elimition of an ox hydrogen reaction that uses photosynthetically generated O_2 and nitrogenase-mediated H_2. Both these effects have been observed in some cyanobacteria (Houchins, 1984). That a loss of reductant does take place by DCMU treatment was first of all proved by Benemann and Weare (1974b) who demonstrated H_2 dependent acetylene reduction in the presence of DCMU. Gogotov *et al.* (1976) have reported that H_2 evolution is not inhibited by DCMU in the presence of pyruvate or formate in *Anabaena variabilis*. Working with isolated heterocysts of *Anabaena* sp. Strain CA, it has been demonstrated that DCMU has no effect on H_2 formation (Kumar and Kumar, 1990b). Xiankong *et al.* (1983) have reported that DCMU addition immediately inhibits H_2 production in *Anabaena* CA. It seems that the effect of DCMU depends on the particular cyanobacterium used; those that produce little H_2 in the absence of DCMU and exhibit a vigorous oxyhydrogen reaction might be stimulated in H_2 formation by DCMU, whereas those in which H_2 and O_2 are not re-utilized until significant tensions accumulate in the environment, such as in the case of uA cylindrical will show no benefit from DCMU addition (Lambert and Smith, 1981b). As a whole it seems that even if DCMU results in stimulation of H_2 formation, the effect is temporary. Furthermore, the effect of DCMU also depends on the level of endogenous reductant present in a particular alga.

(b) Metronidazole (2-methyl-5-nitroimidazole-1-ethanol)

Tetley and Bishop (1979) first of all showed that metronidazole at 1-2 mM levels was a selective inhibitor of nitrogenase activity in *Anabaena* sp. Hydrogenases are insensitive to metronidazole. Although this chemical does not stimulate H_2 production but it offers a model system to distinguish between hydrogenase and nitrogenase activity.

(c) MSO (L-methionine-DL-sulfoximine)

MSO, an inhibitor of glutamine synthetase which derepresses nitrogenase activity, has been found to increase H_2 production for shorter duration (Lambert and Smith, 1981b; Stewart, 1980; Srivastava, 1990). Lambert *et al.* (1979b) found stimulation of H_2 production by treatment of MSO (2µm) in *A cylindrical* B 629. Use of MSO is important especially where cultures are grown at lower concentration of NH_4Cl for longer duration of H_2 production. MSO derepresses nitrogenase even in the presence of NH_4Cl and thus continued production of H_2 occurs.

(d) Organic Carbon Sources

Cyanobacteria are obligate photoautotrophs and as such do not require any organic carbon sources for normal growth (Stewart, 1980). However, there are certain species which show stimulation of growth in light and a few show growth in dark in the presence of certain organic carbon sources. As most of the cyanobacteria show little or no H_2 formation in dark, the role of exogenous organic carbon sources seems important. Chan Van *et al.* (1983) have reported that pyruvate and glucose stimulate photohydrogen production in *Anabaena azollae*. Kumar and Kumar (1990a) did not observe stimulation of H_2 evolution by addition of fructose and erythrose in *Anabaena* CA. Srivastava *et al.* (1989) have reported significant stimulation of H_2 production in a number of N_2-fixing cyanobacteria. They have also demonstrated stimulation of H_2 production by exogenously added carbon sources in the dark (Srivastava, 1990; Srivastava *et al.*, 1989). So far no attempt has been made to stuly the response of added organic carbon sources on H_2 production in any typical heterotrophic cyanobacteria. Differential response of organic carbon sources on H_2 formation seems to be related with transport of carbon sources by specific cyanobacterial species.

(e) Nickel

Nickel has been demonstrated to be an essential metal for uptake hydrogenase activity (Almon and Boger, 1984; Daday and Smith, 1983; Daday *et al.*, 1985; Srivastava, 1990; Xiankong *et al.*, 1984). In almost all the cyanobacteria tested, nickel addition stimulates uptake of H_2 and thus inhibition of H_2 formation takes place. Attempts are being made to isolate nickel-resistant mutant which might be uptake hydrogenase negative (Hup⁻) and therefore may prove better strain in terms of H_2 production.

(f) Na₂S, Thiosulfate and other Reducing Agents

A number of workers have tested the effect of a variety of reducing agents on H_2 production in a few cyanobacteria. It was found that Na_2S addition showed stimulation of H_2 production in *Nostoc muscorum* (Fry *et al.*, 1984). The results obtained by various workers are variable and as such no definite conclusion can be drawn on the role of these substances on H_2 production.

6. H₂ Production by Hdrogenase (Reversible Hydrogenase)

In comparison to nitrogenase-mediated H_2 formation little work has been done on H_2-formation by hydrogenase (Houchins, 1984; Lambert and Smith, 1981b). Whereas nitrogenase catalyzed H_2 formation takes plae solely in N_2-fixing forms, hydrogenase is present both in N_2fixer and non-N_2 fixer forms. There is a lot of confusion about the exact function of reversible hydrogenase. Since this enzyme is found in vegetative cells even when grown in the absence of heterocysts it is unlikely that its function relates to nitrogen fixation. Furthermore there is still ambiquity about the exact number of reversible hydrogenases in cyanobacteria. However, now it is believed that two pools of hydrogenase exist, an oxygen-sensitive activity in vegetative cells and an oxygen-resistant activity in heterocysts.

Reversible hydrogenase activities are usually present in aerobically grown cyanobacteria prior to a dark, anaerobic adaptation period (Daday and Smith, 1979; Houchins and Burris, 1981a, b; Tel-Or *et al.*, 1978). Demonstration of activity requires reductive activation of the enzyme or removal of a tightly bound O_2 molecule. This activation can be achieved in intact cells by addition of dithionite and methyl viologen. The methyl viologen-dependent H_2 evolution activity in cyanobacteria usually increases 2-fold during a dark anaerobic adaptation period (Houchins and Burris, 1981a,b; Spiller *et al.*, 1983).

Little reversible hydrogenase activity develops in the presence of molecular oxygen. By far the most dramatic and sustained increase in activity can be obtained by removal of O_2 during continuous illumination. Increases in activity of one to three orders of magnitude occur over a period of hours to days when cultures are sparged with an anaerobic gas mixture during growth (Houchins, 1984; Tel-Or *et al.*, 1977; Lambert and Smith, 1981b). The enzyme is not catalytically functional *in vivo* under these conditions but is maintained in an inactive state by photosynthetically produced O_2 (Houchins and Burris, 1981a,b; Houchins, 1984).

The presence of combined nitrogen in the growth medium has little effect on the level of reversible hydrogenase activity, but an indirect effect is observed in heterocystous organisms (Houchins, 1984). Though reversible hydrogenase occurs in both heterocysts and vegetative cells, the specific activity in heterocyst during aerobic growth is about 4-times that in vegetative cells (Houchins and Burris, 1981a,b). The micro aerobic environment within the heterocyst probably leads to additional reversible hydrogenase synthesis.

Reversible hydrogenase also shows oxyhydrogen reaction in *Anabaena* 7120 when traces of O_2 are added, although the reaction is rapidly inactivated by O_2 levels as low as 0.1 per cent in contrast to the reaction catalyzed by uptake hydrogenase (Houchins and Burris, 1981a, b). CO_2-dependent H_2 uptake occurs at low light intensity, and higher light intensities leads to deadaptation and reversion to normal O_2-evolving photosynthesis. This reaction has been shown to be strictly light-dependent and is completely inhibited by the uncoupler CCCP, demonstrating a requirement for photophosphorylation.

Hydrogenase dependent H_2 evolution has been demonstrated in a number of species of cyanobacteria (Adams *et al.*, 1980; Asada and Kawamura, 1985; Asada *et al.*, 1987; Belkin and Padan, 1978; Benemann *et al.*, 1982; Daay *et al.*, 1979; Daday and Smith, 1987; Ewart and Smith, 1989a,b,c; Houchins, 1984; Houchins, 1981d; Howarth and Codd, 1985; Laczko, 1980; Laczko and Barabas, 1981; Llama *et al.*, 1979a,b,c,d; Spiller *et al.*, 1983; Tel-Or *et al.*, 1977; Yagi, 1976). A few organic carbon sources have been reported to enhance reversible hydrogenase activity.

7. H_2-Metabolism in Symbiotic Cyanobacteria

Symbiotic cyanobacteria have not been widely employed for studies on H_2 production or uptake. However, a few workers have made attempts to screen the capacity of H_2 production in *Anabaena azollae*, *Anabaena cycadeae* and *Nostoc* sp. (from lichen) Newton, 1976; Peters *et al.*, 1976, 1977; Millbank, 1981; Perraju *et al.*, 1986; Banerjee *et al.*, 1989; Srivastava, 1990). In intact *Azolla* plants grown on N_2, H_2 production and reduction of C_2H_2 was found light dependent and constant for 4days (Peters *et al.*, 1976). An average rate of 21.7 nmol H_2/mg Chl/min was routinely observed in *Azolla caroliniana* Willd. Under Ar + CO_2. This rate of H_2 production was almost 2-times lower than C_2H_2 reduction. H_2 production occurred under Ar but was almost completely inhibited by N_2. The inhibition by N_2 was partially overcome by 2 per cent CO. In contrast to many N_2-fixing organisms, there was no short term effect of nitrate on H_2 production. H_2 production was observed with *Azolla* founds grown for even 1 to 9 months on nitrate supplemented medium. H_2 production was found almost comparable in experiments where *Azolla* was grown under 1 per cent CO_2 in air or under 1 per cent CO_2, 20 per cent O_2 in argon.

Increase of light intensity showed parallel increase in the rate of H_2 production. Thus incubation of *Azolla* under 1200 ft-C showed the highest production of under dark. Experiments done with O_2 under dark showed no production of H_2 suggesting the presence of an "uptake hydrogenase". *Anabaena azollae* freshly isolated from *Azolla* also showed active formation of H_2. A rate of 50.4 nmoles H_2/mg Chl a/min as routinely obtained under Ar + CO_2. Addition of C_2H_2 and CO greatly stimulated H_2 formation. Overall studies conducted by Peters *et al.* (1976) are indicative of nitrogenase mediated H_2 formation by *Azolla* plants or isolated *A. azollae*. Effects of a number of other factors on H_2 production have also been reported (Newton, 1976; Peters *et al.*, 1987).

Recently, another water fern *Azolla pinnata* has been employed for studies on H_2 production by Banerjee *et al.* (1989). With this water fern, highest H_2 evolution (5.40 nmol H_2/mg fresh wt at 24 g) was observed if the fronds were incubated in Allen and Arnon's medium devoid of combined nitrogen sources. H_2 evolution was 4-5 times higher under light-anaerobic medium. Addition of NH_4Cl or KNO_3 (10 mM) did not inhibit H_2 production. H_2 evolution was significantly stimulated by the exogenous addition of phosphate. 60 ppm phosphate addition elicited 6-8 times higher H_2 formation than the control. Furthermore, there were significant differences in the rate of H_2 production between pink and green fronds of *Azolla*. Pink frond always showed higher rate of H_2 formation than the green one. Athough considerable activity (H_2 formation) was detected in the dark but addition of PS-II inhibitor completely suppressed H_2 formation. Possible presence of a soluble hydrogenase has been suggested though mainly based on the data of dark and NO_3^- grown Azolla fronds. A number of workers have recently studied H_2 evolution by immobilized *A. azollae* (Brouers and Hall, 1986; Shi *et al.*, 1987; Srivastaa, 1990). Immobilization elicited multifold stimulation of H_2 production.

Perraju *et al.* (1986) have studied H_2 evolution and uptake in *Cycas-Anabaena cycadeae* association. They demonstrated that H_2 evolution was undetectable in free-living N_2^- fixing cultures of *A cycadeae* but intact coralloid roots showed a rate of 4 μmol H_2/mg Chl a/h. Furthermore, it was found that H_2 uptake was absent in the coralloid root but free living *A. cycadeae* showed an H_2 uptake rate of 28μmol H_2/mg Chl a/h. On the other hand, detectable level of H_2 formation (1.56 nmol H_2/μg Chl a/h) in *A. cycadeae* has recently been reported by Srivastava (1990). Nevertheless, this rate is almost 2-3 fold lower tha the intact coralloid roots.

Other than *Azolla* and *Cycas*, H_2 evolution and utake studies have been performed with three N_2-fixing lichens *viz.*, *Peltigera membranacea*, *P. polydactyla* and *Lobaria pulmonaria* (Millbank, 1981). It was found that H_2 evolved concomitant with N_2 fixation was recycled by means of an uptake hydrogenease, and in general the net evolution was zero or small at 5 and 15°C. At 25°C there as appreciable net evolution of H_2 in the *Peltigera* sp. But *Lobaria* showed no net evolution. Extensive study is essential to understand the mode of H_2 metabolism in symbiotic cyanobacteria in general.

8. H_2-Production by Immobilized Cyanobacteria

Immobilization is a method of conversion of enzyme or whole cells from a water soluble mobile state to a water insoluble immobile state, by trapping them in an insert, insoluble, solid support. In recent years immobilizing action of enzymes, microbial cells and other living systems is being used extensively for the large-scale production of drugs and other compounds (Brodelius, 1985). A number of matrixes are used for immobilization *viz.*, glass beads, agar-agar, alginate etc. although the use of particular system depends upon the nature of investigation. The most common procedure for immobilizing microbes is cell entrapment in natural polymers, in particular agar, shaped into blocks, layers or beads.

Immobilization of cyanobacteria for H_2 production is preferred over liquid culture mainly due to two reasons. First cyanobacteria require continuous agitation in liquid culture to keep cultures uniformly suspended, since without it the organisms settle out and form clumps which exhibit no H_2 formation (Weissman and Benemann, 1977). Second, agitation causes filament breakage and structural degeneration of vegetative cells, both of which are associated with cessation of biophotolysis in liquid cultures.

H_2 production by immobilized cells of cyanobacteria has been studied by a few workers (Weissman and Benemann, 1977; Lambert *et al.*, 1979; Kayano *et al.*, 1981; Ochiai *et al.*, 1983; Brouers and Hall, 1986; Robinson *et al.*, 1986; Kuwada and Ohta, 1987; Shi *et al.*, 1987). In almost all the studies so far conducted, immobilization of cultures showed significant stimulation of H_2 evolution was also prolonged significantly. The immobilizing conditions have been found to vary from strains to strains. Similarly types of matrix used also show significant differences in rate as well as duration of H_2 production. For example, *Anabaena* N-7363 showed maximum hydrogen production (0.52μmole H_2 $h^{-1}g^{-1}$) (wet gel) with 2 per cent agar gel at 30°C and 300 lux light intensity (Kayano *et al.*, 1981). This rate was three times higher than that of free algae. On the other hand *Lyngbya* sp. Strain 108 showed optimum H_2 production when immobilized on 4 per cent alginate; 0.5 M $CaCl_2$ and 0.11 mg dry microbial cells ml^{-1} gel (Kuwada and Ohta, 1987).

The pH, temperature and light intensity were kept at 9.0, 30°C and 1000 lux. On the other hand *A. azollae* showed maximum yield of H_2 production when immobilized on polyvinyl (Brouers and Hall, 1986; Shi *et al.*, 1987). Higher yield with polyvinyl has been shown to be partly related to changes in membranes permeability induced by the immobilization process itself. A similar increase in yield of H_2 production was observed when membrane permeability of alginate immobilized *A. azollae* was increased by an acetone pretreatment (Brouers and Hall, 1986).

The exact reasons pertaining to the enhancement of H_2 formation following immobilization are as yet not clearly8 known (Robinson *et al.*, 1986). However, unlike free-living cultures immobilized cyanobacteria do not settle out and form clump. Moreover, cyanobacteria are protected from filament breakage and

Table 5.9: H₂ Production by Immobilized cyanobacteria

Test Organism	Matrix Used	Increase % or Fold	Duration of H_2 Production	Reference
Anaaena cylindrical	Alginate	Two-fold		Weissman and Benemann (1977)
A. cylindrical B-629	Glass beads	5 per cent	Upto 57 days	Lambert et al. (1979a)
Anabaena N-7363	Agar-agar	3-fold	Upto 7 days	Kayano et al. (1981)
Phormidium sp.	SnO₂ OTE Calcium alginate	ND*	ND	Ochiai et al. (1983)
Anabaena azollae	Polyvinyl polyurethane and alginate	2-fold	-	Brouers and Hall (1986)
Lyngbya sp. Strain 108	Calcium alginate	3.4 fold	5 days	Kuwada and Ohta (1987)
Anabaena azollae	Polyurethane polyvinyl and alginate	2-fold	ND	Shi et al. (1987)

*ND: Not Determined-measurement of photocurrent.

structural degeneration which occurs as a result of agitation. Furthermore, increase of heterocyst frequency, in immobilized cyanobacteria may also account for the increased rate of H_2 production. Increase in heterocyst frequency has been reported in the cyanobacterium immobilized on calcium alginate beads (Kerby *et al.*, 1986).

The prolonged duration of H_2 formation following immobilization has been demonstrated to be a phenomenon related with growth (Srivastava, 1990). The non-growth period is prolonged especially in agar-agar and thus results in sustained H_2 production for a longer period. A potential limitation of immobilization for an organism is the problem of light penetration through the supporting matrix. However, as cyanobacteria themselves stick to the cubes or beads, the large surface area provided would thus ensure a uniform distribution of organisms and thus light penetration. Even if lesser light penetrates, the organism will be benefited as lesser light intensity favours H_2 production (Lambert and Smith, 1981b; Kuwada and Ohta, 1987). Stimulation of H_2 production following immobilization has also been claimed to be due to oxygen tolerance. Although immobilization does not prevent the inhibitory effects of O_2 but most probably it does double the resistance of the system to a moderate level of O_2. Studies so far made no doubt reveal that immobilization improves the rate of H_2 production but if immobilization technology is to advance it seems likely that a greater understanding of the actual effects of immobilization is required. Recently, photoproduction of molecular H_2 by *Rhodospirillum rubrum* immobilized in composite agar layer/microporous membrane has been reported (Planchard *et al.*, 1989). Employing this method the H_2 evolving activity could be maintained over several months by periodically incubating the biocatalytic structures in a rich nutrient broth. It will be a pity if the above technique of immobilization is not employed for H_2 production by cyanobacterial system.

Immobilization and Development of Photochemical Fuel Cell

One of the first biological fuel cells designated to use H_2 was developed by Rohrback *et al.* (1962), who reported on the production of hydrogen from glucose by *Clostridium butyricum*. In this biological fuel cell, glucose was used as a nutrient; however, it is expensive as an energy resource. Solar energy is very attractive for energy production and Berk and Canfield (1964) reported one of the first photochemical fuel cells utilizing microorganisms. In the presence of light, *R. rubrum* produced hydrogen, which was oxidized on the surface of the anode. Later on, coupling of immobilized chloroplasts with *Clostridium butyricum* made possible water-splitting hydrogen evolution (Kayano *et al.*, 1981) and the hydrogen produced was applied to a hydrogen oxygen fuel cell. A photo-current of 0.4 - 1.5 mA was obtained for 4 h from the photochemical fuel cell system using immobilized chloroplasts and *C. butyricum*.

Using the cyanobacterium *Anabaena* No. 7363, a photochemical fuel cell system was first developed by Kayano *et al.* (1981). In this system the alga *Anabaena* N-7363 was immobilized in 2 per cent agar gel and thus produced 3-fold higher H_2 in light (10000 lux). The oxygen evolved was removed by a reactor containing aerobic bacterium *Bacillus subtilis*. The H_2 evolved was passed through soda lime and the flow was regulated by a flow meter before reaching the anode chamber of the fuel cell. The hydrogen oxygen fuel cell consisted of platinized platinum anode (10x50 cm), a porous active carbon cathode (7.5x8.0x2.5 cm) and the electrolyte (0.1 M phosphate buffer solution, pH 8.0). The anode and cathode were separated by an anion exchange membrane. The current, the anode potential and the cell voltage were measured by a millivolt-milliammeter and displayed on a recorder. A photo-current of 15-30 mA was continuously produced for 7days by the photochemical fuel cell consisting of the immobilized *Anabaena* reactor, the O_2-removing reactor and the hydrogen-oxygen fuel cell. The conversion ratio of hydrogen to current was from 80 to 100 per cent.

Ochiai *et al.* (1983) have developed semiconductor electrodes coated with living films of cyanobacteria. They have demonstrated that the intact *Phormidium* sp. Cells immobilized on SnO_2 in a light-dependent reaction. They have also demonstrated that they drying of a "wet" algal electrode at 50°C for 60 min increase photocurrent output capacity by 100-fold.

9. Hydrogen Uptake

Uptake of H_2 occurs via uptake hydrogenase and/or reversible hydrogenase in all the cyanobacteria. Hydrogen functions as an electron donor for both respiratory and light dependent photosynthetic electron flow, the precise nature of which appears to vary from species to species. The characteristics of enzymes involved in uptake of H_2 are described elsewhere (see Figures 5.1 and 5.2 and Tables 5.6–5.8).

Like nitrogenase-mediated H_2 formation, uptake of H_2 is affected by gas atmosphere, culture medium, light intensity and quality and age of the culture (Lambert and Smith, 1981b). Furthermore, the uptake is influenced by incubation conditions (Houchins, 1984). Combined nitrogen sources *viz.*, NO_3^-, NH_4^+ and amino acids completely inhibit uptake hdrogenase activity. Nickel has been demonstrated to be involved in regulating uptake hydrogenase activity (Smith *et al.*, 1985; Xiankong *et al.*, 1984). Friedrich *et al.* (1982) have reported that nickel is a constituent of soluble and particulate hydrogenase of *Alcaligenes eutrophus*. However, till now, nickel as a prosthetic group or co-factor of cyanobacterial hydrogenases has not been demonstrated by any workers (Houchins, 1984; Lambert and Smith, 1981b; Xiankong *et al.*, 1984). However, it has been suggested that nickel may be required for activation of an uptake hydrogenase, or for hydrogenase synthesis, or for synthesis of another protein which is involved in H_2 uptake (Almon and Boger, 1984; Daday and Smith, 1983; Daday *et al.*, 1985; Pederson *et al.*, M 1986; Srivastava, 1990; Xiankong *et al.*, 1984). For cyanobacteria uptake of H_2 has many possible functions, *viz.* (a) as a scavenger of O_2; (b) as a mechanism that prevents H_2 inhibition of nitrogenase; (c) in the production of ATP via H_2 oxidation (Knallgas reaction) and (d) as reductant both under reductant limiting and reductant-rich conditions. All these functions have been demonstrated in one or another cyanobacteria (Houchins, 1984). Though uptake hydrogenase does increase the efficiency of nitrogenase, it adversely affects H_2 activity.

According to Benemann *et al.* (1978) and Hallenbeck and Benemann (1979 vertical glass tubes used in experiments can be replaced by horizontal glas tubes, where gas flows over the surface of the algal culture. This will minimize the energy required to pump the gas phase through the system. There is utmost need to make trails of such system at larger scale for large scale production of H_2.

Transcriptional Regulation of Hydrogenases in Cyanobacteria

Cyanobacteria are microorganisms that can be found in very different environments such as fresh and seawater, in the soil, deserts, polar regions hot water springs, and saline environments (Ward and Castenholz 2000, Vincent 2000, Wynn-Williams 2000). They also have the capacity to form symbiotic relations with other organisms (Adams 2000).

The morphological variation of cyanobacteria is considerable. Both unicellular and filamentous forms are known, and variation -within these morphological types occurs (Rippka *et al.*, 1979). Cyanobacteria are able to perform chlorophyll *a* based, oxygenic photosynthesis using a photosynthetic apparatus similar to that of chloroplasts of algae and higher plants. Some cyanobacteria; are obligate photoautotrophs, while others can grow heterotrophicallv in the dark (Rippka *et al.*, 1979). Many cyanobacteria, both filamentous and unicellular strains, have the capacity' to fix atmospheric nitrogen, a capability only found in prokaryotic organisms. Some filamentous strains can also develop a specialized cell, the heterocyst, where nitrogen fixation takes place. The heterocyst is slightly larger and rounded compared to the vegetative cell, and evenly distributed along the filament (Wolk 1996).

In 1896 Jackson and Ellms for the first time reported that *Anabaena* species from Massachusetts resrvior immediately produce hydrogen when placed in sealed bottled. There was a considerable gap in this area of research since the first report appeared. Studies on hydrogen metabolism by blue green algae (cyanobacteria) restarted only after the unequivocal report of nitrogen fixation by this group of algae. The real hydrogen production mediated by the nitrogenase enzyme complex was first of all reported in Anabaena cylindrical under *in vitro* conditions by the Haystead *et al.* (1970). Subsequently hydrogen production by intact filaments of A cylindrical in the light under an atmosphere of argon and CO_2 was reported by Benemann and Weare (1974a,b). In the same year Russian workers (Oshchepkov *et al.*, 1974)have also reported H_2 production from cyanobacteria. Since then extensive work has beendone on various aspects of H_2 metabolism employing a variety of blue green algal strians (Tamagni *et al.*, 2002; Vyas 1992;Vyas and Kumar 1995;Vyas and Gupta 2003; Vyas 2004)

Enzymes directly involved in cyanobacterial hydrogen metabolism. Three enzymes have been described to be. directly involved in hydrogen metabolism in cyanobacteria; (1) nitrogenase(s), catalysing the production of molecular hudrogen concomitantlv with the reduction of nitrogen to ammonia, (2) an uptake hydrogenase, catalysing the consumption of hydrogen produced bv the nitrogenase, (3) a bidirectional hydrogenase., that has the. capacity to take up and produce hydrogen. The three enzymes can be distributed in different ways. However, multiple copies of single hydrogenases have not been found so far in any cyanobacterium (Wunschiers and Lindblad 2003).

Several nitrogen-fixing strains have all three enzymes, *e.g.* the two filamentous, heterocystous strains *Anabaena* PCC 7120 (Houchins and Burns 1981b,Kaneko *et al.*, 2001) and *Anabaena variabilis* ATCC 29413 (Happe *et al.*, 2000, Schmitz *et al.*, 1995). Others have only an uptake hydrogenase and a nitrogenase, *e.g.* the filamentous, heterocystous strain *Nostoc punctiforme* (also *Nostoc* PCC 73102, *Nostoc* ATCC 29133) (Tamagnini *et al.*, 1997, Meeks *et al.*, 2001) while only a bidirectional hydrogenase is present in *e.g.* in the unicellular strain Synechocystis PCC 6803 (Kaneko *et al.*, 1996, Appel and Schulz 1998).

An uptake hydrogenase is consistently present when a cyanobacterium has the capacity to fix atmospheric nitrogen. An exception might be the unicellular *Synechococcus* PCC 6301 (also *Anacystis nidulans*). It has been reported that it mav possess an uptake hvdrogenase despite it is a non-nitrogen fixing strain(Boison *et al.*, 1996). The distribut.ion of the bidirectional hydrogenase is not as clear as in the case of the uptake hydrogenase and it is missing in several strains (Tamagnini *et al.*, 1997, Tamagnini *et al.*, 2000).

All hydrogenases identified in cyanobacteria belong to [NiFe]-hydrogenases. However, no cyanobacteria] hydrogenase has yet been purified and crystallised.

Nitrogen Fixation and Nitrogenase

Nitrogenase is an enzyme that is directly involved in hydrogen metabolism as a consequence of its production of molecular hydrogen during nitrogen fixation.

The overall reaction of nitrogen fixation can be written as follows:

$N_2 + 8H^- + 16\ ATP — 2NH_3 + H_2 + 16\ ADP + 16P.$

The above reaction is catalysed bv nitrogenase, which consists of two separate protein components: dinitrogenase and dinitrogenase reductase. The dinitrogenase is a heterotetramer consisting of two subunits of NifK and two subanits of NifD. The dinitrogenase reductase is a homodimer of NifH and plays a role in transferring electrons from a ferredoxin, or a flavodoxin, to the dinitrogenase, where the actual reduction of N_2 occurs (Howard and Rees 1996). The reaction requires ATP and low-potential electrons. ATP may be generated by either cyclic photophosporylarion or oxidative phosporylation and the low potential electrons can come from NADPH that may be generated from the degeneration of carbohydrates produced during photosynthesis (Haselkorn and Buikema, 1992). In hetcrocystous cyanobacteria, carbohydrates are imported from vegetative cells and products of nitrogen fixation arc exported to the vegetative cells (Bohme1998). Nitrogen fixation is oxygen sensitive since nitrogenase becomes inactivated by O_2, (Picnkos *et al.*, 1983). Since oxygcn is produced by photosystcm II during photosynthesis, these two processes must be separated either temporally or spatially by cyanobacteria. Heterocystous cyanobacteria separate the oxygen evolution and nitrogen fixation spatially by performing photosynthesis in the vegetative cells and nitrogen fixation in the hetcrocysts. To provide an environment with low oxygen the heterocyst lacks photosystem II activity, is surrounded by a thickened cell wall to reduce the diffusion of oxygen, and has a higher respiration rate (Bohme 1998).

Nitrogenase is subject to strict regulatory controls. Nitrogen fixation is not only inhibited by O_2 but also by ammonium and nitrate (Halbleib and Ludden 2000). A key protein in the control of nitrogen metabolism in cyanobacteria is NtcA, a transcriptional regulator that belongs to the CAP family (Herrero *et al.*, 2001). NtcA is present in both unicellular and filamentous, hetcrocystous cyanobacteria (Frias *et al.*, 1993, Herrero *et al.*, 2001), and has been demonstrated to be essential for heterocyst development (Wei *et al.*, 1994). NtcA binds to a palindromic target motif GTA (N8) TAG upstream of the target gene. Studies of the motif in *Synechococcus* PCC 7942 demonstrated that some variation in the sequence was possible while still maintaining the binding capacity. However, GT(N10)AC was found to be essential for binding NtcA (Vazquez-Bermudez *et al.*, 2002). Examples of known genes regulated by NtcA are *gln*A (glutamine synthetase), nir (nitrate assimilation), NtcA (autoregulatory), and genes important ' for hetcrocyst development, *e.g.* het C and devBC*A* (Herrero *et al.*, 2001). In addition, NtcA has also been demonstrated to bind upstream of xis*A* which is a site-specific recombinase responsible for the excision of an 11.5 kb DNA element located within nifD in *Anabaena* PCC 7120 (Ramasubramanian *et al.*, 1994).

Cyanobacterial Uptake Hydrogenase, Function and Activity

All known cyanobacterial uptake hydrogenascs consist of two subunits, encoded by hupS and hupL, respectively (Figure 5.1). The small subunit, HupS, contains the iron-sulphur [Fe-S] cluster necessarv for electron transfer to the active site, which is located in the large subunit, HupL. HupS also contains cysteins that

The physiological function of the uptake hydrogenase in cvanobactoria is to catalyse the consumption of hydrogen produced by nitrogenase (Happe et al. 2000, Lindberg *et al.*, 2002, Houchins 1984, Oxelfelt *et al.*, 1995, Troshina *et al.*, 1996). A strong correlation between the activities of uptake hydrogenase and nitrogenase has been demonstrated in filamentous cyanobacteria. It is believed that the electrons derived from the hydrogen oxidation catalysed by the uptake hydrogenase recombine, through the respiratory chain, with oxygen in the oxyhydrogen reaction to form water. The advantage would be that energy from the H_2 produced by nitrogenase can be recaptured and oxygen will be consumed and protect nitrogenase from oxygen. The recycling of hydrogen would also supply reducing equivalents to nitrogenase and other cell functions (Wunschiers and Lindblad 2003). However, the uptake hydrogenase is not essential for diazotrophic growth (Happe *et al.*, 2000, Lindberg *et al.*, 2002, Masukawa *et al.*, 2002). Inactivation of the uptake hydrogenase in *Anabaena variabilis* resulted in a lower rate of nitrogen fixation and slightly reduced growth rate compared to the wild type (Happe *et al.*, 2000).

In hetcrocystous cyanobacteria the uptake hydrogenase is located in the hetcrocyst with no activity in the vegetative cells (Peterson and Wolk 1978, Houchins and Burris 1981a, Houchins 1984, Carrasco *et al.*, 1995). The enzyme has been suggested to be membrane bound, located in the thylakoid (Eisbrenner *et al.*, 1978) or in the cytoplasmic membrane (Houchins and Burns 1981b). Since all identified cyanobacterial HupS arc missing the N-terminal signal peptide important for membrane translocation, it has been suggested that the uptake hydrogenase is

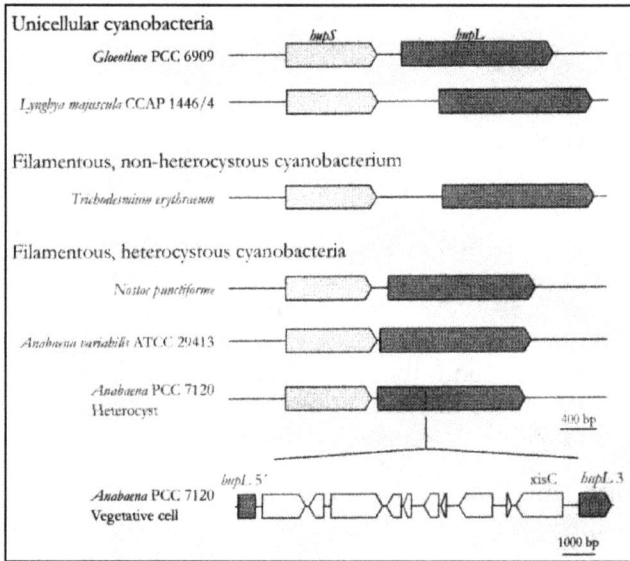

Figure 5.1: Physical map of the genes encoding the uptake hydrogenase (hupSL) in unicellular.filamentous non heterocystous, and filamentous heterocystous cyanobacteria. All genes are contigous except hupL in *Anabaena* PCC 7120. In vegetative cells, a 10.5 kb DNA fragment is located -within hupL. However, this DNA fragment is erased by the site specific recombinase which is XisC located within the fragment. *Gloeothece* PCC 6909 (GenBank AY260103), *Lyngbya majusula* CCAP 1446/4 (GenBank AF368526), *Trichodesmium erythraeum* (http://spider.jgi-psf.org/ JGI_microbial/html/trichodesmium/trichod-homepage. html), *Nostoc punctiforme* (Oxelfelt *et al.*, 1998), *Anabnena variabilis* ATCC 29413 (Happe *et al.*, 2001), and *Anabaena* PCC 7120 (Carrasco *et al.*, 1995) are involved in the coordination of the {Fe-S] clusters. The small subunit in cyanobacteria differs from other small subunits from other microorganisms since the signal peptide in the N-terminus part of HupS is missing (Oxelfelt *et al.*, 1998). HupL contains two putative nickel-binding sites (R x CG x C) necessary for the coordination of the nickel in the active site (Carrasco *et al.*, 1995, Oxelfelt *et al.*, 1998, Happe *et al.*, 2000, Wunschiers and Lindblad, 2003).

located on the cytoplasmic side of the thylakoid or cytoplasm membrane (Appel and Schultz 1998).

The activity of the uptake hydrogenase has been demonstrated to be influenced by different external factors such as nickel, molecular hydrogen, carbon and nitrogen. Addition of extra nickel resulted in an increased hydrogen uptake activity' in *Nostoc punctiforme*, *Anabaena varaiabilis Oscillatorria subbrevis*,and *Anabaena* strains CA and 1 F (Xiankong *et al.*, 1984, Daday *et al.*, 1985, Kumar and Polasa 1991, Oxelfet *et al.*, 1995). However, a nickel concentration above 10 um does not stimulate the hydrogen uptake (Kumar and Polasa 1991, Oxelfelt *et al.*, 1995). A direct dependence of nickel was demonstrated for the induction of the uptake hydrogenase in *Anabaena* strains Cl and IF (Xiankong *et al.*, 1984) and *Anabaena cylindrica* (Daday *et al.*, 1985).

Molecular hydrogen has also been demonstrated to induce higher uptake activities in cyanobacteria. Studies on Anabaena PCC 7120 (Houchins and Burn's

1981a), *Nostoc punctiforme* (Oxelfelt *et al.*, 1995), *Anabaena cylindrica* and *Nostoc muscorum* (Eisbrenner *et al.*, 1978) demonstrated an increase of the hydrogen uptake activity when a fraction of the air was replaced bv molecular hydrogen. However, in *Anabaena variabilis* only a slight stimulatory effect on the hydrogen uptake activity' was observed when exogenous hydrogen was added. It was suggested that the hydrogen produced from the nitrogenase is sufficient for hydrogenase induction (Troshina *et al.*, 1996).

A stimulation of the hydrogen uptake activity, together with nitrogenase activity", could be observed in *Nostoc punctiforme* when organic carbon was added to the medium (Oxelfelt *et al.*, 1995). However, no effect could be observed in *Anabaena variabilis* after addition of carbon (Troshina *et al.*, 1996). Addition of ammonium decreased the activity' of both the uptake hydrogenase and the nitrogenase in *Nostoc punctiforme* (Oxelfelt *et al.*, 1995). A similar observation was made in *Anabaena variabilis* (Troshina *et al.*, 1996).

Transcription of the Genes Encoding the Uptake Hydrogenase

In the two cyanobacteria examined, hupS and hupL are located on a single transcript containing no additional ORFs (Happe *et al.*, 2000, Lindberg *et al.*, 2000). The size of the transcript was determined by Northern blotting in *Anabaena variabilis* ATCC 29413 to be approximately 2.7 kb (Happe *et al.*, 2000). In *Nostoc punctiforme*, hupSL was shown to be a transcript unit and a putative transcriptional terminator could be identified downstream of hupL. The intergenic region contains 7 bp repeats putatively forming a hairpin structure. The function of a hairpin formation in cyanobacterial hupSL is unknown but may be involved in transcript stability' or translational coupling between the structural genes (Lindberg *et al.*, 2000). Studies on the localisation of the hupSL transcript in *Anabaena* PCC 7120 demonstrated the presence, in the heterocysts only (Carrasco *et al.*, 1995). No hupSL-transcript could be detected in vegetative cells of *Anabaena variabilis* using either Northern blot or RT-PCR (Happe *et al.*, 2000). However, another study was able to detect a low level of hupSL,transcript in vegetative cells from a nitrogen-fixing culture of *Anabaena variabilis*. The authors suggested that this could be a result of a basal activity' of the hupSI~ promotor not necessarily resulting in translation (Boison *et al.*, 2000). A low H_2 uptake in ammonia grown cells of *Anabaena variabilis*, an activity thought to be due to the bidirectional hydrogenase has been reported (Troshina *et al.*, 1996). In addition, *Anabaena variabilis* contains an alternative nitrogenase expressed in vegetative cells under anaerobic conditions (Thiel *et a.l* 1995, 2001). In the study where a hupSL transcript was detected in vegetative cells (Boison *et al.*, 2000), no investigations of *e.g.* the presence of an alternative nitrogenase or uptake hydrogenase were performed. Moreover, there is no data on the regulation of the transcription of hupSL in the vegetative cells of *Anabaena variabilis* during anaerobic conditions.

Transfer of non-nitrogen fixing vegetative cells to nitrogen-fixing conditions induces a hupL transcript in Anabama PCC 7120 and *Anabaena variabilis* (Carrasco *et al.*, 1995, Happe *et al.*, 2000, Boison *et al.*, 2000). Prior to expression of hupSL in Anabaena PCC 7120, a 10.5 kb DMA fragment is excised from within hupL. In *Anabaena* PCC 7120, two additional gene rearrangements occur (Golden *et al.*, 1985).

Each excision requires a site-specific recombinase. XisC is responsible for the excision of the 10.5 kb in hupL and its gene is located on the excised fragment. Studies of the upstream region of another site-specific recombinase, xisA, reveaed a binding site of the global nitrogen regulator NtcA. However, no obvious NtcA binding site could be detected upstream of xisC (Carrasco *et al.*, 1995). In contrast to *Anabaena* PCC 7120, no programmed rearrangement occurs in *Anabaena variabilis* ATCC 29413 or *Nostoc punctiforme* (Happe *et al.*, 2000 and Oxelfelt *et al.*, 1998). As in *Anabaena* PCC 7120, a hupL transcript was not detected in non-nitrogen fixing cells but was induced during nitrogen fixing conditions (Happe *et al.*, 2000).

The transcription start site of hupSL has been determined in *Anabaena variabilis* ATCC 29413 and *Nostoc punctiforme* using primer extension and 5 'RACE, respectively (Happe *et al.*, 2000, Lindberg *et al.*, 2000). The transcription start of hupS.L in *Anabaena variabilis* ATCC 29413 was located 103 bp upstream of the translation start site. One half of a putative Fnr binding site was found 144 bp upstream of the translarionat start site (Happe *et al.*, 2000). Fnr is a transcription factor playing a major role in altering the gene expression between aerobic and anaerobic conditions to facilitate changes in energy metabolism (Kiley and Beinert 1999). The transcription start in *Nostoc punctiforme* was located 259 hp upstrearn of the translation start. The promotor region has putative binding sites for NtcA and integration host factor (IHF). NtcA is the global nitrogen regulator in cyanobacteria and IHF is a DNA-binding protein consisting of two subunits that upon binding creates a sharp (more than 160°) bend. This allows proteins that bind further upstream to interact with the promotor region. In addition, IF-IP can also act directly as a represser or activator (Wagner, 2000).

Cyanobacterial bidirectional hydrogenase, function and activity Initially, the bidirectional hydrogenase was suggested to be an enzyme with four subunits, consisting of a diaphorase part, HoxFU, and a hydrogenase part, HoxYH. These subunits are homologous to the heterotetrameric NAD'- reducing hvdrogenase of *Raslastonia eutropha* (Schmitz *et al.*, 1995). Recently, it was suggested that a fifth subunit, HoxE, belongs to the diaphorase part of the bidirectional hydrogenase. Thus, the cyanobacterial bidirectional hydrogenase is encoded by hoxEFUYH (Schmitz *et al.*, 2002).

The physiological function of the bidirectional hydrogenase in cyanobacteria is not completely clear. The bidirectional hydrogenase has been suggested to act as an electron valve during photosynthesis in *Synechocystis* PCC 6803. Inactivation of the bidirectional hydrogenase resulted in a higher fluorescence of photosystem II compared to the wild type (Appel *et al.*, 2000). The enzyme has also been proposed to play a role in fermentation by functioning as a mediator in the release of excess reducing power during anaerobic conditions (Stal and Moezelaar 1997, Troshina *et al.*, 2002). In addition, it has been suggested that the bidirectional hydrogenase is part of the respiratory complex I (Appel and Schulz 1996, Schmitz and Bothe 1996). In cyanobacteria, only 11 subunits out of 14 conserved subunits of a prokaryotic complex I have been identified. Some of the subunits of the bidirectional hydrogenase show sequence similarities with the missing subunits of the respiratory complex I (Schmitz *et al.*, 1995). However, the bidirectional hydrogenase has been demonstrated to be absent from several cyanobacterial strains (Tamagnini *et al.*, 1997, Tamagnini *et al.*,

2000) and studies of the respiration of *Nostoc punctiforme* a strain naturally lacking the bidirectional hydrogenase (Tamagnini *et al.*, 1997), demonstrated rates of respiration comparable to cyanobacteria having a bidirectional hydrogenase (Boison *et al.*, 1999), In addition, mutants of hoxU in *Synchocystis* PCC 6301(Boison *et al.*, 1998) and hoxF in *Synechocystis* PCC 6803 showed non-impaired respirator 02 uptake whilst being affected in H_2 evolution (Howitt and Vermaas 1997). In general, it seems that the bidirectional hydrogenase does not play an essential role in those strains where it is present. Inactivation of hoxH in *Synechocystis* PCC 6803 and *Anabaena* PCC 7120 resulted in a small decrease in growth rate compared to the wild type (Appel *et al.*, 2000, Masukawa *et al.*, 2002).

The bidirectional hydrogenase is present in both vegetative cells and in heterocysts (Hallenbeck and Beneman 1978, Houchins and Burris 1981 a). In *Anabaena* PCC 7120 the bidirectional hydrogenase appeared in the soluble fraction after cell disruption and it was suggested that the enzyme is soluble (Houchins and Burris 1981b, Hallenbeck and Beneman 1978). However, investigations in other cyanobactcria suggest an association with cell membranes. In *Anabaena variabilis* and *Synechocystis* PCC 6803, an association with the thylakoid membrane was demonstrated (Screbryakova *et al.*, 1994, Appel *et al.*, 2000). However, based on immunological data, an association with the cytoplasmic membrane in *Synechocystis* PCC 6301 has been suggested (Kentemich *et al.*, 1989).

The activity of the bidirectional hydrogenase has been examined in both unicellular and filamentous cyanobactcria. *in vivo*, NADH supports H_2 evolution in *Synechocystis* PCC 6301. NADPH also supports H_2 evolution but with less than 50 per cent of the activity obtained using NADH. For H_2 uptake, NAD- is the preferred electron acceptor (Schmitz and Bothe 1996).

The activity of the bidirectional hydrogenase has in several studies been demonstrated to be induced by anaerobic conditions (Schmitz and Bothe 1996, Screbryakova *et al.*, 1994, Houchins and Burris 1981a). The bidirectional hydrogenase in *Anabaena* PCC 7120 is active in both vegetative cells and in heterocysts in aerobically grown filaments, with a several-fold higher activity in hetcrocysts. Transferred to anaerobic conditions, the activity of the bidirectional hydrogenase increased with about two orders of magnitude with approximately the same activities in both cell types (Houchins and Burris 1981a). Similar results have been observed in *Anabaena variabilis* (Screbryakova *et al.*, 1994). In contrast to the filamentous cyanobacteria, the activity of the bidirectiona.l hydrogenase in the unicellular *Gloeocapsa alpicola* is not directly dependent on oxygen. Higher activity is observed under nitrogen starvation and low light, and it was suggested that the bidirectional hydrogenase could act as an alternative electron donor to photosystem I after inactivation of photosystem II due to nitrogen starvation. Under dark anoxic conditions the unicellular cyanobacterium *Gloeocapsa alpicola* produces H_2 catalysed by the bidirectional hydrogenase (Troshina *et al.*, 2002). In addition, the unicellular strain *Chroocociidiopsis thermalis* contains a bidirectional hydrogenase with some catalytic properties more similar to an uptake hydrogenase. It is not inducible under anaerobic conditions or under nitrate starving conditions (Screbryakova *et al.*, 2000).

In contrast to the uptake hydrogenase, the bidirectional, hydrogenase in *Anabaena* PCC 7120 did not respond to added H_2 in aerobically grown cells (Houchins and Burns, 1981a).

Transcription of the genes encoding the bidirectional hydrogenase "The genes encoding the bidirectional hydrogenase, hoxEFUYH, in cyanobacteria are organised in a similar way in many strains. In some strains the genes are not adjacent and must thus be located on at least two operons. It is also possible to identify ORFs that are located between the hox-genes (Figure 5.2).

The information about the transcription and regulation of the hox-genes is limited in cyanobacteria. Transcript(s) of the bidirectional hydrogenase is present in both vegetative cells and heterocysts under nitrogen-fixing conditions and in vegetative cells during non-nitrogen fixing conditions *Anabaena variabilis* ATCC 29413. In addition, hoxFUYH were shown to be transcribed as a transcript unit together with two ORFs with unknown function. These experiments were performed using RT-PCR and do not exclude additional promotors within the operon (Boison *et al.*, 2000). The hox-genes in the unicellular *Synechocystis* PCC 6301 are located on two different transcripts. hoxEF form one transcript and at least 16 kb downstream of hoxF and hoxUYH is forming a second transcript together with hoxW, hypA and hypB (Boison *et al.*, 2000). In *Synechococcus* PCC 7942, hoxEFand hoxUYHW are located on two different transcripts. Using real time PCR and reporter gene

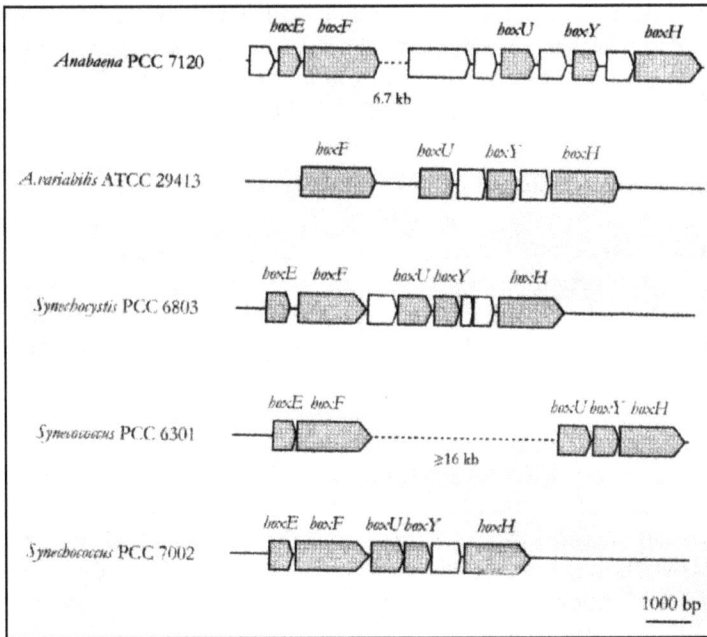

Figure 5.2: Physical map of the genes encoding the bidirectionn.! hydrogenase (hox E'FUYH in cyanobacteria. *Anabaena* PCC 7120 (Kaneko et at 2001), *Anabaena variabilis* ATC(29413 (Schmitz *et al.*, 1995), *Synechocystis* PCC 6803 (Appel and Schulz 1996), *Synechococcus* PCC 6301 (also Anacystis nidulans) (Boison *et al.*, 1996), and Syrnchocystis PCC 700 (GcnBank AF381045).

constructs, it was suggested that a second promotor might be present between hoxH and hoxW. It was also demonstrated that the hox- genes had a circadian clock expression (Schmitz *et al.*, 2001).

Very few regulatory studies have been performed on the transcriptional regulation of the hox-genes in cyanobacteria. Studies of the transcription of hoxY and hoxW *Gloeocapsa alpicola* CALU 743 during nitrogen-limiting growth conditions demonstrated an increase in the enzyme activity but no regulation on the transcriptional level (Sheremetieva *et al.*, 2002). In contrast, transfer to a low level of oxygen in Anabaena variabilis induced both the enzyme activity as well a: the relative amount of hoxH (Sheremerieva *et al.*, 2002).

Maturation of Hydrogenases

"The maturartion of hydrogenases is a complex process in which severalt proteins, encoded by *e.g.* the hyp-genes, are involved. The respective gene involved in the maturation of (NiFe)- hydrogenases have been identified sequenced, and characterized but the corresponding genes for the maturation of [Fe]-hydrogenases. remain to be identified. Very little is known in cyanobacteria, most of the knowledge about hydrogenase maturation is from other microorganisms such as *Eschrichia coli* (Lutz *et al.*, 1991), *Ralastonia eutropha* (Wolf *et al.*, 1998), *Bradirhizobium* japonicum (Olson and Maier 1997), and Rhizobium *leguminosarum* (Rey *et al.*, 1993). Homologues of the hyp-genes have been identified in cyanobacteria but their role in maturation of the hydrogenases remains to be demonstrated. The Hyp-proteins are involved in the insertion of Ni, Fe, and the ligands, CO and CN, into the active site of the large, subunit. Other genes involved in the maturation of the hydrogenases, not belonging to the hyp-genes, encode endopeptidases, which are responsible for a specific cleavage of a C- terminal part of the large subunit of the hydrogenase.

One of the earliest steps in the maturation is the formation of the complex between HypC and the large subunit of the hydrogenase. HypC is assumed to act as a chaperone, maintaining the large subunit in a conformation accessible for metal. insertion. A hypC mutant results in a metal free hydrogenase. The next step is the insertion of the ligands CN and CO into the large subunit. This is performed by HypF and of HypE (Reissman *et al.*, 2003). HypF has been shown bv mutational studies to be absolutely required for hydrogenase maturation, It has been demonstrated that the CN and CO ligands are derived from carbarnoylphosphate. Proteins catalysing O-carbamoylations contain a sequence motif (VxHHxAH) that is also found in HypF. In addition, HypF contains two zinc finger motifs. It is possible that HypF interacts with the large subunit and that the acyl-phosphatase and the carbonoylphosphate domains synthesisc and insert the ligands in the active site. Using a two-hybrid method, it was shown that HypF and HypE interact in *Helocobacter pylori* (Rain *et al.*, 2001). Inactivation of HypE results in a non-mature hydrogenasc, so HypE is also essential for hydrogenase maturation. HypB is suggested to be the main contributor of insertion of nickel.

Deletions of hypB leads to nickel free hydrogenase precursors and an inactive hydrogenase in *E.coli* (Maier *et al.*, 1993). In *R. leguminosarum* (Rey *et al.*, 1994), *B. japonicum* (Olson *et al.*, 1997), HypB has also been shown to have a function in nickel storage., as a result of histidine rich domains in the amino terminus. The function of HypA is to cooperate with HypB during nickel insertion (Hube *et al.*, 2002). In *H. pylori* was demonstrated that HypA and HypB form a heterodimer (Mehta *et al.*, 2003). The role of HypD is unclear and remains to be identified. However, the protein has been demonstrated to form a complex with HupC in *E. coli* (Blokesch and Bock 2002).

The last identified step in the naturation of the large subunit is the proteolytic cleavage of the C-terminus. Recently, two ORFs putarively encoding hydrogenase specific endopeptidases, HoxW and HupW, were identified in.Anabaena PCC 7120. It was suggested that they are responsible for the specific proteolytic cleavage of the C-terminal part of the bidirectional hydrogenase and uptake hydrogenase, respectively. Putative endopeptidases were also found in the unicellular *Synechocystis* PCC 6803, containing only the bidirectional hydrogenase, and the filamentous, hetcrocystous *Nostoc punctiforme*, containing the uptake hydrogenase onlv. Only one, though specific, putative endopeptidase was found in each of these two strains supporting the hypothesis that each hydrogenase has a specific endopeptidase (Wunschiers *et al.*, 2003).

Chapter 6

Role of Fe-Hydrogenase in Biological Hydrogen Production

The potential use of microorganisms for biological production of hydrogen as a future energy resource makes hydrogen metabolism an emerging field of research. Hydrogenase (H_2ase) is the name given to the family of enzymes that catalyse the reversible oxidation of hydrogen into its elementary particle constituents, two protons (H^+) and two electrons:

$$2H^+ + 2e^- \text{ v } H_2.$$

In the light of this reaction, it is reasonable to postulate the bacterial production of hydrogen as a device for disposal of electrons released in metabolic oxidations through the activity of hydrogenases. These are a heterogeneous group of enzymes with different sizes, subunit compositions, metal contents and cellular localizations. On the basis of metal content of catalytic subunit, H_2ase can be grouped into two non-homologous classes – those containing only Fe at the active site, called Fe-H_2ase and those with Ni, Fe and sometimes Se, [Ni–Fe] H_2ase and [Ni–Fe–Se] H_2ase (Malki et al., 1195; Adams and Stielfel 1998). Initially Fe-H_2ase was presumed to be present in a limited number of bacteria and anaerobic living protozoa (Horner et al., 2002). Subsequently, it was revealed that its distribution in eukaryotes is also quite significant. Genes bearing signatures of Fe-H_2ase are found not only in prokaryotes and lower eukaryotes, but also in the genome of higher eukaryotes like mammals, although the physiological activity of these proteins is yet to be found out.

The presence of the enzyme Fe-H_2ase in bacteria has been known for over 70 years (Vignais et al., 2001). The requirement of Fe for its activity was discovered (Peck et al., 1956) in 1950s. The function of the cytoplasmic enzyme is to remove excess reducing equivalents during microbial fermentation and that of the periplasmic enzymes in hydrogen oxidation (Graf *et al.*, 1981). The highly reactive nature of Fe-hydrogenase enzymes is evidenced by their extremely high turnover numbers, 6000 s^{-1} for *Clostridium pasteurianum* and 9000 s^{-1} for *Desulfovibrio* sp. These are almost 1000 times higher than the turnover number of nitrogenases (Benemann and Hallenbeck 2002). The present article updates the role of Fe-hydrogenase in the biological production of molecular hydrogen. Attempts have also been made to highlight some of the salient aspects of classification, structural diversity and biochemical assay of Fe-H$_2$ase and its limitations.

Hydrogen Metabolism

Hydrogen metabolism is probably one of the most fundamental processes of living systems. Transfer of electrons to hydrogenase with subsequent production of hydrogen is intimately related with the primary energy metabolism in different microorganisms. Formation of hydrogen occurs as a major physiological process in some hydrogenase containing organisms, which display anaerobic energy metabolism (Benemann 1996; Nandi and Sengupta 1998). Salient metabolism patterns of different groups of hydrogen-producing organisms have been depicted in Table 6.1. In aerobic metabolism, electrons from substrate oxidation are transferred to oxygen as the ultimate oxidant. But in case of anaerobic metabolism, electrons released from anoxygenic catabolism are used by different terminal oxidants such as nitrate, sulphate and organic compounds derived from carbohydrates (Roy Chowdhury *et al.*, 1988).

Table 6.1: Energy Metabolism Patterns in H$_2$-Producing organisms

Representative Organism	Salient Metabolism Pattern	Energy Source (anaerobic)
Strict anaerobes	Disposal of electrons from energy-yielding oxidations	Fermentation
	Extensive accumulation of reduced organic products Cell yields relatively low	
Facultative anaerobes	Promotion of energy-yielding oxidations through removal of formate Close coupling of H$_2$ evolution with energy-yielding reaction	Fermentation
Photosynthetic organisms Purple non-sulphur bacteria Purple sulphur bacteria Anerobically adapted algae	Cell yields remarkably high Reduced organic byproducts not produced in significant amounts Reduced pyridine nucleotide is the electron donor	Light

Several physiological groups of bacteria can evolve hydrogen under anaerobic conditions. These include fermentative bacteria, sulphate-reducing bacteria and phototrophic bacteria. Under anaerobiosis, fermentative metabolism predominates over others. Most purple and green bacteria can produce hydrogen at high rates either in the dark or in the light. Photoproduction of H$_2$ by the phototrophic bacteria is a nitro-genase-dependent reaction. H$_2$ is only produced under anoxic condition when nitrogen source is limited (Greenbaum 1982; Weare and Shanmugam 1976; Das and Veziroglu 2001; Benemann *et al.*, 1973). Nitrogenase is responsible for nitrogen-fixation (Benemann *et al.*, 1973) and is distributed mainly among prokaryotes, including cyanobacteria (Hall *et al.*, 1995). Nitrogenase-catalysed hydrogen production occurs as a side reaction at a rate of one-third to one-fourth that of nitrogen fixation, even in a 100 per cent nitrogen gas atmosphere. Molecular nitrogen is reduced to ammonia via an irreversible reaction with consumption of electrons released by ferredoxin and ATP:

$N_2 + 8H^+ + 2e^-$ ® $2NH_3 + H_2$,

12ATP

12ADP + 12Pi.

However, nitrogenase catalyses proton reduction in the absence of nitrogen gas with concomitant utilization of ATP.

$2H^+ + 2e^-$ ® H_2,

4ATP

4ADP + 4Pi.

Under normal growth conditions, cyanobacteria (bluegreen algae) undergo photosynthesis (as in plants) in the light and can use water as the primary electron donor (Tamagnini *et al.*, 2002). Both cyanobacteria and green plants have two photosystems. Production of H_2 in these bacteria is solely dependent on nitrogenase. But heterocystous cyanobacteria (*e.g. Anabena* sp. and *Nostoc muscorum*) can produce H_2 in the light without the interference of CO_2 fixation, O_2 evolution and sugar synthesis (Gfeller and Gibbs 1984). In all nitrogen-fixing bacteria, nitrogenase is accompanied with hydrogenase. Benemann and Weare (1974) have demonstrated that a nitrogen-fixing cyanobacterium, *Anabaena cylindrica*, can produce hydrogen and oxygen gas simultaneously in an argon atmosphere for several hours. Such group of bacteria contains two major proteins, hydrogenase and ferredoxin in addition to nitrogenase, to shuttle electrons from the photosynthetic membranes to the hydrogenase (Figure 6.1). However, the efficiency of the process is lowered by high ATP requirement of nitrogenase in these bacteria (Greenbaum 1988; Melis and Happe 2001). Green algae and cyanobacteria both possess reversible hydrogenase with low energy requirements and the manner in which they produce hydrogen is called direct biophotolysis (Gibbs *et al.*, 1986; Wunschiers *et al.*, 2001). The reductant produced from the photosystems is transferred through reduced ferredoxin to hydrogenase.

Classifications of H_2ases

The first isolated and characterized H_2ase was found to be monomeric Fe–S protein (Chen and Mortenson 1974). Initially the classification was only based on the identity of specific electron donors and acceptors, quaternary structures, size and so on. According to the protein sequence homology of thirty sequenced microbial H_2ases, the enzymes can be classified into five groups:

1. The first group contains H_2 uptake membrane bound [Ni–Fe]-H_2ases from aerobic, anaerobic and facultative anaerobic bacteria (Harry *et al.*, 1978).

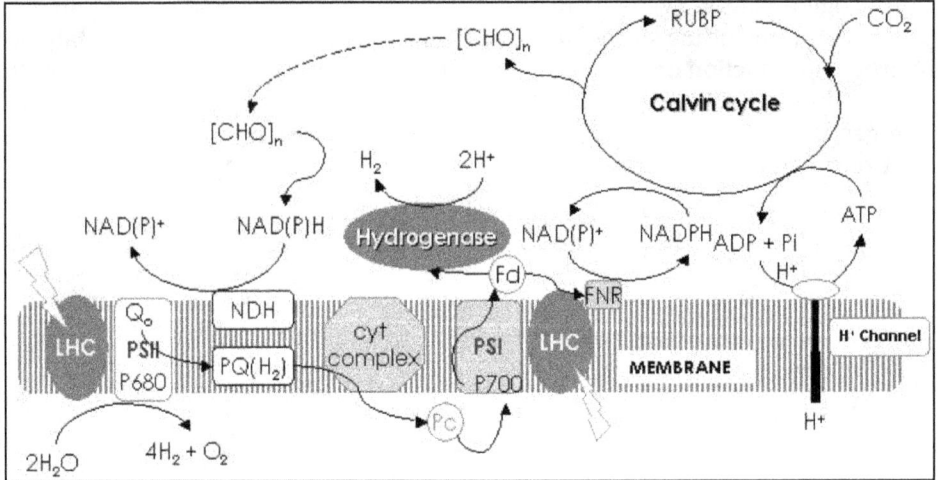

Figure 6.1: Nitrogenase-Mediated Hydrogen Production in Cyanobacteria (Benemann and Weare 1974).

2. The second group comprises membrane bound H_2 uptake [Ni–Fe–Se]-H_2ase from sulphate-reducing bacteria (Gracin *et al.*, 1999).

3. The third consists of periplasmic Fe-H_2ase from strict anaerobic bacteria mainly responsible for hydrogen evolution (Chen and Mortenson 1974).

4. The fourth contains methyl viologen factor [F-420] or NAD-reducing and soluble H_2ases from *Methanobacteria* and *Alcaligenes* (Buurman *et al.*, 2000).

5. The fifth is the labile H_2ases isoenzyme of *Escherichia coli* (Stoker *et al.*, 1988).

The results of the sequence comparisons reveal that these H_2ases share some common sequences. Those of classes (i), (ii), (iv) and (v) are homologous and share the same evolutionary origin. Despite increasingly conspicuous diversity in many respects, H_2ase can be classified broadly into three distinct classes: Ni–Fe H_2ase, Fe-H_2ase and metal-free H_2ase.

The vast majority of known H_2ases belong to the first class (Table 6.2) and over hundreds of these enzymes have been characterized genetically and/or biochemically (Buurman *et al.*, 2000; Stoker *et al.*, 1988; Volbeda *et al.*, 1999). Metal content as well as sequence similarity is a reliable classification criterion. Each of these classes is characterized by a distinctive functional core and is conserved within each group. Such phylogenetically independent classes of H_2ases get support from X-ray crystallography at least in the cases of [Ni–Fe]-H_2ases (Volbeda *et al.*, 1999) and Fe-H_2ases (Peters *et al.*, 1998). Metal-free H_2ase (Buurman *et al.*, 2000) is found to be present in some methanogens (Buurman *et al.*, 2000) but the paucity of sequence data for metal-free H_2ase fails to give detailed information on the enzyme.

Table 6.2: Classification of H$_2$ases

Features Classification Occurrence/source Structure Localization Function

Ni–Fe-H$_2$ase Ni–Fe–Se-H$_2$ase Fe-H$_2$ase

Metal-free H$_2$ase Anaerobic, photosynthetic bacteria, cyanobacteria

Sulphate-reducing bacteria, methanogenes

Photosynthetic bacteria, anaerobic fermentative bacteria, cyanobacteria, green algae, protozoan

Methanogens Heterodimeric, multimeric Oligomeric Monomeric, heteromeric

Monomeric Membrane-bound, cytoplasmic, periplasmic

Membrane-bound, cytoplasmic

Cytoplasmic, mambranebound, periplasmic, chloroplast, hydrogenosomes

Cytoplasmic Uptake of hydrogen Oxidation of hydrogen Production of hydrogen

Formation of hydrogen

H$_2$ase Assay

H$_2$ase activity may be assayed either by the reduction of electron carriers, *e.g.* methylene blue, benzyl viologen and methyl viologen, spectrophotometrically or by the Clark-type electrode method. But a simplified assay of hydrogenase involves an enzyme-catalysed reaction, which includes hydrogen production from common sugar, glucose using two enzymes, glucose dehydrogenase (GDH) and hydrogenase. GDH is an enzyme that oxidizes glucose to gluconic acid. GDH requires NADP$^+$ for its activity. Reduced NADPH is further oxidized to NADP$^+$ by hydrogenase isolated from any microorganism with the evolution of hydrogen. This hydrogen evolution can be detected with the red-ox dye benzyl viologen (BV) that turns purple when reduced. The absorbance is measured spectrophotometrically at a wavelength of 600 nm. The molar extinction coefficient of BV is 7400 at 600 nm. The stoichiometry of the reaction is one mole of glucose reduces one mole of BV; so by measuring absorbance at 600 nm, the amount of hydrogen produced can be estimated (Portman *et al.*, 1998). Hydrogenase activity of the bacterially expressed and purified protein can also be assayed spectrophotometrically using methyl viologen (MV) or NADH substrate (Figure 6.2). One unit of hydrogenase activity is the amount of enzyme that catalyses the reduction of 1 mM of MV or NADH per min, or the production of 1 mM of H$_2$ per min.

Evidences for Involvement of Fe-H$_2$ase in Hydrogen Production

The genes for Fe-H$_2$ase were the first to be cloned and sequenced for *Desulfovibrio vulgaris* and the enzyme was purified to homogeneity (Harry *et al.*, 1978; Pohorelic *et al.*, 2002). Experimental results showed that in *D. vulgaris*, following a transfer of a broad-range host plasmid that constitutively expresses *hydAB* (FeH$_2$ase encoded gene) anti-sense mRNA, causes two to threefold reduced content of Fe-H$_2$ase Pohorelic *et al.*, 2002). The strain with reduced content of Fe-H$_2$ase showed less H$_2$ production compared to the wild-type *D. vulgaris*. Subcellular fractionation, immunochemistry, Western blotting and comparison of the deduced amino acid sequence of H$_2$ase available to date confirmed the presence of Fe-H$_2$ase in the

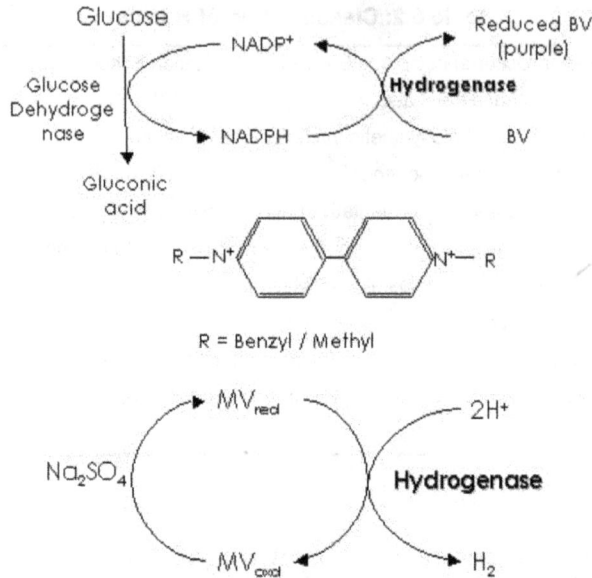

Figure 6.2: Reaction Outline for Measurement of Hydrogen Production in Colorimetric Assay (Portman *et al.*, 1998).

hydrogenosome of anaerobic chytrid (Akhmanova *et al.*, 1998), *Neocallimastix* sp. L2. This is the enzyme responsible for at least 90 per cent of H_2 production of hydrogenosome, and its activity can be blocked by carbon monoxide. Gaffron and Rubin (1942) first reported that a green alga, *Scenedesmus*, produced molecular hydrogen after being kept under anaerobic and dark conditions. Under anaerobic atmosphere, hydrogen metabolism is the only pathway for algae to create high amount of ATP, which is required for survival under stress conditions. Sulphur-deprivation in *Chlamydomonas reinhardtii* brings about prompt degradation of Rubisco (principal enzyme involved in photosynthesis) and substantial accumulation of starch. Thus the organism switches over the photosynthetic pathway to hydrogen metabolism (Tamagnini *et al.*, 2002; Wunschiers *et al.*, 2001). Starch accumulation and subsequent breakdown of Rubisco provide the endogenous substrate that supports H_2-production, both directly by feeding electrons into the plastoquinone pool in chloroplasts, and indirectly by sustaining mitochondrial respiration for the maintenance of anaerobiosis in the cell (Gfeller and Gibbs 1984). Green algae respond to anaerobic stress by switching the oxidative pathway to the fermentative metabolism. Fermentation is mostly associated with hydrogen evolution. It has also been reported that the key enzyme, Fe-H_2ase is synthesized only after an anaerobic adaptation in *C. reinhardtii*. Results of the suppression subtractive hybridization (SSH) approach (Happe and Kaminski, 2002) showed that the Fe-H_2ase gene is differentially regulated under anaerobiosis only. All these compelling evidences from structures, sequences and experimental results suggest that Fe-H_2ases are the distinct class of H_2ases that is the only responsible enzyme in hydrogenesis.

Structural Diversity

X-ray crystallography, spectroscopy and other biochemical studies have contributed significantly to the structural studies of Fe-H$_2$ase. The molecular architecture of the catalytic site of the enzyme has been resolved only with the X-ray crystal structure. Despite the wide occurrence of the Fe-H$_2$ase in prokaryotes and eukaryotes, the structure of only two Fe-H$_2$ases has been solved so far – from cytoplasm of *Clostridium pasteurianum* (CpH$_2$ase) (Peters *et al.*, 1998) and periplasmic space of *Desulfovibrio desulfuricans* (DdH$_2$ase) (Nicolet *et al.*, 1999). The three-dimensional structure of CpH$_2$ase has been determined to 1.8 Å resolution by X-ray crystallography using multi-wavelength anomalous dispersion (MAD) phasing (Peters *et al.*, 1998). These data show that the larger domain of the enzyme possesses the catalytic activity termed as H-cluster and the apoprotein part of the enzyme constitutes some multiple copies of [2Fe-2S/4Fe-4S] clusters, known as the F-cluster. The molecular architecture of the H-cluster was confirmed by FTIR, EPR and IR studies (Figures 6.3 *a–c*). Such spectroscopic data indicate that the H-cluster of the Fe-H$_2$ase consists of a bimetallic centre with two Fe atoms (Fe1–Fe2). Each Fe of the binuclear centre.

The coordination of strong-field diatomic ligands like CO and CN to the metallic centre makes Fe low spin and stabilizes at low oxidation state. Such low oxidation state at the Fe site facilitates the binding of hydrogen (Adams and Mortenson 1984; Zexing and Michael 2001; Pierik *et al.*, 1998) as ligands and acts as a site to donate and accept protons for heterolytic mechanism of hydrogen production. The F-cluster and the non-coordinated Cys residue at the active site accommodate the electron and proton transfer pathways and form a putative channel for the access of hydrogen to the active site (Meyer and Gagnon, 1991), and are responsible for electron transfer path from buried active site to the exterior of the protein. CpH$_2$ase consists of a single polypeptide chain and the structure resembles a mushroom. The large C-terminal domain contains the catalytic centre and makes up the cap of the mushroom and is bridged to the proximal F-cluster via the sulphur atom of the cysteine residue. Fe-H$_2$ase from *D. desulfuricans* possesses two polypeptide chains. The larger subunit contains the catalytic site, H-cluster and is covalently attached to the loops of beta-sheets at the active site domain. The most conserved part of H-cluster domain of Fe-H$_2$ases includes four cysteine ligands at the metal site, with a few residues like methionine or histidine lining the active site. This region exhibits high degree of conservation in sequence alignment of all Fe-only hydrogenases reported so far (Figure 6.3 *d*). Mössbauer and FTIR spectroscopic data (Pereira *et al.*, 2001; Grande *et al.*, 1983) suggest that Fe1 of the bimetallic centre has six ligands in distorted octahedral conformation, whereas Fe$_2$ has five ligands. CpH$_2$ase contains additional water as ligand, but DdH$_2$ase has an empty site. This apparent open site most likely binds hydrogen and is confirmed by the loss of activity of the enzyme by irreversible binding of both CO and CN at this site. Interpretation of EPR data and stretching frequencies of FTIR data explain well this phenomenon of loss of activity of H$_2$ase in the presence of CO/CN and its susceptibility towards oxygen with the plausible conformational changes that might occur at the H-cluster site upon oxidation and on reductive states (Bennett *et al.*, 2000).

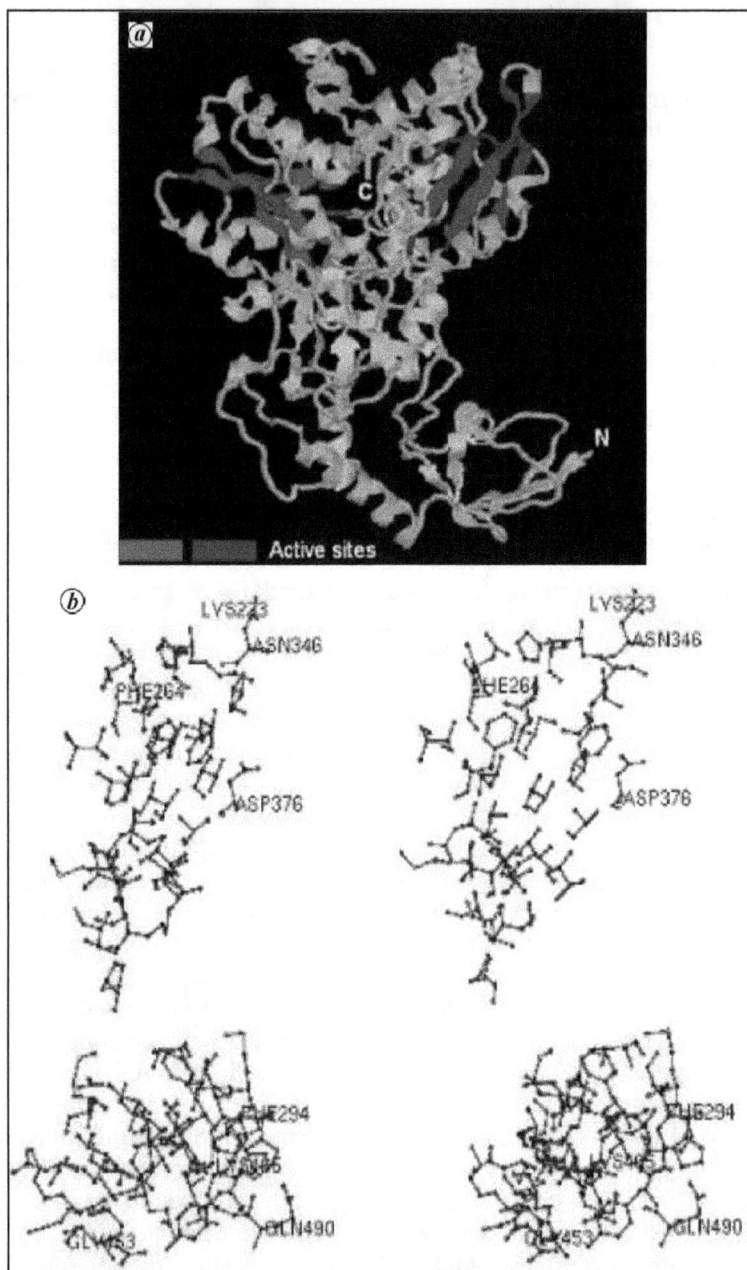

Figure 6.3 *a, b*: *a*, Crystal structure of Fe-H$_2$ase from *Clostridium pasteurianum*. The active site domain of CpH$_2$ase consists of two four-stranded twisted beta sheets, each of which is flanked by a number of alpha helices and forms two equivalent lobes. The left lobe of the active site domain (red-coloured region) possesses four-stranded parallel beta strands and the right lobe (blue-coloured region) of the same contain four-stranded mixed three sheets. The active site cluster is located in the specified cleft (Peters *et al.*, 1998). *b*, Stereo view of active site domain generated by MOLSCRIPT v.2.1.2.

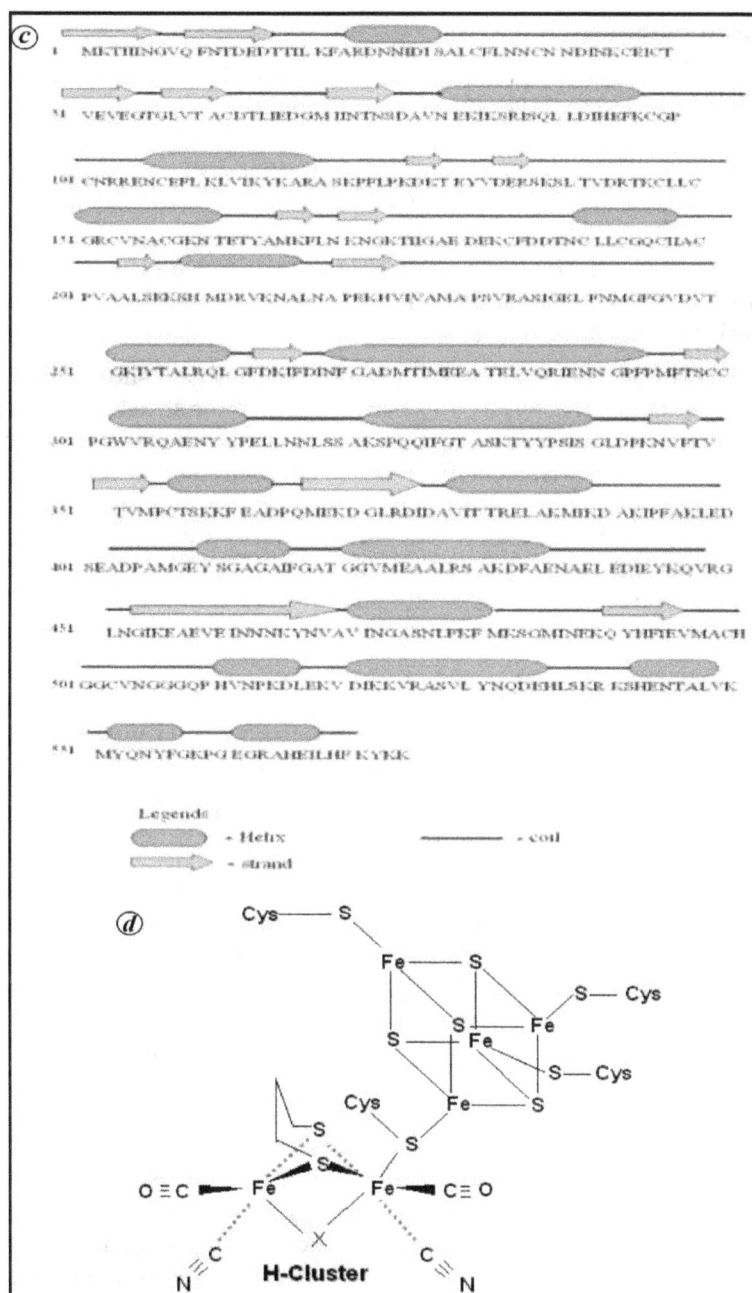

Figure 6.3 *c, d*: *c*, Secondary structure prediction of CpH2ase. *d*, Proposed structure of active site, H-cluster of CpH$_2$ase. H-cluster consists of binuclear Fe atoms bridged to a [4Fe–4S] subcluster through a sulphur residue of cysteine. Fe atoms are coordinated to CO/CN ligands and *X* may be a H$_2$O/CO ligand. is coordinated to diatomic ligands like CO and CN. The two Fe atoms (Fe1–Fe2) in CpH$_2$ase are bridged by CO and in DdH$_2$ase by 1,3-propanedithiol-like bonds (De Lacey *et al.*, 2000; Darensborng *et al.*, 2003).

Accessory Domain(s)

In addition to the H-cluster domain, an N-terminal domain homologous to the bacterial ferredoxin [4Fe–4S] and/or bacterial thioredoxin [2Fe–2S] is also present in many Fe-H$_2$ases. *Trichomonas vaginalis* and *Megasphaera elsdenii* possess two monomeric subunits as accessory domain. *Clostridial*-type H$_2$ase is of 64 kDa, with three domains in addition to the H-cluster domain. *Desulfovibrio fructosovorans* H$_2$ase resembles *Clostridial*-type. The catalytic subunit of *Thermotaga maritima* is large (73 kDa), but its C-terminal domain is homologous to Nuo E subunit of NADH-ubiquinone oxido-reductases. The largest catalytic subunit so far reported is the monomeric H$_2$ase of anaerobic eukaryote, *Nycotherus ovalis* (Malki *et al.*, 1995).

On the contrary, N-terminal part of eukaryotic FeH$_2$ase shows heterogeneity, which is lesser among bacterial Fe-H$_2$ases. Regarding the structural configuration, in most bacteria four motifs are implicated in the coordination of one [2Fe–2S] and three [4Fe–4S] accessory clusters (Peters *et al.*, 1998). In contrast among eukaryotes, *T. vaginalis* and *N. ovalis* contain a coordination of four putative accessory [Fe–S] clusters (Riualt and Klausner 1996), surprisingly, green algae like *Chlorella*, *Chlamydomonas* and *Scenedesmus* (Wunschiers *et al.*, 2001; Florin *et al.*, 2001) lack accessory domains though H$_2$ evolution and biochemical results are similar to *C. pasteurianum*.

Red-ox Partners of Fe-H$_2$ase

Fe-H$_2$ases are efficient users of a wide range electron donors or acceptors. However, they differ greatly in size and number of accessory domains, which determine the selectivity of enzymes towards the red-ox partner. Structural information on the recognition of its red-ox partners is essential to understand the structure–function relationships of the enzyme. Considerable divergence is reached at the level of natural red-ox partners of H$_2$ases. Flavodoxins, ferredoxin, rubredoxins, monoheme cytochromes, and multi-heme cytochromes NADH, NADPH have all been implicated as putative partners of H$_2$ase (Table 6.3). This multitude defines electron transfer to and from H$_2$ase as a research area in itself. The red-ox partner can change in response to nutrient variation (the flavodoxin/ferredoxin switch depending on available Fe levels) or to a changing function (high-potential/low-potential partner depending on whether H$_2$ should be consumed or produced; Wunschiers *et al.*, 2001; Riualt and Klausner 1996). Different H$_2$ases in the same cell can use different red-ox partners, and one particular H$_2$ase can be red-ox connected to different streams of metabolism. All this switching and branching of electron transfer requires research for enzymological, physiological and technological implications. Their occurrence in diverse organisms is endowed with a wide range of metabolic capabilities. Fe-H$_2$ase is versatile with respect to electron donors and acceptors. The metabolism of *M. elsdenii* is closely related to the *C. pasteurianum*. These bacteria use 4Fe–4S ferrodoxin under rich iron condition but flavodoxin during iron starvation (Firzgerald *et al.*, 1980). In green algae, *e.g.* *Scendusmus obliquus*, *C. reinhardtii* and *Chlorella fusca*, the electrons for H$_2$ evolution are provided by the fermentative metabolism via plastoquinone pool into the photosystem-I (PS I) which in turn reduces plant-type [2Fe–2S] (Florin *et al.*,

2001). In *D. fructosovorans* and *T. maritima*, NADP plays a physiological role in the functioning of the enzyme. The only correlation between accessory domains and red-ox partner specificity appears to be with the presence of Nuo-E and Nuo-F-like domains (subunits of NADP ubiquinone: oxidoreductase) in FeH_2ases interacting with NADP (Jeong and Bowien 1999).

Table 6.3: Characteristics of Fe-Hydrogenase

Organism	Size (No. of Amino acid residues)	Red-ox partners	Approximate Mol. Wt. of Fe Hydrogenase (kDa)
Clostridium pasteurianum 574	Fd, Fv, plant Fd	60–64	
Megasphaera elsdenii 484	Fd, Fv	57–59	
Desulfovibrio vulgaris Hildenborough	421–606	Fd, Fv, plant Fd	50
D. desulfuriacans	422	Fd, Fv, plant Fd	50
D. fructosovorans	421–585	Fd, Fv, plant Fd	NA
Escherichia coli	NA	Fd, Fv	64
Pyrococcus furiosus	NA	NAD(P)H, Fd	65
Thermotaga maritima	300–645	Fd, Hfd, Fv	NA
Chlamydomonus reinhardtii	497	Plant Fd, plastoquinone	53
Chlorella fusca	436	Plant Fd, plastoquinone	NA
Scendusmus obliquus	448	Plant Fd, plastoquinone	49
Nyctotherus ovalis	1206	Fd, Fv, Hfd, NE, NF	130
Neocallimastix sp. L2	NA	Fd, Fv, plant Fd	66.5
Trichomonas vaginalis	450–590	Fd	51.5–53
Giardia intestinalis	468	Fd	49

NA, Not available; Fd, Ferredoxin; Fv, Flavodoxin; NADP(P)H, Nicotinamide adenine dinucleotide phosphate (reduced form); Hfd, His ligand ferradoxin.

Distribution and localization of Fe-H$_2$ase

Extensive genomic sequencing efforts have currently revealed unexpected evolutionary connections among phylogenetically distant organisms. One of the most surprising examples is the presence of Fe-H$_2$ase in eukaryotic organisms like yeast, few anaerobic fungi, some ciliates, parabasalid flagellates, some micro aerophillic protozoa, parasite, diplomonad spiro nucleus, green algae, plants and even in mammals, including humans (Horner *et al.*, 2002). However, there are differences of opinion over the existence of H$_2$ases in various eukaryotes. The putative gene encoding H$_2$ase from each of these groups is cloned and characterized at the molecular level. The degree of the amino acid sequence homology between H$_2$ases and several homologous eukaryotic putative proteins is much higher than those previously reported. However, there is now evidence that sequences bearing common ancestry with Fe-H$_2$ases are located in aerobic eukaryotes, including humans. These genes are termed as NARF (nuclear prelamin A recognition factor) and display extensive similarity to Fe-H$_2$ase, specially with respect to the conservation of residues implicating in the coordination of the unique active H-cluster, as reported from two hybrid techniques. Human NARF protein interacts with prelamin A, the precursor form of a protein involved in the maintenance of the structural integrity of the nucleus (Lutz *et al.*, 1992; Barron and Worman 1999). In yeast, deletion of NARF-like gene is lethal in the haploid background (Winzeler *et al.*, 2000). Nuclear localization of NARF as well as the absence of reports on hydrogen production by higher eukaryotes suggests that these proteins are involved in energy metabolism. But clusters of Fe–S can facilitate electron transfer and contribute to the catalytic activity or help in maintaining the structural integrity of proteins. Proteins containing Fe–S are found to be involved in the sensing of oxidative stress, and also in red-ox dependent regulation of gene expression in prokaryotes. Processing and cleavage of nuclear lamins play a key role in programmed cell death (apoptosis) of higher eukaryotes (Zheng and Storz 2000). Thus it can be speculated that such NARF-like protein may be involved in a variety of eukaryotic cellular processes, including cell cycle. Thus the proposition that Fe-H$_2$ases play a crucial role in eukaryogenesis, of both the new premises, the hydrogen hypothesis (Martin and Muller 1998) and the syntrophic hypothesis (Moretra and Gracia 1998) is found to be true. Furthermore, in eukaryotic fungi, *e.g. Neocallimastix frontalis, Trichomonads* (Bui and Johnson 1996; Kulda 1999) and in anaerobic ciliates, *e.g. Nyctothermus ovalis*, the enzyme Fe-H$_2$ase is present in the organelle called hydrogenosome, a peculiar organelle that supplies ATP and also produces molecular hydrogen by a similar mechanism like some eubacteria (Akhmanova *et al.*, 1998). Hydrogenosome is generally derived from mitochondria, which in turn is originated from the free-swimming proteobacteria. But *Giardia intestinalis*, a eukaryote without hydrogenosome, produces hydrogen (Akhmanova *et al.*, 1998; Lloyd *et al.*, 2002). Under strictly anaerobic conditions, a mass spectrometric investigation of gas production indicates a low level of generation of hydrogen gas about tenfold lower than that in *T. vaginalis*, under similar conditions. In green algae, *C. fusca, S. obliqus, C. reinhardtii* (Lloyd *et al.*, 2002). the enzymes are located in the chloroplast stroma and are linked via ferredoxin to the photosynthetic electron transport chain (Happe and Kaminski 2002; Florin *et al.*, 2001).

Improvement of Hydrogenase Activity–Some Approaches

Hydrogen metabolism in different microorganisms involves a coordinated action of two enzymes: nitrogenase and H$_2$ase. However, the yield of hydrogen in hydrogenase-catalysed reaction is much higher than that of nitro-genase-catalysed reaction (Benemann and Hallenbeck 2002). Genetic studies on fermentative microorganisms have markedly increased, but relatively few genetic engineering studies have focused on altering the characteristics of these microorganisms, particularly with respect to enhancing the hydrogen-producing capabilities compared to photosynthetic bacteria and cyanobacteria. Different strategies can be followed for the amelioration of biological hydrogen production.

Development of Oxygen-Tolerant H$_2$ase

Among the H$_2$ases, Fe-H$_2$ase is an extremely oxygen-sensitive enzyme. In algal systems, H_2 photo-production ceases abruptly if oxygen is present. Oxygen inactivation is thought to occur by the direct binding of O_2 to one of the iron species (with an unoccupied coordination site) located at the catalytic centre (Hall *et al.*, 1995). Thus several precautions are required during purification of H_2ases. Sustained hydrogen generation is only possible if the enzyme H_2ase remains active in all physiological conditions, aerobic and anaerobic. Different strategies for surmounting the 2-sensitivity problem include: (i) molecular engineering of the hydrogenase to remove O_2 sensitivity, and (ii) development of physiological means to separate O_2 and H_2 production. The approach to overcome oxygen sensitivity is only through classical genetics, either by site-directed mutagenesis or point mutation at *hyd*A (encoding FeH$_2$ase) gene. The focus of this approach is to identify the region, the particular amino acid residue, where the oxygen irreversibly binds and substitution of the residue may result in O_2 tolerance. Mutant prokaryotic organism, *Azotobacter vinelandii* (Mctavish *et al.*, 1995) containing H_2ase with increased oxygen tolerance has been reported, which suggests that the enzyme is amenable to manipulations.

Repression of Uptake Hydrogenase

The presence of multiple forms of H_2ases in a single organism has been reported (Pedro *et al.*, 2000). Hydrogen production via indirect photolysis using cyanobacteria can be improved by screening for wild-type strains possessing highly active hydrogen evolving enzymes (nitogenases and/or H_2ases) in combination with high heterocyst formation (Hagen *et al.*, 1986). Genetic modification of strains to eliminate uptake H_2ases and increase levels of bidirectional hydrogenase activity may yield significant increases in H_2 production. For example, a mutant strain of Anabaena (AMC 414), in which the large subunit of the uptake hydrogenase (hupL) was inactivated by a deletion event (Zhang *et al.*, 1983), produced H_2 at a rate that was more than twice that of the parent wild-type strain, Anabaena PCC 7120. The uptake hydrogenase mostly contains Ni as the prosthetic group and is required for the assembly of the holo enzyme and also for its catalytic activity. If the cells are allowed to grow in Ni-deficient medium, synthesis of the uptake H_2ase will be blocked resulting in comparatively greater hydrogen evolution (Hall *et al.*, 1995).

Metabolic Shift

Disposal of excess reducing equivalents generated during fermentation is one of the major bottlenecks in facultative anaerobes which produce hydrogen (Lay *et al.*, 1999). These excess reducing equivalents could be disposed of via proton-reduction, facilitated by hydrogenase and electron carriers, leading to the formation of hydrogen in organisms such as *Enterobacter aerogenes*, *Enterobacter cloacae*, etc. In addition to volatile fatty acids, anaerobic fermentation also leads to formation of alcohols. These reduced end-products such as ethanol, butanol and lactate contain additional H atom that has not been liberated as gas (Levin *et al.*, 2003). Thus alcohol production gives correspondingly lower hydrogen yield. Therefore, to maximize the yield of hydrogen, the metabolism of bacteria must be directed away from alcohols (ethanol, butanol) and reduced acids (lactate) towards volatile fatty acids (Kumar *et al.*, 2001). H_2ase, which consists of two subunits, interacts with NADH (reducing equivalent) on the cytoplasmic side and with protons on the periplasmic side.

NADH is usually generated by catabolism of glucose to pyruvate via glycolysis. The conversion of pyruvate to ethanol, butanediol, lactic acid and butyric acid involves oxidation of NADH. The concentration of NADH would be increased if the formation of these alcoholic and acidic metabolites could be blocked (Kumar *et al.*, 2001). This, in turn would augment the yield of hydrogen through the oxidation of NADH. The yields are reportedly (Kumar *et al.*, 2001) increased to 3.8 mol (mol glu-cose)$^{-1}$ by blocking the pathways of alcohol and organic acid formation by allyl alcohol and proton-suicide technique using NaBr and NaBrO$_3$. Similar enhancement of hydrogen yield using *E. aerogenes* HU-101 is reported by blocking the formation of alcoholic and acidic metabolites by allyl alcohol and also by proton suicidal technique (Hawkes *et al.*, 2002; Mahyudin *et al.*, 1997).

Over-expression of Fe-H$_2$ase Gene

Cloning and sequence analysis of Fe-H$_2$ase encoded gene (*hydA*) of many different fermentative bacteria and cyanobacteria is now under study. It is now possible to over-express the *hydA* gene from any hydrogen-producing organism into fast growing bacteria like *E. coli* under strong promoters (Mishra *et al.*, 2002; Jameson *et al.*, 1986). Genetic manipulation with mutation at the transcription regulatory site can result in the constitutive expression of *hydA* gene, which was previously found to be active under derepressed condition only (Mishra *et al.*, 2004).

Conclusion

Hydrogenase research has been strengthened with X-ray crystallography and consequent simulations of other investigations. Complete genome sequences together with genetic and biochemical data indicate that Fe-H$_2$ases occur in bacteria and eukarya, but having distinct nature regarding accessory domains, size, red-ox partner specificity, charge distribution, etc. Discovery of H$_2$ase-like sequences in genomes of aerobic eukarya, including mammals implies the involvement of the enzyme in the evolution of eukaryotes, the hydrogenosomes and in the formation of eukaryotic cells, which can be an emerging track of H$_2$ase research. H$_2$ases of prokaryotic cells as well as of some eukaryotes are oxygen-sensitive. Thereby another future scope for the improvement of biological hydrogen production lies in the development of oxygen tolerant mutants.

Chapter 7

Hydrogen Production by Symbiotic Systems

1. Introduction

Symbiosis is perhaps best defined as a state of equilibrated physiological interdependence of two or more organisms involving no permanent stimulation of defensive reaction mechanism'. Practically all symbionts systems are characterized by the fact that one of the symbionts is the morphologically dominant parther of the association. When one sysmbiont is involved in intracellular association then it is termed as 'endozoic' association but when the association is endocellular, then it is called as ' syncyanoses' involving blue-green algae (cyanobacteria). A detailed discussion on symbiotic systems involving cyanobacteria has described in Chapter IV. therefore, no further attempts would be made to elucidate it again.

The nitrogen-fixing bacterial symbiosis may be broadly categorized into two main types : Those involving interaction between legumes and bacteria of the genus 'Rhizobium' (however, some rhizobia have been shown to form symbiotic association with the non-leguminous angiosperms and the actinomycete Frankia spp. (Table 7.II). Thus a common denominator to legume and non-legume symbiosis is the fact that bacteria are harboured in specialized structures called nodules. Although considerable amount of work has been conducted on rhizobial symbiosis, knowledge of Frankia and non-symbiotic association is much less extensive.

In this section, we briefly summarize those works where an attempt has been made to understand hydrogen evolution through symbictic forms. Altohugh hydrogen evolution in these forms has been reported, as an alternative source of energy.

Leguminous Plant and Rhizobium Associations

Several reviews have been devoted to H_2 metabolism and H_2 recycling in legume nodules (Dixon, 1978; Evans *et al.,* 1980; phillips, 1980; Evans *et al.,* 1982; Eisbrenner and Evans, 1983). Although rhizobia contain an active hydrogenase, the rate of hydrogen production by nitrogenase may exceed the rate of hydrogen consumption. In many cases, the overproduction of hydrogen gives rise to escape from the cell. Quite substantial amounts of hydrogen are released from the excised root nodules as well as from the root nodules in situ (Evans and Barber, 1977; Schubert and Evans, 1976), about 30 per cent of the reducing equivalents provided to nitrogen ended as hydrogen that escaped to soil. For example 75 per cent of the leguminous species studied by Schubert and Evans (1976) spend 40 to 60 per cent of their ATP and reductant on nitrogen catalysed H_2 production under an atmosphere of air. Under such condition hydrogen evolution may be observed from soils also. This hydrogen evolution affects the surrounding ' microenvironment' and influences the number and diversity of bacteria occupying the region, including those which will utilize hydrogen (Cuningham *et al.,* 1988) such as in the case of free-living *Rhizobium meliloti* (Ruiz-Argueso *et al.,* 1979).

Many strains of *R. japonicum* possess a membrane bound uptake hydrogenase which enables them to reoxidize the formed hydrogen and to regain the part of energy lost (Carter *et al.,* 1978; Emerich *et al.,* 1979). A significant increase in dry weights and nitrogen contents has been observed in soja plants incoulated with Hup + strains of *R. japonicum* compared with Hup *R. japonicum* strains (Albrecht *et al.,* 1979 ; Schubert *et al.,* 1978). Most strains of the fast growing Rhizobia have no hydrogenase activity at all or too little activity to recycle all the hydrogen evolved by nitrogenase (Nelson and Salminen, 1982).

Whilst certain strains of *Rhizobium* possess the ability to consume H_2, both outside their host and in symbiotic association, others, including commercially important strains, do not. This has prompted considerable interest in the genetics of H_2 uptake by *Rhizobium* species, with the aim of transferring this property to other strains and thereby improving the efficiency of N_2-fixation in legumes infected with these transformed bacteria.

The genes encoding H_2 uptake (hup) have been isolated from *Bradyrhizobium japonicum* and have also been identified on sym. Plasmid. Furthermore, transfer of *hup* genes from *hup* + to *hup* strains of *B. japonicum*, and also to *hup*- strains of *R. meliloti* and *R. trifolii*, resulted in low but detectable rates of H_2 consumption in nodules of infected soybeans, lucerne and clover, respectively. Transfer of *hup* genes within *Rhizobium* and *Bradyrhizobium* is therefore possible (Lambert *et al.,* 1985).

Unfortunately, the cosmic used to transfer the hup genes was rather unstable, so it was not possible to assess any benefit that such transfer might have for N_2-fixation. Nevertheless, in general, legumes infected with *hup* + strains of *B. japonicum* show greater rates of N_2- fixation than do *hup*- strains (Evans *et al.,* 1980; Lambert *et al.,* 1985) though the same may not be true of legumes infected with *R. leguminosarum* (Cunningham *et al.,* 1985).

Actinorhizal Symbiosis

It is known since long that many non-leguminous trees and shrubs form N_2-fixing root nodules in association with bacteria. Such root nodules are found in over 200 species of woody. Dicotyledonous trees and shrubs (Table 7.1) throughout temperate and tropical climates (Torrey, 1985). The bacterial counter parts of these root nodules are not rhizobia but members of the group *Streptomyces*, and are classified under the generic name Frankia. These associations between non-legumes and *Frankia* have been given the name actinorhizal associations. Initially attempts were made to describe different species of Frankia based on host specificity (Becking, 1970) but this has now been largely abandoned. Current practice is to refer to the generic name only (*i.e.* Frankia) without attempting to define the species.

Table 7.1: Symbiotic Systems Involving Bacteria and Actinomycete

Host-plant	Symbiont	Type of Symbiosis	References
Soybean	*Rhizobium japonicum*	Root-nodule	Arp and Burris (1979, 1981)
Pea	*R. leguminosarum*	Root-nodule	Ruiz-Argueso *et al.* (1978), Edie (1983)
Trifolium sp.	*Rhizobium* sp.	Root-nodule	Van Kessel and Burris (1983)
Cow-pea	*Rhizobium* sp.	Root-nodule	Rainbird *et al.* (1981)
Chick-pea	*Rhizobium* sp.	Root-nodule	Sindhu *et al.* (1986)
Parasponia parviflora	*Rhizobium* sp.	Root-nodule	Becking (1983)
Alnus incana	*Frankia* sp.	Plant-actinomycete	Sellstedt (1986)

The successful isolation and culture of Frankia by Callaham *et al.* (1978) made it possible to study the microsymbiont outside the host plant. This was followed by successful demonstration that Frankia, in pure culture could reduce acetylene (Tjepkema *et al.*, 1980). Since actinorhizal plants are good, often aggressive, colonizers, they may be used to regenerate poor soils. In this connection, N_2-fixation rates as high as 260 kg/ha/yr have been reported for *Casuarina equisetifolia* (Dommergues *et al.*, 1984).

Uptake hydrogenase activity (O_2-depent activity) in root nodules from *Alnus glutionsoa, A. rhombifolia, A. rubra* and *Myrica pennsylvanica* has been demonstrated from different sources may have different patterns of coupling to various electron acceptors has been described (Benson *et al.*, 1980). Sellstedt (1986,1988) has shown in *Frankia-Alnus incana* association that the root nodules evolve hydrogen. Strain specific differences in H_2 uptake and evolution have been demonstrated in various strains of *Frankia*. For example a local *Frankia* strain showed no measure able H_2 uptake but had high H_2 evolution rate (Sellstedt *et al.*, 1986).

The genus *Frankia* is closely related to *Streptomyces*, an organism in which cell fusion has proved a promising method for generating recombinant strains (Beringer, 1981). However, most of the studies made so far have involved isolating *Frankia* DNA, digestion with restriction endonucleases and hybridization with appropriate genes from *Rhizobium* species and from *Klebsiella pneumoniae*. Using these techniques, the nif H, nif D and nif K have been identified (Ligon and Nakos, 1985). The latter two genes appear to be contiguous, whereas nif H is located some distance away. In this respect organization of the nif genes in *Frankia* resembles that in *Bradyrhizobium japonicum*. A region of the Frankia DNA hybridizing with the Rhizobium nod C gene has also been found (Drake *et al.*, 1985).

Table 7.2: Nodulated Genera of Non-Leguminous Angiosperms Forming Root Nodules. Incidence refers to the ratio of species bearing nodules to the total number of known species (from Silver, 1971)

Genus	Incidence
Alnus	25/35
Arctostaphylos	1/40
Casuarinas	14/45
Ceanothus	30/55
Cercocarpus	1/20
Coriaria	12/15
Discaria	1/10
Dryas	3/4
Elaeagnus	9/45
Hippophae	1/1
Myrica	12/55
Purshia	2/2
Shepherdia	2/3

Chapter 8

Hydrogen Production by Chloroplasts

For majority of investigations, chloroplasts from *Pisum sativum, Spinacia oleracea, Lactuca sativa, Nicotiana* spp. and *Zea mays* have been used (Personov *et al.*, 1977; Hoffmann *et al.*, 1977; Krasnovskii *et al.*, 1980; Mactsev and Krasnovskii, 1982, 1983; Mactsev *et al.*, 1986; Rao *et al.*, 1976). Chloroplasts from eukaryotic algae such as *Chlamydomonas reinhardii* or *Euglena gracilis* have also been employed (King *et al.*, 1977). On comparative grounds, the most unstable component of the photosynthetic hydrogen production system is the chloroplast. This has been attributed to the fact that photosynthetic electron transport activity decays in dark as well as in light. Dark inhibition has been observed to be partially prevented by addition of Ficol 400 or such reagents as bovine serum albumin (Cocquempot *et al.*, 1982). These agents absorb the fatty acids liberated by the action of galactolipases on membrane lipids during aging (Morris *et al.*, 1982). Light induced inhibition of chloroplast function is more pronounced in photosystem II, where water splitting occurs, than in photosystem I have a strong reductant is the end product. Under continuous illumination of chloroplasts, there seems to be loss of a number of proteins which are integral components of light harvesting reaction centers, and electron transfer complexes of the photosynthetic apparatus, which are essential for the activities of the thylakoid membrane. Santos and Hall (1982) studied chloroplast ageing in the dark and in the light and observed a correlation between the light induced decay of photosystem II activity and decrease in the content of membrane polypeptides. These polypeptides had apparent molecular weight of 36, 48, 50 KD. They may probably be associated with the photosystem II reaction centre.

Oxygen is the first product released during photosynthesis; therefore, reactivity of the catalysts with oxygen is another major draw-back for continuous hydrogen production by water photolysis. In a closed homogeneous system, the oxygen released by photosystem II dissolves in aqueous medium. This reacts with the reduced electron carriers (*e.g.*, ferredoxin or methylviologen) generated by photosystem I and produces superoxide or hydrogen peroxide (Foyer and Hall, 1980). Since the reduced electron carrier reacts more rapidly with the oxygen present in the immediate environment than with hydrogenase, the rate of hydrogen evolution is decreased markedly. Cuendet and Gratzel (1982) have described a number a number of photoreducible, low potential, organic and inorganic relays which can couple with hydrogenase. However, most of them are not suitable catalysts for the simultaneous H_2 and O_2 evolution. It is due to these reasons that simultaneous hydrogen and oxygen evolution is observed from illuminated single stage systems containing isolated chloroplasts, hydrogenase and an electron mediator. The hydrogen which is evolved is produced at the expense of oxidation of water to H_2O_2. Good rates of H_2 production from water were obtained in presence of oxygen and peroxide scavengers. The scavengers are usually a observed a correlation between the light induced decay of photosystem II activity and decrease in the content of membrane polypeptides. These polypeptides had apparent molecular weight of 36, 48, and 50 KD. They may probably be associated with the photosystem II reaction centre.

Oxygen is the first product released during photosynthesis; therefore, reactivity of the catalysts with oxygen is another major draw-back for continuous hydrogen production by water photolysis. In a closed homogeneous system, the oxygen released by photosystem II dissolves in aqueous medium. This reacts with the reduced electron carriers (*e.g.*, ferredoxin or methylviologen) generated by photosystem I and produces superoxide or hydrogen peroxide (Foyer and Hall, 1980). Since the reduced electron carrier reacts more rapidly with the oxygen present in the immediate environment than with hydrogenase, the rate of hydrogen evolution is decreased markedly. Cuendet and Gratzel (1982) have described a number of photreducible, low potential, organic and inorganic relays which can couple with hydrogenase. However, most of them are not suitable catalysts for the simultaneous H_2 and O_2 evolution. It is due to these reasons that simultaneous hydrogen and oxygen evolution is observed from illuminated single stage systems containing isolated chloroplasts, hydrogenase and an electron mediator. The hydrogen which is evolved is produced at the expense of oxidation of water to H_2O_2. Good rates of H_2 production from water were obtained in presence of oxygen and peroxide scavengers. The scavengers are usually a mixture of glucose, glucose-oxidase, catalase and ethanol (Person *et al.*, 19077; Packer and Cullingford, 1978; Mactsev *et al.*, 1986).

Two important factors which limit the rate and duration of biophotolytic hydrogen evolution from water are: (1) The relative instability of the oxygen-evolving water splitting) complex of photosystem II, and (2) The oxygen sensitivity of the catalysts used. If water is not used as electron donor, the effect of above-mentioned factors on hydrogen production can be eliminated. This approach then utilizes the photosystem I only of the thylakoid membranes from chloroplasts for light harvesting and electron transport. The systems are, therefore, comparable to the electron transport chain encountered in photosynthetic bacteria. Extraneously added ascorbate, glutathione, cystein or dithiothreitol in conjunction with a mediator dye (2,6-dichlorophenolindophenol or N, N, N', N'-tetramethyl-p-phenylamine) donates electrons prior to photosystem I (at or near cytochrome f/plastocyanin complex). The possibility of water splitting reaction (O_2-evolution) is eliminated by addition of DCMU or by controlled heat treatment of the chloroplast which does not affect the activities associated with photosystem I (Personov *et al.*, 1976; Rao *et al.*, 1976). Methylviologen, which is thus reduced at photsystem I, is coupled to bacterial hydrogenase for hydrogen evolution.

Rao and Hall (1983) defined conditions for optimal hydrogen production from ascorbate in chloroplasts methylviologen-hydrogenase catalytic systems which are given below:

1. The presence of intact thylakoid vesicles to decrease internal cycling of electrons from reduced methylviologen, which is produced on the outside of the vesicles, beck to plastocoyanin which is located inside (Hoffmann *et al.*, 1977).

2. The addition of dithlothreitol or glutethione to reduce any dehydroiascorbate to ascorbate and also to keep the chloroplasts and possibly the hydrogenase in a reducing environment.

3. pH of the reaction medium is kept neutral (Muallem and Hall, 1982).

Good rates of hydrogen production are usually obtained for a period of 3-5 h in the system; however, the reaction slows down subsequently. This has been attributed to accumulation of products in the reaction media which inhibit the coupling of electron transport from ascorbate to hydrogen without causing permanent inactivation of any of the components of the reaction (Muallem and Hall, 1982). The H_2-production could reach a level of 700 umol H_2/mg Chl/h (Aristarkhov at *et al.*, 1986).

If one considers the fact that relatively expensive molecules have to be used as an electron source and the life-time of H_2-production is reduced by product inhibition, the advantages of hydrogen production with photosystem I alone are limited as compared to photolytic system catalysed by photosystem II and. I. Nevertheless, it is interesting to note that in system using organic electron, donors, there is only one gaseous products-hydrogen-and there is no need to separate oxygen from hydrogen as in the case with water photolysis.

A number of studies as have been conducted to stabilize the photosynthetic activity of isolatests chloroplasts by endogenous or exogenous immobilization of photosynthetic apparatus (Bherezin and Varfolomeev, 1979; Rao and Hall, 1983). Such attempts are summarized in Table 8.1. As mentioned above, addition of certain reagents which may alter the environment of chylorophasts, *e.g.*, serum albimin, glycerol or polyethylene glycol, or improve the stability of chlorophasts. It has been reported that electron transport activity of isolated pea chlorophasts could be extended from 2 h to 360 h by incubation in medium containing 40 per cent (v/v) glycerol and 0.1 mM bovine serum albumin.

Table 8.1: Hydrogen Evolution by Immobilized Chloroplasts

Organelle	Immobilization Agents and Conditions	Stability and Photosynthetic Activity after Immobilization, Compared to Free	References
Chlorplast	0.05 per cent glutaraldehyde at 4°C	Higher stability in storage. No improvement in light stability. H_2 evolution with ferredoxin and hydrogenase.	Rao et al. (1976)
Chloroplast-thylakioids	Encapsulation in protamine + toluene diisocyanate	Loss of PS-II activity. PS-I activity maintained.	Kitajima and Butler (1976)
Chloroplast-thylakoids	0.37 per cent glutaraldehyde + serum albumin at −30°C	70 per cent O_2 evolution activity, 27 per cent ATP synthesis, stability in light for 7d.	Cocquempot et al. (1979)
Chloroplast thylakoids	Agar in hollow fiber reactor	NADP reduction for 2h, 31 per cent activity for one week at 5°C.	Karube et al. (1980)
Chloroplasts-thylakoids	2 per cent (w/v) calcium alginate ad films	100 per cent O_2 exchange, upto 67 per cent H_2 evolution. Light stability increased to 7h in comparison to 4h for non-immobilized	Gisby and Hall (1980)
Chloroplast-thylakoid	0.063 per cent glutaraldehyde + 5 per cent (w/v) gelatin	45 per cent activity in ferricyanide reduction	Cocquempot et al. (1981)
Chloroplast-thylakoids	2 per cent (w/v) agar + 5 per cent (w/v) serum albumin	50 per cent ferricyanide reduction activity. stability upto 400 h.	
Chloroplast-thylakoids	2 per cent (w/v) agar + 0.5 per cent serum albumin	PS-II and PS-I stabilized without loss of activity. Better H_2 evolution with other immobilized catalysts	Rao et al. (1982)
Chloroplast	2 per cent (w/v) agar + Clostridium butyricum, ferredoxin and benzylviologen.	H_2 evolution for 6h, subsequently reduced to 30 per cent.	Karube et al. (1981)

Gisby *et al*. (1982) used agar (2 per cent, w/v) and calcium alginate (2 per cent, w/v) as entrapment gels for the endogenous immobilization of thylakoid membranes. The rates of hydrogen evolution observed for agar immobilized chloroplasts were comparable to those of free chloroplast. Also the kinetics of O_2 evolution (measured as ferricyanide reduction) by agar immobilized chloroplasts parallel the kinetics of H_2-evolution. The rates of H_2 evolution catalyzed by alginate entrapped chloroplasts were lower than in free chloroplasts; nevertheless, in the former case, the H_2-evolution. The rates of H_2 evolution catalyzed by alginate entraped chloroplasts were lower than in free chloroplasts; nevertheless, in the former case, the H_2-evolution was sutained for a longer period. Entrapment in polyacrylamide gels had limited success in the case of isolated chloroplast fragments because of the inactivation of photosynthetic electron transport by the monomers and cofactors of polymerization (Ochial *et al*., 1978; Weetal and Kranmpitz, 1980). Entrapment of chloroplast fragments in polyvinyl alcohol gave more encouraging results (Ochial *et al*., b). Thus it is evgi8dent that these gel immobilized cholorophasts may preview useful in construction of heterogeneous photoplasts (with solid catalysts) for H_2 production.

Chapter 9

Bio-Hydrogen Production from Waste Materials

1. Introduction

The worldwide energy need has been increasing exponentially, the reserves of fossil fuels have been decreasing, and the combustion of fossil fuels has serious negative effects on environment because of CO_2 emission. For these reasons, many researchers have been working on the exploration of new sustainable energy sources that could substitute fossil fuels. Hydrogen is considered as a viable alternative fuel and "energy carrier" of future. Hydrogen gas is clean fuel with no CO_2 emissions and can easily be used in fuel cells for generation of electricity. Besides, hydrogen has a high energy yield of 122 kJ/g, which is 2.75 times greater than hydrocarbon fuels. The major problem in utilization of hydrogen gas as a fuel is its inavailability in nature and the need for inexpensive production methods.

Demand on hydrogen is not limited to utilization as a source of energy. Hydrogen gas is a widely used feedstock for the production of chemicals, hydrogenation of fats and oils in food industry, production of electronic devices, processing steel and also for desulfurization and re-formulation of gasoline in refineries.

About 50 million tonnes of hydrogen are traded annually globally with a growth rate of nearly 10 per cent per year for the time being (Winter 2005). The contribution of hydrogen to total energy market will be 8–10 per cent by 2025 (2). Due to increasing need for hydrogen energy, development of cost-effective and efficient hydrogen production technologies has gained significant attention in recent years.

Conventional hydrogen gas production methods are steam reforming of methane (SRM), and other hydrocarbons (SRH), non-catalytic partial oxidation of fossil fuels (POX) and auto-thermal reforming which combines SRM and POX. Those methods are all energy intensive processes requiring high temperatures (>85 0 ~C). Among other methods developed to improve the existing technologies are the membrane processes, selective oxidation of methane and oxidative dehydrogenation (Armor 1999).

Biomass and water can be used as renewable resources for hydrogen gas production. Utilization of wide variety of gaseous, liquid and solid carbonaceous wastes was investigated by Kim (2003) as renewable sources for formation ofhydrogen gas by steam reforming. Despite the low cost of waste materials used, high temperature requirement (T= 1200°C) is still the major limitation for this process. Electrolysis of water may be the cleanest technology for hydrogen gas production. However, electrolysis should be used in areas where electricity is inexpensive since electricity costs account for 80 per cent of the operating cost of H_2 production. In addition, feed water has to be demineralized to avoid deposits on the electrodes and corrosion (Armor, 1999).

Biological hydrogen production is a viable alternative to the aforementioned methods for hydrogen gas production. In accordance with sustainable development and waste minimization issues, bio-hydrogen gas production from renewable sources, also known as "green technology" has received considerable attention in recent years. Bio-hydrogen production can be realized by anaerobic and photosynthetic microorganisms using carbohydrate rich and non-toxic raw materials. Under anaerobic conditions, hydrogen is produced as a by-product during conversion of organic wastes into organic acids which are then used for methane generation. Acidogenic phase of anaerobic digestion of wastes can be manipulated to improve hydrogen production. Photosynthetic processes include algae which use CO_2 and H_2O for hydrogen gas production. Some photo-heterotrophic bacteria utilize organic acids such as acetic, lactic and butyric acids to produce H_2 and CO_2. The advantages of the later method are higher H_2 gas production and utilization of waste materials for the production. However, the rate of H_2 production is low and the technology for this process needs further development (Levin *et al.*, 2004).

Production of clean energy source and utilization of waste materials make biological hydrogen production a novel and promising approach to meet the increasing energy needs as a substitute for fossil fuels. Considering the above mentioned facts, this articles focuses on potential use of carbohydrate rich wastes as the raw material, microbial cultures, bio-processing strategies and the recent developments on bio-hydrogen production.

2. Types of Waste Materials

The major criteria for the selection of waste materials to be used in bio-hydrogen production are the availability, cost, carbohydrate content and biodegradability. Simple sugars such as glucose, sucrose and lactose are readily biodegradable and preferred substrates for hydrogen production. However, pure carbohydrate sources are expensive raw materials for hydrogen production. Major waste materials which can be used for hydrogen gas production may be summarized as follows.

2.1. Starch and Cellulose Containing Agricultural or Food Industry Wastes

Many agricultural and food industry wastes contain starch and/or cellulose which are rich in terms of carbohydrate contents. Complex nature of these wastes may adversely affect the biodegradability. Starch containing solid wastes is easier to process for carbohydrate and hydrogen gas formation. Starch can be hydrolyzed to glucose and maltose by acid or enzymatic hydrolysis followed by conversion of carbohydrates to organic acids and then to hydrogen gas. Cellulose containing agricultural wastes requires further pre-treatment. Agricultural wastes should be ground and then delignifiedby mechanical or chemical means before fermentation. Cellulose and hemicellulose content of such wastes can be hydrolyzed to carbohydrates which are further processed for organic acid and hydrogen gas production. It was reported that there is an inverse relationship between lignin content and the efficiency of enzymatic hydrolysis of agricultural wastes (De Vrije *et al.*, 2002). Figure 9.1 depicts a schematic diagram for bio-hydrogen production from cellulose and starch containing agricultural wastes by two stage anaerobic dark and photo-fermentations.

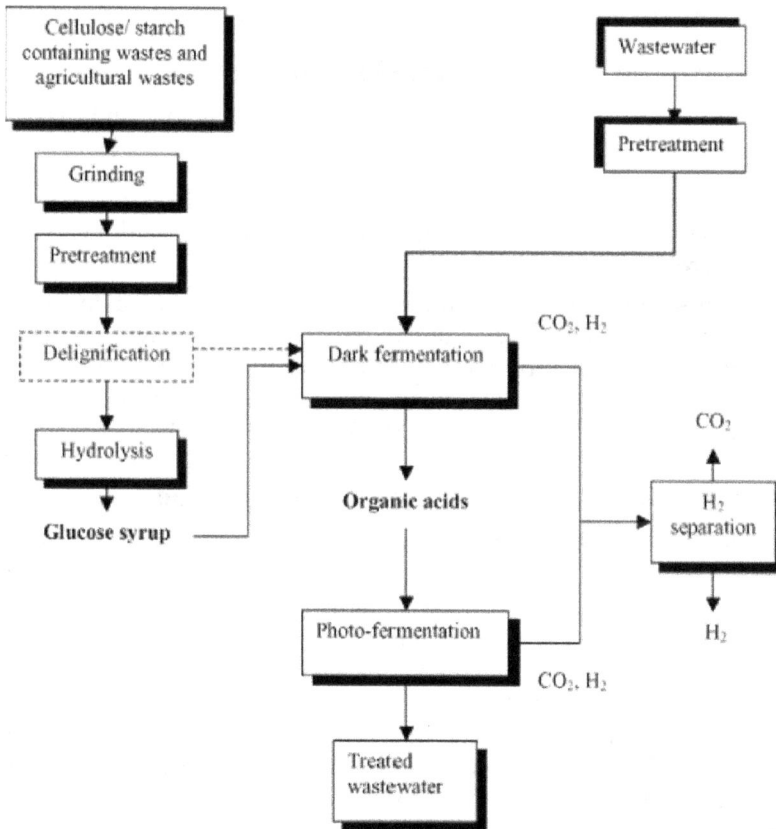

Figure 9.1: A Schematic Diagram for Bio-hydrogen Production from Cellulose/Starch Containing Agricultural Wastes and Food Industry Wastewaters.

2.2. Carbohydrate Rich Industrial Wastewaters

Some biodegradable carbohydrate containing and non-toxic industrial effluents such as dairy industry, olive mill, baker's yeast and brewery wastewaters can be used as raw material for bio-hydrogen production. Those wastewaters may require pre-treatment to remove undesirable components and for nutritional balancing. Carbohydrate rich food industry effluents may be further processed to convert the carbohydrate content to organic acids and then to hydrogen gas by using proper bio-processing technologies. Figure 9.1 shows schematic diagram for bio-hydrogen production from food industry wastewaters by two stage anaerobic dark and photo-fermentations.

2.3. Waste Sludge from Wastewater Treatment Plants

The waste sludge generated in wastewater treatment plants contains large quantities of carbohydrate and proteins which can be used for energy production such as methane or hydrogen gas. Anaerobic digestion of excess sludge can be realized in two steps. Organic matter will be converted to organic acids in the first step (acidogenic phase) and the organic acids will be used for hydrogen gas production in the second step by using photo-heterotrophic bacteria.

3. Bio-processes for Hydrogen Gas Production

Major bio-processes utilized for hydrogen gas production can be classified in three categories:

1. Bio-photolysis of water by algae.
2. Dark-fermentative hydrogen production during acidogenic phase of anaerobic digestion of organic matter.
3. Two stage dark/photo-fermentative production of hydrogen.The metabolic pathways, types and function of enzymes involved in biological hydrogen production for different microbial processes are summarized in details in some recent review articles (Das and Veziro 2002; Asada and Miyake 1999; Ghirardi *et al.*, 2000; Koku *et al.*, 2002).

3.1. Hydrogen Gas Production from Water by Algae

Algae split water molecules to hydrogen ion and oxygen via photosynthesis. The generated hydrogen ions are converted into hydrogen gas by hydrogenase enzyme. *Chlamydomonas reinhardtii* is one of the well-known hydrogen producing algae (Melis 2002). Hydrogenase activity has been detected in green algae, *Scenedesmus obliquus* (Florin *et al.*, 2001), in marine green algae *Chlorococcum littorale* (Ueno *et al.*, 1999; Schnackenberg et al1996), *Playtmonas subcordiformis* (Guan *et al.*, 2004) and in *Chlorella fusca* (Winkler *et al.*, 2002). However, no hydrogenase activity was observed in *C. vulgaris* and *Duneliella salina* (Winkler *et al.*, 2002; Cao *et al.*, 2001). The hydrogenase activity of different algae species was compared by Winkler *et al.*, (Winkler *et al.*, 2002) and it was reported that enzyme activity of the *Scenedesmus* sp. (150 nmol/ µg Chl a.h) is lower than *C. reindhartii* (200 nmol/µg Chl a.h).

Cyanobacterial hydrogen gas evolution involves nitrogen fixing cultures such as non-marine*Anabaena* sp., marine cyanobacter *Oscillatoria* sp., *Calothrix* sp. and non-nitrogen fixing organ-isms such as *Synechococcus* sp., *Gloebacter* sp. and it was reported that *Anabaena* sp. have higher hydrogen evolution potential over the other cyanobacter species (Pinto *et al.*, 2002). The growth conditions for *Anabaena* include nitrogen free media, illumination, CO_2 and O_2. Since nitrogenase enzyme is inhibited by oxygen, hydrogen production is realized under anaerobic conditions. CO_2 is required for some cultures during hydrogen evolution phase (Pinto *et al.*, 2002) although inhibition effects of CO_2 on photo-production of H_2 was also observed (Tsygankov *et al.*, 1998). Four to 18 per cent CO_2 concentrations were reported to increase cell density during growth phase resulting in higher hydrogen evolution in the later stage (Yoon *et al.*, 2002). The use of simple sugars as supplement was reported to promote hydrogen evolution (Shah *et al.*, 2001). Recent studies are con-centrated on development of hydrogenase and bi-directional hydrogenase deficient mutant of *Anabaena* sp. in order to increase the rate of hydrogen production. There are a number of cyanobacterial species other then Anabena reported by Vyas (2003) which have ability to produces H_2 Vyas (1992) also reported N_2 are actively inhibited by UV B radiation resulting inhibitory effect on H_2 production.

The algal hydrogen production could be considered as an economical and sustainable method in terms of water utilization as a renewable resource and CO_2 consumption as one of the air pollutants. However, strong inhibition effect of generated oxygen on hydrogenase enzyme is the major limitation for the process. Inhibition of the hydrogenase enzyme by oxygen can be alleviated by cultivation of algae under sulfur deprivation for 2–3 days to provide anaerobic conditions in the light (Guan *et al.*, 2004, Winkler *et al.*, 2002, Melis *et al.*, 2000, Laurinavichene *et al.*, 2002). Low hydrogen production potential and no waste utilization are the other disadvantages of hydrogen production by algae. Therefore, dark and photo-fermentations are considered to be more advantageous due to simultaneous waste treatment and hydrogen gas production.

3.2. Hydrogen Gas Production by Dark Fermentation

3.2.1. Type of Organisms and Conditions

Many anaerobic organisms can produce hydrogen from carbohydrate containing organic wastes. The organisms belonging to genus *Clostridium* such as *C. buytricum* (Yokoi *et al.*, 2001), *C. thermolacticum* (Collet *et al.*, 2004), *C. pasteurianum* (Liu and Shen 2004; Lin and Lay 2004), *C. paraputrificum* M-21 (Evvyernie *et al.*, 2001) and *C. bifermentants* (Wang *et al.*, 2003) are obligate anaerobes and spore forming organisms. *Clostrida* species produce hydrogen gas during the exponential growth phase. In batch growth of *Clostridia* the metabolism shifts from a hydrogen/acid production phase to a solvent production phase, when the population reaches to the stationary growth phase. Investigations on microbial diversity of a mesophilic hydrogen producing sludge indicated the presence of *Clostridia* species as 64.6 per cent (Fang *et al.*, 2002). The dominant culture of *Clostridia* can be easily obtained by heat treatment of biological sludge. The spores formed at high temperatures can be activated when required environmental conditions are provided for hydrogen gas production.

The species of the genus enterobactericeae have the ability to metabolize glucose by mixed acid or the 2–3 butanediol fermentation. In both patterns, CO_2 and H_2 are produced from formic acid in addition to ethanol and the 2–3 butanediol (Podest *et al.*, 1997). Hydrogen production capacity of anaerobic facultative bacterial culture *Enterobacter aerogenes* has been widely studied (Yokoi *et al.*, 2001; Palazzi *et al.*, 2002; Nakashimada *et al.*, 2002; Yokoi *et al.*, 1997 and 1998; Tanisho and Ishiwata 1994 and 1995; Fabino *et al.*, 2002) *Enterobacter cloacae* ITT-BY 08 produced 2.2 mol $H_2/$ mol glucose (Kumar and Das 2005). Hydrogen production from glucose by *E. coli* and *Hafnia alvei* was studied by Podest *et al.* (1997) and trace amount of hydrogen yield was detected.

Recently, hydrogen producing aerobic cultures such as *Aeromonos* spp., *Pseudomonos* spp. and *Vibrio* spp. were identified. Anaerobic cultures like *Actinomyces* spp., *Porphyromonos* spp. beside to *Clostridium* spp. have been detected in anaerobic granular sludge. The hydrogen yield varied between 1 and 1.2 mmol/mol glucose when the cultures were cultivated under anaerobic conditions (Oh *et al.*, 2003). Hydrogen production by *Thermo-to gales* species and *Bacillus* sp. were detected in mesophilic acidogenic cultures (Shin *et al.*, 2004).

Hydrogen gas production capacity of some anaerobic thermophilic organisms belonging to the genus *Thermoanaerobacterium* has also been investigated (Shin *et al.*, 2004; Zhang *et al.*, 2003; Liu *et al.*, 2003; Ueno *et al.*, 2001). Shin reported *T. thermosaccharolyticum* and *Desulfotomaculum geothermicum* strains producing hydrogen gas in thermophilic acidogenic culture (Shin *et al.*, 2004). A hyperthermophilic archeon, *Thermococcus kodakaraensis* KOD1 with 85~C optimum growth temperature was isolated from a geothermal spring in Japan and identified as a hydrogen producing bacteria (Kanai *et al.*, 2005). *Clostridium thermolacticum* can produce hydrogen from lactose at 58~C (Collet *et al.*, 2004). Recently, a hydrogen producing bacterial strain *Klebisalle oxytoca* HP1 was isolated from hot springs with maximal hydrogen production rate at 35 ~C (Minnan *et al.*, 2005).

Environmental conditions are the major parameters to be controlled in hydrogen production. Medium pH affects hydrogen production yield, biogas content, type of the organic acids produced and the specific hydrogen production rate. The reported pH range for the maximum hydrogen yield or specific hydrogen production rate is between pH 5.0 and 6.0 (Fang *et al.*, 2002; Lay *et al.*, 1999; Lay 2000; Khanal *et al.*, 2004; Chen *et al.*, 2001). However, some investigators report the optimum pH range between 6.8 and 8.0 (Collet *et al.*, 2004, Liu and Shen 2004, Zhang *et al.*, 2003, Kanai *et al.*, 2005, Lay *et al.*, 2001) and around pH 4.5 for the thermophilic culture (Shin *et al.*, 2004). Most of the studies indicated that final pH in anaerobic hydrogen production is around 4.0–4.8 regardless of initial pH (Yokoi *et al.*, 2001;Liu and Shen 2004; Zhang *et al.*, 2003; Liu *et al.*, 2003;Lay 2001, Morimoto *et al.*, 2004). The decrease in pH is due to production of organic acids which depletes the buffering capacity of the medium resulting in low final pH (Khanal *et al.*, 2004). Gradual decreases in pH inhibit hydrogen production since pH affects the activity of iron containing hydrogenase enzyme (Dabrock *et al.*, 1992). Therefore, control of pH at the optimum level is required. Initial pH also influences the extent of lag phase in batch hydrogen production. Composition of the substrate, media composition, temperature and the type of microbial culture are also important parameters affecting the duration of lag phase. Some studies reported that low initial pH of 4.0–4.5 causes longer lag periods such as 20 h (Liu and Shen 2004, Khanal *et al.*, 2004). High initial pH levels such as 9.0 decrease lag time; however, lower the yield of hydrogen production (Zhang *et al.*, 2003).

The majorproducts in hydrogen production by anaerobic dark fermentation of carbohydrates are acetic, butyric and propionic acids. Formation of lactic acid was observed when lactose and molasses (sucrose) were used as the substrates (Collet *et al.*, 2004;Tanisho and Ishiwata 1994 and 1995). pH also affects the type of organic acids produced. More butyric acid is produced at pH 4.0–6.0. Concentration of acetate and butyrate could be almost equal at pH 6.5–7.0 (Fang *et al.*, 2002). Ethanol production was observed depending on the environmental conditions (Collet *et al.*, 2004; Oh *et al.*, 2003; Zhang *et al.*, 2003; Liu *et al.*, 2003; Ueno *et al.*, 2001; Yu *et al.*, 2002). Methane was not detected in most of the hydrogen production studies because of elimination of methane producers by heat digestion of sludge (Liu and Shen 2004; Lin and Lay 2003; Yu *et al.*, 2002). However, long retention times may cause methane formation by the mesophilic cultures (Shin *et al.*, 2004). Methane production was also observed when sewage sludge was used as the substrate (Wang *et al.*, 2003; Kim *et al.*, 2004).

Since the hydrogenase enzyme present in anaerobic organ-isms oxidizes reduced ferrodoxin to produce molecular hydrogen, external iron addition is required for hydrogen production. Liu reported that high iron concentrations (100 mg/L) increases lag phase in batch operations and also composition volatile fatty acids (VFA) may vary as a result of metabolic shift in anaerobic digestion. Ten milligram per liter iron concentration was deter-mined to be the optimum in batch hydrogen production by *C. pasteurianum* from starch (Liu *et al.*, 2004).

Nitrogen is an essential nutrient for hydrogen production by dark fermentation under anaerobic conditions. Yokoi reported that the highest level of hydrogen (2.4 mol/mol glucose) could be obtained from starch in the presence of 0.1 per cent polypepton. But no hydrogen production was observed when urea or other nitrogen salts were used as nitrogen source (Yokoi *et al.*, 2001). Maximum specific hydrogen production rate was obtained as 178 mL/g VSS d in the presence of 5.64 g/L $(NH4)_2 HCO_3$ (Liu and Shen 2004). Corn-steep liquor which is a waste of corn starch manufacturing process could be used as nitrogen source (Yokoi *et al.*, 2002). Lin reported that the C/N ratio affected hydrogen productivity more than the specific hydrogen production rate (Lin and Lay 2004).

Hydrogen gas producing organisms are strict anaerobes. Therefore, reducing agents such as argon, nitrogen, hydrogengas and l-cystine·HCl are used to remove trace amounts of oxygen present in the medium. However, the use of such reducing agents is relatively expensive, and therefore uneconomical for industrial production of hydrogen gas. *Enterobacter aero genes* is a facultative anaerobe and the amount of hydrogen produced by this culture is comparable to *Clostridum* sp. (Nakashimada *et al.*, 2002; Yokoi *et al.*, 1998; Yokoi *et al.*, 1997; Tanisho and Ishiwata, 1994; Tanisho and

Ishiwata, 1995; Fabiano and Perego, 2002). The culture has the ability to survive in the presence of slight amount of oxygen generated during anaerobic biodegradation. There-fore, utilization of *E. aerogenes* along with *Clostridum* instead of expensive chemical reducing agents was suggested by Yokoi for effective hydrogen gas production by dark fermentation.

3.2.2. Type of Substrates

3.2.2.1. Use of Simple Sugars

Glucose is an easily biodegradable carbon source, present in most of the industrial effluents and can be obtained abundantly from agricultural wastes. Theoretically bioconversion of 1 mol of glucose yields 12 mol of hydrogen gas (H_2). According to reaction stoichiometry, bioconversion of 1 mol of glucose into acetate yields 4 mol H_2/mol glucose, but only 2 mol H_2/mol glucose is formed when butyrate is the end product. The highest hydrogen yield obtained from glucose is around 2.0–2.4 mol/mol (Ueno *et al.*, 2001;Fang and Liu 2002; Morimoto *et al.*, 2004). Production of butyrate rather than acetate may be one of the reasons for deviations from the theoretical yield. Fang suggested that partial biodegradation of glucose could be another reason for lower yields (Fang and Liu 2002). However, even when more than 95 per cent glucose was degraded, the yield could be less than 1.7 mol H_2/ mol glucose (Lin and Chang 2004). Therefore, utilization of substrate as an energy source for bacterial growth is the main reason for obtaining the yields lower than theoretical estimations.

Batch and continuous hydrogen gas production from sucrose has been widely studied (Tables 9.1 and 9.2). Chen obtained a yield of 4.52 mol H_2/mol sucrose in a CSTR with 8 h hydraulic residence time (Chen and Lin 2004). This yield is higher than the other reported studies such as 3.47 mol H_2/mol sucrose in CSTR (Chen *et al.*, 2001) and 1.5 mol H_2/mol sucrose in UASB (Chang and Lin 2004) at the same HRT. How-ever, the yield from glucose was only 0.91 mol H_2/mole glucose under the same operating conditions in CSTR (Chen and Lin 2000). Optimization of C/N ratio at 47 provided efficient conversion of sucrose to hydrogen gas with a yield of 4.8 mol H_2/mol sucrose (Lin and lay 2004). Similarly, cumulative hydrogen production from sucrose was 300 mL while it was only 140 mL from starch (Khanal *et al.*, 2004). *Enterobacter cloacae* ITT-BY 08 produced 6 mol H_2/mol sucrose which is the highest yield among the other tested carbon sources (Kumar and Das 2000). Collet reported maximum hydrogen yield of 3 mol H_2/mol lactose although theoretical yield is 8 mol H_2/mol lactose (Collet *et al.*, 2004). The results of these studies indicated that the higher hydrogen yields could be obtained from sucrose compared to other simple sugars. However, the yield per mole of hexose remains almost the same for all types of the disaccharides.

3.2.2.2. Use of Starch Containing Wastes

Starch containing materials are abundant in nature and have great potential to be used as a carbohydrate source for hydrogen production. Tables 9.1 and 9.2 summarize the yields and the rates of hydrogen production for batch and continuous operations when starch was used as the substrate. According to the reaction

Table 9.1: Yields and Rates of Bio-hydrogen Production from Pure Carbohydrates by Batch Dark Fermentations

Organism	Carbon Source	H₂ Yield	Per cent H₂ Yield	H₂ Content in Gas Mixture (per cent)	Reference
E. coli	Glucose (20 g/L)	4.73×10^{-8} mol/molglucose			Podestai *et al.*, 1997
Sludge compost	Glucose (10 g/L)	2.1 mol/mol glucose			Morimoto *et al.*, 2004
Mixed culture	Glucose (1 g COD/L)	0.9mol/*mol* glucose	23	60	Logan *et al.*, 2002
Mixed culture	Sucrose (6 g/L)	300 mL/g COD		40	Khanal *et al.*, 2004
C. pasteurium (dominant)	Sucrose (20 g COD/L)	4.8 mol/mol sucrose		55	Lin and Chang, 2004
Mixed culture	Sucrose (1 g COD/L)	1.8 mol/mol sucrose	23		Logan *et al.*, 2002
The rmoanaero ba cteri um	Cellulose (5 g/L)	102 mL/g cellulose	18		Liu *et al.*, 2003
Clostridium sp.	Microcristalline cellulose (25 g/L)	2.18 mmol/g cellulose		60	Lay, 2001
E. aerogenes	Starch[a] (20 g glucose/L)	1.09 mol/mol glucose			Fabiano and Perego, 2002
The rmoanaero ba cteri um	Starch (4.6 g/L)	92 mL/g strach	17	60	Zhang *et al.*, 2003
C. pasteurium	Starch (24 g/L)	106 mL/g starch	19		Liu and Shen, 2004
Mixed culture	Potato starch (1 g COD/L)	0.59 mol/mol starch	15		Logan *et al.*, 2002
Mixed culture	Sugar beet juice	1.7 mol H₂/mol hexose			Hussy *et al.*, 2005

a: Hydrolysate; SHPR: Specific hydrogen production rate; VHPR: Volumetric hydrogen production rate.

stoichiometry, a maximum of 553 mL hydrogen gas is produced from one gram of starch with acetate as a by-product (Zhang *et al.*, 2003). However, the yield may be lower than the theoretical value because of utilization of substrate for cell synthesis.

Table 9.2: Yields and Rates of Bio-hydrogen Production from Pure Carbohydrates by Continuous Dark Fermentations

Organism	Carbon	H_2 Yield	Reactor	Reference
C. acetobutyricum	Glucose	2 mol/mol glucose	Fed-batch	Chin *et al.*, 2003
Mixed culture	Glucose (20 g COD/L)	1.1 mol/mol glucose	CSTR	Chen and Lin, 2000
Mixed culture	Glucose (13.7g/L)	1.2 mol/mol glucose	Trickling biofilter	Oh *et al.*, 2004
Clostridia sp.	Glucose (20 g COD/L)	1.7 mol/mol glucose	CSTR	Lin and Chang, 2004
Mixed culture	Glucose (7 g/L)	2.1 mol/mol glucose	CSTR	Fang and Liu, 2002
Mixed culture	Glucose (20 g/L)		UASB	Kim *et al.*, 2005
Clostridium sp.	Glucose (10g/L)		AMBR[a]	Oh *et al.*, 2004
E. aerogenes HO39	Glucose (10g/L)		Fixed film	Yokoi *et al.*, 1997
Mixed culture	Sucrose (20 g COD/L)	3.47 mol/mol sucrose	CSTR	Chen *et al.*, 2001
Mixed culture	Sucrose	2.1 mol/mol sucrose	CIGSBR[b]	Lee *et al.*, 2004
Mixed culture	Sucrose (20 g COD/L)	1.5 mol/mol sucrose	UASB	Chang and Lin, 2004
Mixed culture	Sucrose (20 g COD/L)	2.6 mol/mol glucose	SBR	Lin and Jo, 2003
Klebsiella oxytocaHP1	Sucrose (50mM)	3.6 mol/mol sucrose	CSTR	Minnan *et al.*, 2005
Mixed culture	Sucrose (20g COD/L)	1.48 mol/mol sucrose	CSTR	Yokoi *et al.*, 2002
C. butyricum+	Starch (2 per cent)	2.5 mol/mol glucose	CSTR	Yokoi *et al.*, 1998
C. butyricum + E.aerogenes	Starch (2 per cent)	2.6mol/mol glucose	Immobilized[c]	Yokoi *et al.*, 1998
Thermococcus kodakaraensis KOD1	Starch (5 g/L)	3.33 mol/mol starch	Gas-liftfermenter	Kanai *et al.*, 2005
Mixed culture	Wheat starch (10 g/L)	0.83 mol/mol starch d	CSTR	Hussy *et al.*, 2000
Mixed culture	Starch (6kg starch/m³)	1.29L/g starch COD	CSTR	Lay, 2000
C. termolacticum	Lactose (29mmol/L)	3 mol/mol lactose	CSTR	Collet *et al.*, 2004

a: Anaerobic membrane bioreactor; b: CIGBR, carrier induced granular bed reactor; c: Immobilization on porous glass beads; SHPR, specific hydrogen production rate; VHPR, volumetric hydrogen production rate.

The maximum specific hydrogen production rate was 237 mL H_2/g VSS d when 24 g/L edible corn starch was used as the substrate by *C. pasteurianum* (Liu and Shen 2004). Zang obtained higher specific yield of 480 mL H_2/gVSS d with 4.6 g/L starch concentration at 37~C using a mixed sludge. Thermophilic conditions did not improve the production rate yielding 365 mL H_2/g VSS d with *Thermoanaerobacterium* at 55°C (Zhang *et al.*, 2003).

Yokoi used dried sweet potato starch residue for hydrogen production by the mixed culture of *C. butyricum* and *E. aerogenes*. Hydrogen yield obtained in long term repeated batch operations was 2.4 mol H_2/mol glucose from 2.0 per cent starch residue containing wastewater (Yokoi *et al.*, 2001).

3.2.2.3. Use of Cellulose Containing Wastes

Cellulose is the major constitute of plant biomass and highly available in agricultural wastes and industrial effluents such as pulp/paper and food industry. Hydrogen gas production potential from micro-crystalline cellulose at mesophilic conditions with heat-digested sludge was investigated by Lay (Lay 2001). Increasing cellulose concentration resulted in lower yields with the maximum value of 2.18 mol H_2/mol cellulose with 12.5 g/L cellulose concentration. However, 25 g/L cellulose concentration provided the highest specific hydrogen production rate of 11.16 mmol/gVSS d. Liu reported that cellulose is converted to hydrogen with a higher rate at 37°C, but more hydrogen was accumulated at thermophilic range. The maximum hydrogen yield obtained in this study was 10^2 mL/g cellulose which is only 18 per cent of the theoretical yield (Liu *et al.*, 2003). Low yield was explained as partial hydrolysis of cellulose. Taguchi hydrolyzed the cellulose and used the hydrolysate for fermentation by a *Clostridium* sp. During an 81 h period of stationary culture, the organisms consumed 0.92 mmol glucose/h and produced 4.10 mmol H_2/h (Taguchi *et al.*, 1996). The same culture was also used for hydrogen production from pure xylose or glucose and enzymatic hydrolysate of Avicel cellulose or xylan. The hydrogen yield from the hydrolysate was higher than that of carbohydrates as 19.6 and 18.6 mmol H_2 per gram of substrate consumed, respectively (Taguchi *et al.*, 1995).

3.2.2.4. Use of Food Industry Wastes and Wastewater

Food industry wastes constitute a major fraction of the municipal solid wastes. Landifilling, composting and incineration are the conventional approaches for the solid waste management. How-ever, high carbohydrate content in form of simple sugars, starch and cellulose makes the solid food wastes a potential feedstock for biological hydrogen production. The problem with the food waste is the variations in carbohydrate and protein types and concentrations in the mixture. Each component requires different environmental and bio-processing conditions for hydrogen gas production. Table 9.3 summarizes hydrogen gas production from different wastewaters and solid wastes.

The feasibility of biological hydrogen production from organic fraction of municipal solid wastes (OFMSW) was investigated by Lay *et al.* (1999). under mesophilic conditions by using mixed anaerobic bacterial flora. At high F/M ratio (0.4 g OFMSW/g biomass), the pre-treated digested sludge had a high hydrogenic activity with 43 mL/g VSS h specific production rate and 125 mL/g TVS h production potential (Lay *et al.*, 1999). Kim obtained 111.2 mL H_2/g VSS h when food waste was used as sole substrate. Addition of sewage sludge onto food waste as a rich protein source did not improve the production rate (Kim *et al.*, 2004). Similarly, hydrogen production potential of carbohydrate rich high solid organic waste (HSOW) was 20 times larger than those of fat rich HSOW and protein rich HSOW. This is probably because of the consumption of hydrogen gas to form ammonium using nitrogen generated from biodegradation of protein rich solid wastes (Lay *et al.*, 2003). Shin reported higher production potential and specific H_2 production rates from food wastes under thermophilic conditions as compared to the mesophilic processes (Shin et al., 2004). The effect of dilution rate (HRT) on hydrogen production from food wastes was studied by Han and 58 per cent COD reduction, 70 per cent hydrogen formation efficiency, over 100 L cumulative H_2 gas were obtained at optimum dilution rate of 4.5 d-1 or HRT of 5.3h (Han and Shin 2004).

Table 9.3

Organism	Carbon Source	Per cent H₂ Content	Reference
Mixed culture	OFMSW	66	Lay et al., 1999
The rmoanaero ba cteri um	Food waste (6 gVS/L)	55	Shin et al., 2004
Mesophilic mixed culture	Food waste (3 per cent VS)	1	Shin et al., 2004
Mixed culture	Food waste (3 per cent VS)		Kim et al., 2004
Mixed culture	Potato Ind. WW (21 g COD/L)	60	Ginkel et al., 2005
Mixed culture	Apple (9 g COD/L)	60	Ginkel et al., 2005
Mixed culture	Domestic WW	23	Ginkel et al., 2005
E. aerogenes	Molasses (2 per cent sucrose)	60	Yokoi et al., 1997
Mixed culture	Rice winery WW (36 g COD/L)	53–61	Yu et al., 2002
Mixed culture	Biosolid		Wang et al., 2003
Mixed culture	Filtrate		Wang et al., 2003
C. butyricum + E. aerogenes	Sweet potato starch residue (0.5 per cent)		Yokoi et al., 2001
C. butyricum + E. aerogenes	Sweet potato starch residue (2 per cent)		Yokoi et al., 2002

OFMSW, organic fraction of solid waste; SHPR, specific hydrogen production rate; VHPR, volumetric hydrogen production rate.

Food processing industrial wastewaters are carbohydrate rich effluents. Ginkel studied hydrogen production from confectioners, apple and potato processor industrial effluents and also from domestic wastewater. The highest production yield was obtained as 0.21 L H_2/g COD from potato processing wastewater (Ginkel et al., 2005). Molasses is another carbohydrate rich substrate and it is a good source of sucrose. The maximum and available rate of hydrogen production in continuous operation with E. aerogenes strain E.82005 was 36 and 20 mmol H_2/L h, respectively. The available yield was 1.5 mol H_2/mol sugar expressed in terms of sucrose (Tashino and Ishiwata 1994). Immobilization of cultures on polyurethane foam increased the yield to 2.2 mol H_2/mol sugar (Tashino and Ishiwata 1995).

3.2.2.5. Use of Waste Sludge

Biosolids (sludges) from wastewater treatment plants contain large amounts of polysaccharides and proteins. Hydrogen yields of 1.2mg H_2/g COD (Wang et al., 2003) and 0.6 mol/kg CODi (Wang et al., 2004) were reported when sludge was used as the raw material. However, higher hydrogen yields (15 mg H_2/g COD) were obtained from the filtrate (Wang et al., 2003). Pre-treatment of the sludge increased the soluble COD enhancing the hydrogen yield (0.9 mmol/g dried sludge) (Wang *et al.*, 2003).

3.3. Hydrogen Gas Production by Photo-Fermentations

3.3.1. Types of Organisms and the Conditions

Some photo-heterotrophic bacteria are capable of converting organic acids (acetic, lactic and butyric) to hydrogen (H_2) and carbon dioxide (CO_2) under anaerobic conditions in the presence of light. Therefore, the organic acids produced during the acidogenic phase of anaerobic digestion of organic wastes can be converted to H_2 and CO_2 by those photosynthetic anaerobic bacteria. Hydrogen gas production capabilities of some purple photosynthetic bacteria such as *Rhodobacterspheroides* (Koku *et al.*, 2002), *Rhodobacter capsulatus* (He *et al.*, 2005), *Rhodovulum sulfidophilum* W-1S (95,96) and *Rhodopseudomonaspalustris* (97) have been investigated to some extent. Photoproduction of hydrogen from CO or other organic acids by carbon-monoxide dependent dehydrogenase (CODH) enzyme containing cultures such as *Rhodospirillum rubrum* and *Rodopseudomonospalsutris* P4 has also been reported (Najafpour *et al.*, 2004; Oh *et al.*, 2004; Oh *et al.*, 2002).

The optimum growth temperature and pH for the photo-synthetic bacteria is in the range of 30–35 ~C and pHopt 7.0, respectively (Kim *et al.*, 2006; He *et al.*, 2005). The hydrogen production takes place under anaerobic conditions with light illumination. The organisms prefer organic acids as carbon source such as acetic (Fang *et al.*, 2005), butyric (Fang *et al.*, 2005), propionic (Shi and Yu, 2004), lac-tic (Federov *et al.*, 1998; Kondo *et al.*, 2002; Fascetti and Todini, 1995; He *et al.*, 2005; Miyake *et al.*, 1999) and malic acid (Eroglu *et al.*, 1992). However, other carbohydrates (Maeda *et al.*, 2003; Ike *et al.*, 1999) and industrial effluents may also be used for hydrogen gas production by photosynthetic bacteria (Eroglu *et al.*, 2004; Singh *et al.*, 1994). Table 9.4 summarizes the yields and the rates of hydrogen production from different organic acids by the photo-fermentative organisms. Hydrogen production rates vary depending on the light intensity, carbon source and the type of microbial culture. On the basis of available literature the highest conversion efficiency was obtained using lactic acid as the carbon source (Federo *et al.*, 1998; Fascetti and Todini, 1995; He *et al.*, 2005).

Nitrogenase is the key enzyme that catalyzes hydrogen gas production by photosynthetic bacteria. The activity of the enzyme is inhibited in the presence of oxygen, ammonia or at high N/C ratios (Koku *et al.*, 2003). Therefore, the process requires ammonium limited and oxygen free conditions (Zhu *et al.*, 2001; Takabatake *et al.*, 2004). Hydrogen production by R *sphaeroides* is completely inhibited at ammonia concentrations above 2mM (Yokoi *et al.*, 1998). Hydrogen gas production was lower in the presence of ammonia salts, while proteins such as albumin, glutamate and yeast extract as a nitrogen source enhanced the production (Oh *et al.*, 204; Takabatake *et al.*, 2001). The metabolism shifts to utilization of organic substance for cell synthesis rather than hydrogen production in the presence of high nitrogen concentrations resulting in excess biomass growth and reduction in light diffusion (Fascetti and Todini, 1995; Oh *et al.*, 2004). However, hydrogen production activity can be recovered after ammonia is consumed. It was reported

Table 9.4: Yields and Rates of Bio-hydrogen Production from Organic Acids by Photo-fermentations

Organic Acid	Organism	Concentration	Light intensity	Conversion Efficiency (per cent)	Process	Reference
Acetate	Rhodopseudomonas	22mM	680 ~molphotons/m² s	72.8	Batch	Barbosa et al., 2001
	R. palustris	22mM	480 ~molphotons/m² s	14.8	Batch	Barbosa et al., 2001
	R. palustris		2500 lux	60–70	Batch	Oh et al., 2004
	R. capsulata	4g/L	200W/m²	76.5	Batch	Fang et al., 2005
	R. capsulata	1.8g/L	4170 lux	t32.6	Batch	Shi and Yu, 2005
Lactate	Rhodopseudomonas	50mM	680 ~molphotons/m² s	9.6	Batch	Barbosa et al., 2001
	R. palustris	50mM	480 ~molphotons/m² s	12.6	Batch	Barbosa et al., 2001
	R. sphaeroides RV	100mM	3klx	80	CSTR	Fascetti et al., 1995
	R. capsulatus IR3	30 mmol	120W	84.8	Batch	He et al., 2005
	R. sphaeroides GL-1	20mM	300W/m²	86	c	Federov et al., 1998
Butyrate	Rhodopseudomonas	27mM	680 ~molphotons/m² s	8.4	Batch	Barbosa et al., 2001
	R. capsulata	1 g/L	200W/m²	67.6	Batch	Fang et al., 2005
Malate	Rhodopseudomonas	15mM	680 ~molphotons/m² s	6.6	Batch	Barbosa et al., 2001
	R. palustris	15mM	480 ~molphotons/m² s	36	Batch	Barbosa et al., 2001
	R. sphaeroides	15mM	200W/m²		Batch	Eroglu et al., 1999
	R. sphaeroides	7.5mM	150–250W/m²	35–45	Batch	Koku et al., 2002
PHB[d]	R. sulfidophilum		190W/m²		Batch	Maeda et al., 2003
Succinate	R. sulfidophilum	50mM	190W/m²		Batch	Maeda et al., 2003

a: Light conversion efficiency; b: H_2 yield mol/mol substrate; c: Immobilized on polyurethane foam; d: PHB, poly-hydroxy butyrate; 210 ~mol photons/m² s = 190W/m².

that presence of carbonate enhanced ammonia removal and stimulated hydrogen production (Takabatake *et al.*, 2004). Two stage ammonia removal and hydrogen production process has been suggested for hydrogen production from high level ammonia containing wastewater (Fascetti and Todini, 1995).

Hydrogenase enzyme in photo-fermentative bacteria is an uptake hydrogenase which utilizes hydrogen gas and therefore is antagonistic to nitrogenase activity (Koku *et al.*, 2002). Uptake hydrogenase activity should be limited for enhanced hydrogen gas production. Hydrogenase deficient mutant cultures of photo-fermentative bacteria could produce 2–3 times more hydrogen (Kim *et al.*, 2006).

One of the parameters affecting the performance of photo-fermentation is the light intensity. Increasing light intensity has a stimulatory affect on hydrogen yield and production rate, but has an adverse effect on the light conversion efficiency (Shi and Yu, 2005; Barbosa *et al.*, 2001). Kondo found that the reduced pigment mutant of *R. sphaeroides* MTP4 produces hydrogen more efficiently under high light intensity as compared to the wild type (Kondo *et al.*, 2002; Kondo *et al.*, 2002). Light intensity might also affect the consumption rates of organic acids. Shi stated that butyrate consumption requires higher light intensities (4000 lux) as compared to acetate and propionate (Shi and Yu *et al.*, 2005).

Hydrogen production under dark conditions is usually lower than that of the illuminated conditions (Koku *et al.*, 2003; Oh *et al.*, 2002). However, alternating 14 h light/10 h dark cycles yielded slightly higher hydrogen production rates and cell concentrations as compared to continuous illumination (Koku *et al.*, 2003). Similarly, Wakayama reported that hydrogen production rate during 30 min dark/light cycle was 22 L/m^2 d which was twice as much as that obtained by illuminated culture during a 12 h cycle under the same conditions (Wakayama *et al.*, 2000).

3.3.2. Types of Substrates

Utilization of industrial effluents for hydrogen gas production by photosynthetic bacteria is possible although, these cultures prefer organic acids as carbon sources. One of the major problems in hydrogen gas production from industrial effluents is the color of wastewaters, which could reduce the light penetration. High ammonia concentration is another problem which inhibits the nitrogenase enzyme reducing the hydrogen productivity. High organic matter content (COD) and presence of some toxic compounds (heavy metals, phenolics and PAH) in industrial effluents may require pre-treatment before hydrogen gas production.

Table 9.5 summarizes hydrogen production studies from some food industry wastewaters by photo-fermentation. Photo-production of hydrogen from pre-treated sugar refinery wastewater (SRWW) was studied by Yetis in a column photo-bioreactor using *R. sphaeroides* OU 001 (Yetis *et al.*, 2000). The hydrogen production rate was 3.8mL/Lh at 32~C in batch operation with 20 per cent diluted SRWW. Addition of malic acid (20 g/L) into SRWW enhanced the production rate to 5 mL/L h. Eroglu reported that high dilutions (3–4 per cent) of olive mill waste (OMW) are necessary to alter the inhibitory effects of high organic content and dark color of OMW. Two percent dilution resulted in the highest hydrogen production potential of 13.9 L H_2/L WW at 32~C with *R. sphaeroides* OU 001 and around 35 per cent COD reduction was observed (Eroglu *et al.*, 2004).

Tofu wastewater is a carbohydrate and protein rich effluent. Hydrogen yield from tofu wastewater (1.9 L/L wastewater at 30~C) was comparable to the yield from glucose (3.6 L/L wastewater) by immobilized *R. sphaeroides* RV on agar gels. No ammonia inhibition (2 mM) was observed and 41 per cent of TOC was removed (Zhu *et al.*, 1999). Fifty percent dilution of the wastewater increased the yield to 4.32 L/L wastewater and TOC removal efficiency to 66 per cent (Zhu *et al.*, 2002).

Singh investigated the hydrogen gas production from potato starch, sugar cane, juice and whey by using *Rhodopseudomonas* sp. Among the three substrates sugar cane juice yielded the maximum level of hydrogen production (45 mL/mg DW h) as compared to potato (30 mL/mg DW h) and whey (25 mL/mg DW h) (Singh *et al.*, 1994). On the contrary, no hydrogen production from raw starch was observed by salt-tolerant photosynthetic bacteria *Rhodobium marinum* (Ike *et al.*, 1999).

3.3.3. Photo-bioreactors for Bio-hydrogen Production

The major types of photo-bioreactors developed for hydrogen production are tubular, flat panel and bubble column reactors.The features of these photo-bioreactors have been reviewed by Akkerman *et al.* (2002) and the importance of photochemical efficiency (theoretical maximum 10 per cent) in hydrogen production was strongly emphasized. High illuminations cause lower light conversion yields, but higher hydrogen production rates. How-ever, excess light could also cause photo-inhibition resulting in decreases in hydrogen production rate. A mutant type photosynthetic bacteria have been developed to increase the light conversion efficiency and hence hydrogen production rate (Miyake *et al.*, 1999). Although an improvement was observed by mutant type, the light conversion efficiency was around 6 per cent which is still less than theoretical efficiency. El-Shishtawy reported 9.23 per cent maximum light conversion efficiency by using light-induced and diffused photo-bioreactor (IDPBR) at 300 W/m^2 light intensity (El-Shishtawy *et al.*, 1997). The width of the culture significantly affected the productivity which reached to 7577 mL H_2/m^2 h or 50 mL/L culture h hydrogen production rate with 1cm culture width. Similar results were observed by Nakada in a photo-bioreactor composed of four compartments aligned along the light penetration axis (Nakada *et al.*, 1995). The efficiency of the conversion of light to hydrogen increased with the depth in the reactor and 1 cm depth showed the highest efficiency.

In relating to solar hydrogen production, the light conversion efficiency could be less during mid-day because of high light intensity (1.0 kW/m^2). In addition, a delay of 2–4 h was observed in maximum hydrogen production rate (3.4 L H_2/m^2 h) after the highest light intensity at noon with an average light conversion efficiency of 1.4 per cent (Miyake *et al.*, 1999). Wakayama developed a light shade bands photo-bioreactor system to improve the solar hydrogen production efficiency. 3.5 per cent light conversion efficiency at mid-day with over 0.8 kW/m^2 light intensity was obtained in photobioreactors with light shade bands whereas photo-inhibition was observed at 0.4 kW/m^2 in the ones without shade bands (Wakayama and Miyake, 2002).

The other important parameter to be controlled in photobioreactors is mixing. Argon gas was commonly used for mixing and providing anaerobic conditions in photo-bioreactors although not cost-effective. It was observed that continuous argon sparging inhibited the growth of *Rhodopseudomonas* in a pneumatically agitated photo-bioreactor (Figure 9.2a) because of CO_2 loss whereas recirculation provided better growth of the culture (Hoekemann *et al.*, 2002). A novel flat-panel airlift photo-bioreactor with baffles (Figure 9.2b) was developed by Degen *et al.* (2001). It was observed that both installation of baffles for better mixing and reduction in the light path provides a significant increase in the biomass productivity. Although this photo-biorecator was used for cultivation of *Chlorella vulgaris*, it may also be used inhydrogen gas production.

Tredici type multi-tubular photo-bioreactors (Figure 9.2c) was used for hydrogen gas production in the presence and absence of light by using *Spirulina* (Benemann, 2000).Tubular reactors are made up of parallel transparent tubes filled with water. The system is inclined with a 10–30 per cent slope to allow gas bubbles to rise. A modification of tubular reactor was developed by Modigell as a modular outdoor photo-bioreactor (Figure 9.2d) (Modigell and Holle, 1998). The hydrogen production rate from lactate reached 2 L/m² h with light conversion efficiency of 2 per cent in outdoor experiments.

3.4. Hydrogen Gas Production by Sequential Dark and Photo-fermentation

Higher hydrogen production yields can be obtained when two systems are combined (Yokoi *et al.*, 2002; Yokoi *et al.*, 1998). Further utilization of organic acids by photo-fermentative bacteria could provide bet-ter effluent quality in terms of COD. However, the system should be well-controlled to provide optimum media composition and environmental conditions for the two microbial components of the process (Yokoi *et al.*, 2001; Yokoi *et al.*, 2002; Yokoi *et al.*, 1998; Fascetti *et al.*, 1998). The ammonia concentration and C/N ratio in the effluent of anaerobic fermentation should not be at the inhibitory level for the photosynthetic bacteria (Lee *et al.*, 2002; Fascetti *et al.*, 1998). Dilution and neutralization of dark fermentation effluents are required before photo-fermentation to adjust the organic acid concentration and the pH7 for the optimal performance of photosynthetic bacteria (Fascetti *et al.*, 1998; Kawaguchi *et al.*, 2001).

Bio-hydrogen production by co-culture of anaerobic and photosynthetic bacteria in single stage has also been studied. Yokoi obtained higher hydrogen production yield (4.5 mol/mol glucose) by co-culture of *C. butryricum* and *Rhodobacter* sp. as compared to single stage dark fermentation (1.9 mol/mol glucose) and sequential two step fermentation (3.7 mol/mol glucose) of starch (Yokoi *et al.*, 1998). Similarly, higher hydrogen yields from different substrates were reported by co-cultures of *R. marinum* and *V. fluvialis* compared to *R. marinum* alone (Ike *et al.*, 1999). Better hydrogen yield (60 per cent), but lower production rate was observed in combined fermentation and hydrogen gas production by *Lactobacillus amylovous* and *R. marinum* from starch accumulating algae in comparison to sequential two stage fermentation (45 per cent) (Kawaguchi *et al.*, 2001). In addition, pH of the mixed fermentation remained around pH 7 which is considered as an advantage over the two stage fermentation process. However, the differences in organic acid production/consumption rates, and therefore potential accumulation of organic acids in the media, decreases in light penetration because of suspended solids are the major problems in mixed fermentation processes.

Hydrogen is considered as the 'energy for future' since it is a clean energy source with high energy content as compared to hydrocarbon fuels. Unlike fossil fuels, petroleum, natural gas and biomass, hydrogen is not readily available in nature. Therefore, new processes need to be developed for cost-effective production of hydrogen. Chemical methods such as steam reforming of hydrocarbons and partial oxidation of fossil fuels operate at high temperatures, and therefore are energy intensive and expensive. Biological methods offer distinct advantages for hydrogen production such as operation under mild conditions and specific conversions. However, raw material cost is one of the major limitations for bio-hydrogen production. Utilization of some carbohydrate rich, starch or cellulose containing solid wastes and/or some food industry wastewaters is an attractive approach for bio-hydrogen production.

Among the various methods used for bio-hydrogen production are: (a) water splitting by photosynthetic algae, (b) dark fermentation of carbohydrate rich wastes and (c) photo-fermentation of organic acid rich wastewaters. Algal production of hydrogen is rather slow, requires sunlight and is inhibited by oxygen. Hydrogen is produced as a by-product during acidogenic phase of anaerobic digestion of organic wastes which is known as the dark fermentation. The yield of hydrogen production by dark fermentation is low and the rate is slow. Organic acids produced during the dark fermentation of carbohydrate rich wastes may be converted to hydrogen and CO_2 by photoheterotrophic bacteria. The process requires special organisms, light and strict control of the environmental conditions. Sequential or combined bioprocesses of dark and photo-fermentations seem to be the most attractive approach resulting in high hydrogen yields for hydrogen production from carbohydrate rich wastes.

The major problems in bio-hydrogen production from wastes are the low rates and yields of hydrogen formation. Large reactor volumes are required for bio-hydrogen production due to low hydrogen production rates. Low yields and the rates of hydrogen formation may be overcome by selecting and using more effective organisms or mixed cultures, developing more efficient processing schemes, optimizing the environmental conditions, improving the light utilization efficiency and developing more efficient photo-bioreactors. Due to inhibition of bio-hydrogen production by oxygen and ammonium-nitrogen, microbial growth and hydrogen formation steps may need to be separated in order to improve the hydrogen productivity. Considerable research and development studies are needed to improve the 'state of the art' in bio-hydrogen production.

Chapter 10

Application of Electrochemical Technology for Bioproduction of Hydrogen and Sustainable Development

1. Introduction

As the depletion of fossil fuel generated a great concern for the use of renewable resources as the supplement for fuel. Solar, hydroelectric power and microbial system are known to be abundant renewable resources for fuel production. Hydrogen has a energy yield of 122 kJ/g, which is 2.75 times greater than hydrocarbon fuels (Kapdan and Kargi, 2006). Hydrogen together with oxygen is the key element in the biological energy cycle on the earth. In all organic matter hydrogen atoms are bound to carbon, nitrogen, sulphur and other elements.

Biological processes for the production of hydrogen, which are environment-friendly and less energy intensive, may be categorized into bio-photolysis, photo-fermentation and dark fermentation. Bio-photolysis occurs in organisms such as green algae or cyanobacteria, which carry out plant-type photosynthesis, using captured solar energy to split water. Non-sulphur purple photosynthetic bacteria undergo photo-fermentation to perform an anaerobic photosynthesis. By dark fermentation, a variety of different microbes anaerobically breaks down carbohydrate rich substrates to hydrogen and by-products (Das and Veziroglu, 2001; Hallenbeck and Benemann 2002). Gaseous hydrogen is produced as well as consumed by living microorganisms in the presence or absence of oxygen (under both oxic and anoxic conditions). The anoxic condition is observed during dark fermentation of microbes.

Among the processes, dark fermentation presents a high rate of hydrogen production, using fermentative bacteria, such as *Enterobacter* species (Palazzi *et al.*, 2000; Kumar and Das, 2000; Kumar and Das, 2001; Nakashimada *et al.*, 2002; Kurokawa and Tanisho, 2005; Zhang *et al.*, 2005; Shin *et al.*, 2007), *Clostridium* species (Chin *et al.*, 2003; Lee *et al.*, 2004; Levin *et al.*, 2006; Jo *et al.*, 2008) and *Escherichia coli* (Yoshida *et al.*, 2005). Hydrogen production through bacterial fermentation is currently limited to a maximum of 4 moles of hydrogen per mole of glucose, and under these conditions results in a fermentation end product (acetate; 2 mol/mol glucose) that bacteria were unable to further convert to hydrogen. Thermophiles produced up to 60–80 per cent of the theoretical maximum, demonstrating that higher hydrogen yields can be reached by extremophiles than using mesophilic anaerobes (Chin *et al.*, 2003).

Several problems still remain for the commercial scale production of bio-hydrogen including low hydrogen yield. Alternatively the by-products are to be used by microorganisms or by bioelectrochemical technology, so that higher moles of hydrogen may be produced.

Electrolysis is a method of separating bonded elements and compounds by passing an electric current through them. One important use of electrolysis is to produce hydrogen, which has been suggested as an energy carrier for powering electric motors and internal combustion engines. All electrolysers work according to a principle of two electrodes separated by an electrolyte. A so-called half cell reaction resulting in the formation of hydrogen and oxygen respectively takes place at each electrode. The role of the electrolyte is to close the electrical circuit by allowing ions (but not electrons) to move between the electrodes.

Bioelectrochemically assisted microbial system has the potential to produce 8-9 mol H_2/mol glucose (Liu *et al.*, 2005). Hence, the hybrid technology is an alternative for the production of hydrogen with higher efficiency.

2. Hydrogen Production using Microbial Systems

Different microorganisms participate in the biological hydrogen generation by using photofermentation or dark fermentation such as green algae, microalgae and bacteria, as shown in Table 10.1.

2.1 Major Enzymes for Metabolizing and Producing Hydrogen

The enzymes catalyzing the formation and the oxidation of hydrogen are collectively called hydrogenases. The enzyme reaction is represented by Equation 1:

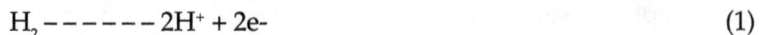

$$H_2 ------ 2H^+ + 2e- \tag{1}$$

In spite of many similarities between the hydrogenases their catalytic and physicochemical properties vary widely. There are three fundamentally different hydrogen producing and metabolizing enzymes found in algae and bacteria (Schlegel and Schneider, 1978):

☆ Reversible or classical hydrogenases,

☆ Membrane-bound hydrogenases, and

☆ Nitrogenase enzymes

Hydrogenase can be differentiated with respect to their position in electron transport systems and their location in the cell. The natural electron donor/acceptor is known only for the soluble, cytoplasmic or loosely bound periplasmic enzymes. For the membrane-bound hydrogenases this information is incomplete or lacking. Details of types and properties of hydrogenases are presented in the literature (Schlegel and Schneider, 1978; Adams *et al.*, 1981). A compilation of papers on function and structure of hydrogenases has been published (Yagi, 1981). Nitrogenase is also responsible for hydrogen evolution by many bacteria. Hence, hydrogenases and nitrogenases possessing microbes can produce hydrogen by their metabolic pathways (Schlegel and Schneider, 1978).

2.1.1 Reversible Hydrogenases

The reversible hydrogenase is located at the cytoplasmic membrane (Kentemich, 1991). It has the dual function of catalysing hydrogen evolution and hydrogen uptake (Lambert and Smith, 1981). It has been suggested that this enzyme functions as a valve for low potential electrons generated during the light reaction of photosynthesis, thus preventing the slowing down of the electron transport chain (Appel, 2000). It is available in the majority of the nitrogen-and non-nitrogen-fixing cyanobacteria (Eisbrenner, 1978). Reversible hydrogenase is a heterotetrameric, NAD-reducing enzyme, consisting of a hydrogenase (encoded by hoxY and hoxH genes) and a diaphorase part (encoded by hoxF and hoxU genes).

Broad Classification	Microorganisms	Enzymes Involved
Green algae	*Scenedesmus obliquus*	Hydrogenase
	Chlamydomonas reinhardii	
	C. moewusii	
Cyanobacteria Heterocystous	*Anabaena azollae*	Nitrogenase
	Anabaena CA	
	A. variabilis	
	A. cylindrical	
	Nostoc muscorum	
	N. spongiaeforme	
	Westiellopsis prolifica	
Cyanobacteria Nonheterocystous	*Plectonema boryanum*	Nitrogenase
	Oscillotoria Miami BG7	
	O. limnetica	Nitrogenase, Membrane-bound hydrogenase
	Synechococcus sp.	Nitrogenase
	Aphanothece halophytico	

Contd...

Contd...

Broad Classification	Microorganisms	Enzymes Involved
	Mastidocladus laminosus	
	Phormidium valderianum	
Photosynthetic bacteria	*Rhodobater sphaeroides*	Nitrogenase, Membrane-bound hydrogenase
	R. capsulatus	
	R. sulidophilus	
	Rhodopseudomonas sphaeroides	
	R. palustris	
	R. capsulate	
	Rhodospirillum rubnum	
	Chromatium sp. Miami PSB	
	Chlorobium limicola	
	Chloroexu aurantiacus	
	Thiocapsa roseopersicina	
	Halobacterium halobium	
Fermentative bacteria	*Enterobacter aero genes*	Hydrogenase
	E. cloacae	
	Clostridium butyricum	
	C. pasteurianum	
	Desulfovibrio vulgaris	
	Magashaera elsdenii	
	Citrobacter intermedius	
	Escherichia coli	

2.1.2 Uptake Hydrogenases

Uptake hydrogenase is located at the cytoplasmic face of the cell membrane or thylakoid membrane, where it uses hydrogen evolved by nitrogenase. There is a considerable loss of energy through the production of hydrogen during nitrogen fixation. Some of this energy can be regained through the action of uptake hydrogenase. This enzyme splits the hydrogen and feeds the electrons back into the electron-transport chain. The reduction of a substrate with a relatively high redox potential like cytochrome through this hydrogenase seems to be a wasteful process. But since nitrogen-fixing cells maintain a highly reducing environment, it seems necessary to use part of the reductive power of hydrogen and saving reducing equivalents. Hydrogen-using uptake hydrogenase has several functions:

☆ It serves as one of the mechanisms to protect oxygen-sensitive nitrogenase (Robson and Postgate, 1980).

☆ It generates ATP in the hydrogen-dependent respiratory oxygen uptake (Knallgas or oxyhydrogen reaction) and

☆ It provides additional reducing equivalents to photosystem-I.

Uptake hydrogenase has been found in all heterocystous cyanobacteria and in some non-heterocystous cyanobacteria (Peschek, 1979). The structural genes encoding cyanobacterial uptake hydrogenases have been sequenced and characterized in only a few strains (Axelsson, 1999). The large subunit of the enzyme is encoded by hupL genes and small subunit is encoded by hupS genes. In the organisms studied so far, there is a high degree of homology in the gene sequence of hupSL (Tamagnini, 1997). However, the mode or rearrangement of the genes varies from one organism to another (Axelsson, 1999).

2.1.3 Nitrogenase

All nitrogenases studied so far are catalysts for H_2 production as they liberate H_2 during the reduction of nitrogen to ammonia. A minimum of 25 per cent of the electron flux through nitrogenase is used in the reduction of protons to H_2.

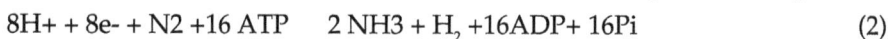

$$8H+ + 8e- + N2 +16\ ATP \quad 2\ NH3 + H_2 +16ADP+ 16Pi \tag{2}$$

ATP, reductant and electrons are provided by photosynthesis or by degradation of sugars in cyanobacteria. Nitrogenase is a metalloenzyme complex consisting of dinitrogenase (MoFe protein: ~2â2) and dinitrogenase reductase (Fe protein: â2). The Mo-Fe protein or component-I is a larger component is responsible for the catalytic reduction of substrate molecules. The Mo-Fe protein from all sources examined are O_2 labile, have molecular weights of approximately 220,000 daltons. Approximately 2 mol of molybdenum and 24±32 mol of iron and sulphide are found per mol of protein (Kim and Rees, 1994). The second protein dinitrogenase reductase or component II accepts electrons from donors such as ferredoxin or flavodoxin, or dithionite and transfers these electrons to dinitrogenase with the concomitant hydrolysis of two molecules of ATP per electron transferred. The six electron reduction of N_2 to $2NH_3$, therefore requires a minimum of 12 ATP molecules making nitrogen fixation an energetically expensive process. The Fe protein is also O_2 labile and has an average molecular weight of about 60,000 daltons. The protein consists of two subunits of equal weight (Kim and Rees, 1994). In addition to reducing nitrogen to ammonia, dinitrogenase can reduce a number of substrates such as protons, acetylene, cyanide, nitrous oxide and azide. Apart from the conventional molybdenum-based nitrogenase, an alternative vanadium-based nitrogenase has also been reported (Kentemich, 1988). *A. variabilis* can express a third nitrogenase when grown under vanadium and molybdenum deficiency (Kentemich, 1991). This nitrogenase contains vanadium in the prosthetic group. A novel mutant of Azotobacter which has a tungsten-based nitrogenase has also been isolated (Kajii, 1994). In photosynthetic bacteria and cyanobacteria, photohydrogen production is mainly associated with nitrogenase rather than hydrogenase and coupled with ferredoxin or flavodoxin (Kosaric and Lyng, 1988). It requires ATP and is inhibited by N2 or NH_4. In this case, ferredoxin is reduced (1) directly by a light-driven reaction, (2) indirectly by ATP-driven reversed electron transport, or (3) by dehydrogenation or oxidative de carboxylation reactions of intermediary metabolism not involving electron transport chains (Kosaric and Lyng, 1988). Nitrogenase is an extremely common, if not universal, enzyme in photosynthetic bacteria (Stewart, 1973). It is difficult to ascertain its prevalence in cyanobacteria since oxygenic photosynthesis in these microbes is inherently incompatible with the nitrogenase protein. Cyanobacteria have evolved several mechanisms to overcome the O_2 incompatibility of nitrogenase.

2.1.4 Genetic Engineering Aspects of Biohydrogen Production

Genetic engineering is the transfer of genes of interest from one organism into other known organism for its ease of culturing and its efficient metabolic activity. Usually *E.coli* is considered as the universal host and it is consequently well characterized for harbouring the foreign genes. Especially for hydrogen, *E. coli* possesses different membrane-bound hydrogenases under specific conditions: the two enzymes are hydrogenase 3(Hyd-3) and hydrogenase 4(Hyd-4) responsible for hydrogen gas production as well as hydrogenase 1 (Hyd-1) and hydrogenase 2 (Hyd-2) responsible for hydrogen uptake. The entire gene regulation in *E.coli* for hydrogen production is shown in Figure 10.1. *E. coli* cells convert glucose to various organic acids (such as succinate, pyruvate, lactate, formate, and acetate) to synthesize energy and hydrogen from formate by the formate hydrogen-lyase (FHL) system that consists of hydrogenase 3 and formate dehydrogenase-H. Bacterial strain, *E.cloacae* IIT-BT 08 was isolated and characterized shown enhancement in biohydrogen production (Kumar and Das, 2000). The gene [Fe]-hydrogenase encoding gene isolated from *E.cloacae* IIT-BT 08 has been over-expressed in fast growing non-hydrogen producing *E.coli* BL-21 using pGEX 4T-1 vector (Mishra, 2004). Hence genetic engineering helps in the effective production of hydrogen.

2.1.5 Biohydrogen Production using Phototrophic Microorganisms

Photosynthetic bacteria can use small-chain organic acids as electron donors for the production of hydrogen at the expense of light energy. In such a system, anaerobic fermentation of carbohydrates (or organic wastes) produces intermediates such as low-molecular hydrogen by photosynthetic bacteria in the second step using a photobioreactor (Nath and Das, 2004). Complete degradation of glucose to hydrogen and carbon dioxide is impossible by anaerobic digestion. However, photosynthetic bacteria could use light energy to overcome the positive free energy of the reaction (bacteria can utilize organic acids for hydrogen production) (Das and Veziroglu, 2001). The conversion of malate and lactate to hydrogen by photosynthetic bacteria (mainly purple non-sulphur bacteria) has been documented (Koku, 2002; Kondo, 2002). Cyanobacteria are using two sets of enzymes to generate hydrogen gas (nitrogenase and hydrogenase). Hydrogen photo evolution catalyzed by nitrogenases or hydrogenases (Wünschiers *et al.*, 2003) can only function under anaerobic conditions due to their extreme sensitivity to oxygen. Since oxygen is a by-product of photosynthesis, organisms have developed the following spatial and temporal strategies to protect the enzyme from inactivation by oxygen (Linus Pauling, 1970; Lopes Pinto, 2002). These factors can be arranged into two categories: environmental factors (light, temperature, atmosphere, nutrient availability) and intrinsic factors (genetic or certain sensitive proteins) (Beral and Zapan, 1977).

Genetic engineering has made possible in cyanobacteria for effective hydrogen production (Theil, 1994). The strategies and regulatory studies of enzymes responsible for biohydrogen production in cyanobacteria was well characterized (Hansel and Lindblad, 1998). Cyanobacterial hydrogen production is not rapid which can be circumvented by combining electrochemical technology for higher efficiency of hydrogen production.

3. Hydrogen Production by Fermentative Bacteria using Acids

Clostridium diolis JPCC H-3 was obtained from soil and it is capable of producing hydrogen from slurry solution having acetic and lactic acid at higher rates compared with other isolated *clostridium* spp. Maximum hydrogen production by *C.diolis* JPCC H-3 of 6.03±0.15 ml from 5 ml of slurry solution was achieved at pH 6.8 and 40°C (Matsumoto and Nishimura, 2007). *E. coli* produces hydrogen from formic acid with high productivity. Formic acid can be derived from biomass or carbon monoxide plus methanol. Bio-hydrogen production from formic acid by facultative anaerobe is catalyzed by formate hydrogen lyase (FHL) (Das and Veziroglu, 2001; Bagramyan and Trchounian, 2003; Sawers, 2005; Vardar-Schara, 2008). The direct decomposition of formic acid into H_2 and CO_2 by FHL would provide a high hydrogen production rate without the generation of by-products except CO_2. *Enterobacter* species have a higher potential for hydrogen production than *E. coli* (Das and Veziroglu, 2001). However, hydrogen production from formic acid by *Enterobacter* species has not been studied. It was reported that the hydrogen production by FHL-1 system in *E. coli* was also active only at acidic pH and high formic acid concentration (Bagramyan *et al.*, 2002). Although acids are used by bacteria, hydrogen production is not same as that of stoichiometric yield. Hence, this biochemical barrier can be overcome by generating hydrogen gas from acids using electrochemical technology.

4. Electrochemical Technology

4.1 Electrolysis

Many different types of electrolysis cells have been proposed and constructed. The different electrolysis cells can be divided into groups based on the electrolyte which capable of using H_2O as reactant to produce H_2. However, only the solid oxide cell is capable of using CO_2 to produce CO (Table 10.2).

Table 10.2: Types of Electrolysis Cells (Vendt, 1990)

Types	Alkaline	Acid	Polymer Electrolyte	Solid oxide
Charge carrier	OH ·	H+	H+	O_2
Reactant	Water	Water	Water	Water, CO_2
Electrolyte	Sodium or Potassium hydroxide	Sulphuric or Phosphoric acid	Polymer	Ceramic
Electrodes	Nickel	Graphite with Pt, polymer	Graphite with Pt, polymer	Nickel, ceramics

Generally, the electrolysis cell consists of two electrodes and an electrolyte. The electrolyte may be a liquid (alkaline or acid) or a solid (polymer electrolyte or solid oxide). It serves to conduct ions (the charge carrier) produced at one electrode to the other. There has been a great deal of research in splitting water to make hydrogen and oxygen; in fact its commercial uses date back to the 1890s (Norbeck *et al.*, 1996). Water splitting in its simplest form uses an electrical current passing through two electrodes to break water into hydrogen and oxygen. Commercial low temperature electrolyzers have system efficiencies of 56–73 per cent (70.1–53.4 kWh/kg H_2 at 1 atm and 25°C) (Turner *et al.*, 2008). It is essentially the conversion of electrical energy to chemical energy in the form of hydrogen, with oxygen as a useful by-product using proton exchange membrane (PEM) (Grigoriev *et al.*, 2006; Norbeck et., 1996; Pettersson *et al.*, 2006). Currently, electrolysis is more expensive therefore if non-renewable power generation is used to make the electricity for electrolysis, and results in higher emissions compared to natural gas reforming (Bradley, 2000; Janssen *et al.*, 2004). Several approaches have been addressed these shortcomings. These include using renewable sources of energy such as solar, wind, and hydro, to produce the electricity (Janssen *et al.*, 2004; Koroneos *et al.*, 2004) or excess power from existing generators to produce hydrogen during off-peak times (Yumurtaci and Bilgen, 2004). Since water needs high electrical energy f or its electrolysis, use of weak acids or dilute acids which are obtained from wastes or by-products can be electrolyzed for supplementing hydrogen demands using low electrical appliances.

4.2 Electrohydrogenesis

Electrohydrogenesis is a recently developed electrolysis method for directly converting biodegradable material, organic acids into hydrogen using modified microbial fuel cells (MFCs) (Liu *et al.*, 2005; Rozendal *et al.*, 2006; Ditzig *et al.*, 2007; Cheng and Logan, 2007; Rozendal *et al.*, 2008;). In fact, these types of cells are rather versatile and have been shown to be able to generate hydrogen from a variety of substrates, including some wastewaters (Ditzig *et al.*, 2007). The open circuit potential of ~ -300mV is needed for the electrolysis of acetate, if hydrogen is produced at the cathode; the half reactions occurring at the anode and cathode are as follows:

Anode:

C2H4O2 +2H2O   2CO$_2$ +8e⁻ +8H⁺   (3) Cathode:

8H+ + 8e-   4H2   (4)

Producing hydrogen at the cathode requires a potential of at least $E° = -410mV$ (NHE) at pH 7.0. This voltage is substantially lower than that needed for hydrogen derived from the electrolysis of water, which is theoretically 1210mV at neutral pH. In practice, 1800-2000mV is needed for water hydrolysis (under alkaline solution conditions) due to overpotential at the electrodes (Liu *et al.*, 2005). Hence electrolysis of acids requires less electrical energy compared to electrolysis of water.

4.3 Types of Ion Exchange Membranes

A thin sheet or film of ion-exchange material which may be used to separate ions by allowing the preferential transport or either cations (in the case of a cation-exchange membrane) or anions (in the case of an anion exchange membrane). If the membrane material is made from only ion-exchanging material, it is called a homogeneous ion-exchange membrane. If the ion-exchange material is embedded in an inert binder, it is called a heterogeneous ion-exchange membrane. The difference between anion and cation exchange membrane are summarized in Table 10.3. The cation exchange membrane based on fluorinated polymer and sulfonic acid group is used as major membrane for PEMFC because of the excellent proton conductivity and durability. On the other hand, AEM based on quaternary ammonium group and hydrocarbon polymer backbone has been considered to have low thermal durability and low OH- conductivity under the condition of fuel cell (Gasteiger *et al.*, 2008).

Table 10.3: Differences between Ion Exchange Membranes

Anion Exchange Membrane	Cation Exchange Membrane
OH-conductive	H+ conductive
	-SO3-, ~-PO4-, -CO$_2$-~
Pt free catalyst available	High ion conductivity
Advantage for cathode O$_2$ reduction	Excellent ionomer solution
Low ion conductivity	High cost materials
Low thermostability	Fuel crossover
Influence of CO$_2$	

4.4 Cation Exchange Membrane Electrolyser

PEM electrolyser is a recent advancement in PEM fuel cell technology. PEM-based electrolysers typically use platinum black, iridium, ruthenium, and rhodium for electrode catalysts and a Nafion membrane as the proton exchanger (Pettersson *et al.*, 2006; Turner *et al.*, 2008). The performance that is the hydrogen generation rate can be increased by using efficient electrodes, proton exchange membranes and by reducing electrode spacing (Liu *et al.*, 2005). Proton exchange membranes (PEMs) are one of the most important components in microbial fuel cells (MFCs), since PEMs physically separate the anode and cathode compartments while allowing protons to transport to the cathode in order to sustain an electrical current. The Nafion 117 membrane used in this study is generally regarded as having excellent proton conductivity. Nafion, a sulfonated tetrafluorethylene, consists of a hydrophobic fluorocarbon backbone (-CF^2-CF^2-) to which hydrophilic sulfonate groups (SO3-) are attached. The presence of negatively charged sulfonate groups in the membrane explains the high level of proton conductivity of Nafion, while also showing a significant undesirable affinity for other cations rather than protons (Chae *et al.*, 2008). Most MFCs are operated at a neutral pH in order to optimize bacterial growth in the anode chamber, while other cations (Na$^+$, K$^+$, Ca$^+$, MG^{2+} and NH$_4^+$) contained in growth medium are typically present at a 105 times higher concentration than protons (Rozendal *et al.*, 2006). Consequently, these cations combine with the sulfonate groups of Nafion and inhibit the migration of protons produced during substrate degradation, causing a decrease in the MFC performance due to the pH reduction in the anode chamber. In addition, the frequent replacement of the buffer solution as a catholyte reduced the economic viability of MFCs. Nafion operated over a period of 50 days was contaminated with biofilm causing adverse effects on mass transport through the membrane (Chae *et al.*, 2008).

4.5 Anion Exchange Membrane Electrolyser

Anion exchange membrane fuel cells (AEMFCs) are a viable alternative to PEMFCs and are currently gaining renewed attention. In an AEMFC, an anion exchange membrane (AEM) conducts hydroxide (or carbonate) anions (as opposed to protons) during current flow, which results in several advantages: (1) The oxygen reduction reaction (ORR) is much more facile in alkaline environments than in acidic environments. This could potentially facilitate the use of less expensive non-PGM catalysts with high stability in alkaline environments. (2) The electro-oxidation kinetics for many liquid fuels (including non-conventional choices of importance to the military, such as sodium borohydride) is enhanced in an alkaline environment. (3) The electroosmotic drag associated with ion transport opposes the crossover of liquid fuel in AEMFCs, thereby permitting the use of more concentrated liquid fuels. This is an advantage for portable applications. (4) The flexibility in terms of fuel and ORR catalyst choice also expands the parameter space for the discovery of highly selective catalysts that are tolerant to crossover fuel. These potential advantages make AEMFCs an attractive future proposition (Christopher *et al.*, 2010).

For a traditional AEMFC with hydrogen fuel and air/oxygen as the oxidant, the half cell and overall chemical reactions are as follows: (Varcoe and Slade, 2005)

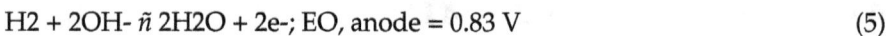

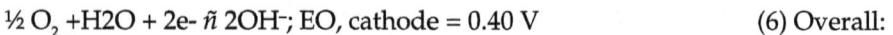

$$H2 + 2OH\text{-} \text{ñ } 2H2O + 2e\text{-}; EO, anode = 0.83 \text{ V} \tag{5}$$

$$\tfrac{1}{2} O_2 + H2O + 2e\text{-} \text{ñ } 2OH^-; EO, cathode = 0.40 \text{ V} \tag{6} \text{ Overall:}$$

$$H2 + \tfrac{1}{2} O_2 \text{ ñ } H_2O; EO, cell = 1.23 \text{ V} \tag{7}$$

In an AEMFC, hydroxide ions are generated during electrochemical oxygen reduction at the cathode. They are transported from the cathode to the anode through the anion conducting (but electronically insulating) polymer electrolyte, wherein they combine with hydrogen to form water. The electrons generated during H_2 oxidation pass through the external circuit to the cathode, where they participate in the electrochemical reduction of oxygen to produce - OH. Note that in practice, the ideal thermodynamic cell voltage of 1.23 V (at standard conditions) is not realized even at open circuit (zero current) due to myriad irreversibilities that arise during AEMFC operation. The phenomenological sources of irreversibility are very similar to those in PEMFCs and include oxygen and water activities that are less than unity, and gas crossover at open circuit leading to mixed potentials, and activation, ohmic, and mass transfer losses (overpotentials) during current flow. Hence, AEM may be a suitable membrane for electrolysis of acid wastes, waste waters and biomass.

Chapter 11

Bioreactors for Biohydrogen Production

Introduction

Hydrogen (H_2) is a promising alternative to conventional fossil fuels, because it release energy explosively in heat engines or generate electricity quietly in fuel cells while producing water as the only by- product. H_2 is also raw material for synthesis of ammonia, alcohols, and aldehydes, and for hydrogenation of various petroleum and edible oils, coal, and shale oil (Fang *et al.*, 2002). H_2 is proposed as ultimate transport fuel for vehicles because of its non- polluting characteristics, it enables use of highly efficient fuel cells to convert chemical energy to electricity (Forsberg *et al.*, 2007).

Most of H_2 is being generated from fossil fuels through thermochemical processes (hydrocarbon reforming, coal gasification and partial oxidationof heavier hydrocarbons) (Das and Veziroglu, 2001; Levin *et al.*, 2004). Biohydrogen production studies have focused on biophotolysis of water using algae and cyanobacteria, photo-decomposition of organic compounds by photosynthetic bacteria and dark fermentation of organic compounds with anaerobes. Under anaerobic fermentation, H_2 is produced in first stage as an intermediate product, which at second stage is used as an electron donor by methanogens. Microbial consortia, mainly methanogenic archaea, acetogenic bacteria and sulphate- reducing bacteria, utilize H_2. It is possible to harvest H_2 at acidification stage of anaerobic fermentation, leaving remaining acidification products for further methanogenic treatment. A possible approach is by promoting acidogens to produce H_2, CO_2 and volatile fatty acids (VFAS) in first stage, while final stage or methanogenesis and other H_2- consuming biochemical reactions are inhibited. This can be achieved through regulating biohydrogen cultures at low pH, and/or short hydraulic retention time (HRT) (Kim *et al.*, 2004; Mizuno *et al.*, 2000), or through inactivating H_2 consumers by heat treatment (Lay, 2001; Logan *et al.*, 2002) and chemical inhibitors (Sparling *et al.*, 1997; Wang *et al.*, 2003). Anaerobic H_2 fermentation, which has several positive features (high production rate, low energy demand, easy operation znd sustainability) has to compete with commercial H_2 production processes from fossil fuels in terms of cost, efficiency and reliability. Bioreactor is required for large scale hydrogen production. Some off the bioreactor are discussed below.

Batch and Semi- continuous

Han and Shin (2004) developed a semi- continuous mode for anaerobic H_2 production from food waste. Pretreated seed sludge and food waste were loaded into anaerobic leaching- bed reactor, and dilution water was continuously fed to reactor by a peristaltic pump at different dilution rates. Microbial reaction was considered accomplished as biogas production ceases, which generally took around 7 days. Appropriate control of dilution rate could enhance H_2 fermentation efficiency by improving degradation of readily degradable matters. Also, dilution rate might delay shift of predominant metabolic flow from H_2 and acid forming pathway to solvent forming pathway.

A high- rate anaerobic sequencing bath reactor (ASBR) has been used to evaluate H_2 productivity of an acid- enriched sewage microflora at 35°C (Lin and Chou, 2004) A 4 h cycle, including feed reaction, settle, was operated on 5-1 ASBR. Sucrose substrate concentration was kept at 20 g COD/I, and HRT was maintained initially at 12-120 h and thereafter at 4-12 h. Reaction/settle period ratio was maintained at 1.7. Hydrogenic activity of sludge microflora was found HRT dependent, and that proper pH control was necessary for a stable operation of bioreactor. Peak hydrogenic activity was noted at an HRT of 8 h and an organic loading rate (OLR) of 80 Kg COD/m^3.day. Each mole of sucrose in reactor produced 2.8 mol of H_2 and each gram of biomass produced 39 mmol of H_2 per day. Very low HRT might deteriorate H_2 productivity. Concentration ratios of butyric acid to acetic acid, as well as VFA and soluble microbial products to alkalinity can be used as monitoring indicators for hydrogenic bioreactor.

Continuous Stirred Tank Reactor (CSTR)

In CSTR which is frequently used for H_2 production (Chen and Lin, 2003; Majizat *et al.*, 1997; Yu *et al.*, 2003), H_2- producing bacteria are well suspended in mixed liquor and less suffered from mass transfer resistance. Because of its intrinsic structure and operating pattern, a CSTR is unable o maintain high levels of biomass inventory. Depending on operating HRTs, biomass measured in terms of volatile suspended solids (1-4-g- VSS/I) is commonly reported (Chen and Lin, 2003; Horiuchi *et al.*, 2002; Lin and Chang, 1999; Zhang *et al.*, 2006). Washout of biomass may occur at short HRTs, and thus H_2 production rates are considerably restricted. Highest H_2 production rate of CSTR culture fermenting sucrose with a mixed H_2- producing culture was reported as 1.12/1.h (Chen and Lin, 2003).

Accordingly Vanderhaegen *et al.*,(1992) found that granular sludge disappeared within three weeks when CSTRs were incubated statically instead of being shaken. Spontaneous granulation of H_2- producing bacteria can occur with reduced HRT in CSTR (Fang *et al.*, 2002; Oh *et al.*, 2004; Yu and Mu, 2006). In such a conventional system, H_2- producing bacteria are well suspended in mixed liquor and less suffered from mass transfer resistance, but washout of biomass may occur at shorter HRTs. H_2 production rates are thus restricted considerably by a low CSTR biomass retention and low hydraulic loading (Lay *et al.*, 1999; Yu *et al.*, 2002). Show *et al.*,(2007) and Zhang *et al.*,(2007) found that formation of granular sludge significantly increased overall reactor biomass to as much as 16.0 g- VSS/1, which enabled CSTR to operate at an OLR of up to 20 g-glucose/1.h and hence enhanced performance in H_2 production.

Memberane Bioreactor

One method for increasing reactor biomass concentration is the use of a membrane in a chemostat to control biomass concentration. At a HRT of 3.3 h, Oh *et al.*,(2004) demonstrated that biomass concentration increased from 2.2g/l in a control reactor (no membrane chemostat) to 5.8 g/l in an anaerobic membrane bioreactor (MBR). This was achieved by controlling sludge retention time (SRT) at 12h, corresponding to a slight increase in H_2 production rate from 0.50 to 0.64 1/1.h. Icreasing SRT can further enhance biomass retention, which favors substrate utilization, but may result in a decrease in H_2 production ate. By summarizing several studies of H_2 production by MBR, According to Li and Fang (2007) H_2 production rates between 0.25-0.691/1.h in MBR systems. However, this process may achieved advantage in contrast to other high efficiency H_2 production systems. In addition, membrane fouling and high operating cost would limit the use of MBR process in anaerobic H_2 fermentation.

Immobilized-Cell Reactor

Immobilized cell systems, are capable of maintaining higher biomass concentration can be operated at high dilution rates without biomass washout (Oh *et al.*, 2004). Biomass immobilization can be achieved through forming granules, biofilm or gel- entrapped bioparticles. Many reasearchers immobilized pure or mixed cultures of H_2 producing bacteria by gel entrapped in a form of biogels such as *C. butyricum* strain in agar gel (Yokoi *et al.*, 1997) *E. aerogenes* strain HO-39 in K- carrageenan, calcium alginate or agar gel (Yokoi *et al.*, 1997b), sewage sludge in calcium alginate beds, or alginate bed with adding activated carbon powder polyurethane and acrylic latex/silicone (Wu *et al.*, 2002), and sewage sludge and activated carbon powder fixed by ethylene- vinyl acetate copolymer (Wu *et al.*, 2005). Peak H_2 production rates obtained by continuous gel- immobilized sludge ranged from 0.090 1/1.h (Yokoi *et al.*, 1997b) in a chemostat with stirring to 0.93 1/1.h (Wu *et al.*, 2003) in a fluidized bed reactor.

Biofilm attachment on solid and porous support carriers seems to be superior to gel- entrapped bioparticles in continuous H_2 production. In continuous cultures without any pH control, H_2 production and glucose consumption rates (Yokoi *et al.*, 1997) with *C. butyricum* immobilized on porous glass beads were higher than corresponding values with cells immobilized in agar gel at HRTs of 3h and 5h.

Granular sludge has some advantages over biofilm sludge in continuous dark H_2 fermentation. Firstly, fast growing characteristics of H_2 producing cultures might cause system upset of fixed- bed biofilm processes. Maximum specific growth rate (Horiuchi *et al.*, 2002; Chen *et al.*, 2001; Van Ginkel and Logan, 2005) (0.17-$0.5h^{-1}$) and biomass growth yield (Fang *et al.*, 2002; Yu& Mu,2006; Chen *et al.*, 2001; Kim *et al.*, 2006) (0.08-0.33g- VSS/g-COD) of H_2 producing bacteria indicated that H_2 producing bacteria would increase rapidly if higher OLR was employed. OLR for immobilized sludge H_2 production was reported as high as 80 g- glucose/1.h (Wu *et al.*, 2006). Nicolella (2000) mentioned that biofilm reactors are not particularly useful when dealing with fast- growing organisms with a maximum growth rate faster than 0.1 h^{-1}. Rapid buildup of H_2 _ producing biofilms could result in system upset due to mass transfer limitation. Oh *et al.*,(2004) reported microbial growth of H_2 producing bacteria too excessive under mesophilic condition, causing system upset just after one week of operation. On the other hand, a packed- bed reactor using cylindrical activated carbon as support matrix exhibited steady and efficient H_2 production. Fed with sucrose at 20 g COD/1, system was operated at 0.5-5h HRT and 35°C for 15 days (Lee *et al.*, 2003). Reduction of bed porosity from 90 per cent to 70 per cent would result in decrease in H_2 production performance, and pressure drop was higher when bed porosity was lower. System stability of such a biofilm- based process may be challenged by long term operation. System upset might occur interstitial void space in pack- bed reactor are clogged with biomass.

Washout of support carriers might be an intrinsic drawback of biofilm processes. Zhang *et al.*,(2008) investigated H_2 production by granular sludge and biofilm sludge growing on granular activated carbon in two fluidized bed reactors at a pH of 5.5 and an OLR of 40g- glucose/1.h. A similar performance in H_2 production was observed with two immobilized structures; both were tested at different HRTs (0.125-3h) and influent substrate concentrations (5-120 g/l). Biofilm sludge was washed out substantially and reactor biomass was replaced by granular sludge after 50 days of operation. But H_2 production was not affected during transition. Severe washout of support carriers is presumably attributed to fast- growing characteristics of H_2 _producing bacteria, wherein maximum specific growth rate and cell yield coefficient were determined to be 0.5 h^{-1} and 0.12g- VSS/g-glucose, respectively (Zhang *et al.*, 2008). A large amount of support carriers is normally required to support microorganism growth in biofilm processes. Carriers occupy a considerable space in reactor and reduce – effective volume for biomass- substrate interactions, resulting in lower reactor performance and efficiency. Supporting carriers need to be replaced periodically due to wear and tear. Cost of material replacement could be major economic consideration in maintenance.

Granular sludge processes generally exhibit long startup. A complete development of H_2 _ producing granules may take several months (Fang *et al.*, 2002; Yu and Mu, 2006). During startup of an UASB H_2 producing reactor, Mu and Yu (2006) found that small granules (diam 400-500 1/4m) were formed at reactor bottom after 140 days of operation. Granules developed further to sizes larger than 2.0mm upon 200 days. Although, reactor reached steady- state H_2 production and substrate degradation after 5 months of startup operation, development and accumulation of mature and stable granular sludge were only completed beyond 8 month of operation. Chang and Lin (2007) noted that a USAB reactor took 39 days to achieve constant gas production at a HRT of 24h and granules become visible after 120 days of operation. A longer period (180 days), however, was required for further development of granules.

Granules of H_2 producing cultures can be markedly accelerated. Packing of a small quantity of carrier matrices significantly accomplished sludge granulation within 80-290h in a novel carrier- induced granular sludge bed (CIGSB) bioreactor (Lee *et al.*, 2006).Column reactors were initially packed with cylindrical activated carbon, spherical activated carbon, sand or filter sponge at a bed height of 4-8 cm and with bed porosities of 90-99 per cent. Granulation of seed sludge could take place in all carrier- packed reactors as HRTs were shortened to 4-8h, dependent on carrier type.

By adding cationic polymer (cationic polyacrylamide) and anionic polymer (silica sol), rapid granulation of H_2 producing culture could be accomplished within 5 min (Kim *et al.*, 2005). As sludge has a negative charge of -26mV, high molecular weight cationic polymer (MW, 15,000,000) with 0.7 per cent (w/w) of dry sewage digester sludge was added and stirred at 200 rpm for 2 min to neutralize sludge.

Since residual cations may cause detrimental effect on microorganisms, anionic silica gel of 0.7 per cent (w/w) of dry sewage digester sludge were added and stirred at 200 rpm for 2 min. Total time required for granulation was about 5 min. When granular sludge was maintained stably its size ranged from.0 to 3.0 mm and maximum concentration sludge was found to be approx. 7 g/l. Zhang *et al.*,(2008) developed an approach of acid incubation to initiating formation of H_2 producing granules rapidly in a CSTR. H_2 producing granules were formed rapidly within 114h as seed microbial culture was subjected to a 24 h period of acid incubation at a pH of 2.0. Changing culture pH would result in improvement in surface physiochemical properties of culture favoring microbial granulation

Reactor Type

Fixed Bed-Reactor

Fixed- or packed-bed reactor is operated under lesser extent of hydraulic turbulence, thus its microbial cultures usually encounter mass transfer resistance resulting in lowered rates of substrate conversion and H_2 production. Kumar and Das (2001) investigated H_2 production by *Enterobacter cloacae* attaching on coir in packed- bed reactors at HRT of 1.08h, and found that rhomboid bioreactor with convergent – divergent configuration gave maximum H_2 production (1.60l/l.h) as compared with tapered reactor (1.46 l/l.h) and tubular reactor (1.40l/l.h), attributed to higher turbulence created by reactor geometry favoring mass transfer and reduced gas hold-up.

Rachman *et al.*,(1998) found that high H_2 molar yield could not be maintained consistently in a packed- bed reactor, although pH in effluent was controlled at >

6.0. This is because pH gradient distribution along reactor column resulted in a heterogeneous distribution of microbial activity. In order to overcome mass transfer resistance and pH heterogeneous distribution, fluidized- bed or expended bed reactor system with recirculation flow was recommended to be more appropriate in further enhancing H_2 production rate and yield. Increasing slurry recycle ratio can alleviate mass transfer resistance in a packed- bed reactor. Kumar and Das (2001) observed that both H_2 production and substrate conversion rates of a packed- bed reactor increased with recycling ratio. Maximum H_2 production rate (1.69l/l.h) was noted at a recirculation ratio of 6.4.

Support materials have important effect on biomass retention and consequently H_2 production in fixed- bed reactors. Chang et al.,(2002) immobilized acclimated sewage sludge on surfaces of porous supports using loofah sponge, expanded clay, and activated carbon for continuous H_2 fermentation in fixed bed reactor. Besides loofah sponge, other carriers exhibited better biomass yields. By comparing two support carriers favoring biomass yield, activated carbon was found a better choice of support carriers used in H_2 – producing fixed- bed reactors, with which maximum H_2 production rate (1.32l/l.h) was reached at a HRT of 1h and a sucrose concentration of 20g/l.

Fluidized Bed Reactor (FBR)

In gel- immobilized sewage sludge (Wu et al., 2005) immobilized culture was able to produce H_2 efficiently in a three phase FBR operated at a HRT between 1-6h with a maximal steady- state rate (0.93 l/l.h) and an optimal yield of H_2 (2.67 mol/mol sucrose), which was highest value reported in gel- immobilized culture systems. Zhang et al.,(2007) obtained higher H_2 production by biofilm culture (pH4.0) growing on granular activated carbon in an anaerobic FBR at HRTs of 0.5-4h and influent glucose concentrations of 10-30g/l. At operating pH biofilm sludge concentration was retained up to 21.5g- VSS/l. H_2 might be produced efficiently in an anaerobic FBR as H_2 production rate reached a maximum rate of 2.36l/l.h.

UASB Reactor

UASB reactor system has been applied in H_2 production due to its potential of high biomass concentration and treatment efficiency. Chang and Lin (2007) found that H_2 yield stabilized at 1.5 mol H_2/mol sucrose at HRT of 8-20h in a UASB granular reactor. The yield drastically decreased at a HRT of 4 or 24h. H_2 production rate (0.25l/l.h) and specific H_2 production rate (53.5 mmol H_2/g-VSS.day) peaked a a HRT of 8h. Biomass concentration reached maximum value of 7.2g- VSS/l at a HRT of 24h, but decreased to 5.0 g/l at optimum HRT of 8h. Yu and Mu (2006) studied H_2 production (yield 0.49-1.44mol- H_2/mol-glucose) from synthetic sucrose wastewater in UASB reactor with granular sludge operated at 38°C and a pH of 4.4±0.1 for over 3 years. H_2 production rate increased with increasing substrate concentration from 5.33 to 28.07 g- COD/l, but decreased with increasing HRT from 3 to 30h. However, optimum operating conditions only gave rise to a low H_2 production rate (0.21/l.h).

CSTR Granular Sludge Reactor

Fang *et al.,*(2002) demonstrated that H_2 producing acidogenic sludge could agglutinate into granules in a well- mixed CSTR reactor treating a synthetic sucrose-containing wastewater at 26°C, pH5.5 and HRT of 6h.

Formation of granular sludge enhanced biomass growth up to 20g/l and consequently H_2 production rate up to 0.54l/l.h with 97 per cent sucrose being degraded. In a similar CSTR system with granular sludge fermenting glucose waste water (10g/l) at a pH of 5.5 and 37°C, a maximum H_2 yeild (1.81mol- H_2/mol-glucose) and a maximum H_2 production rate (3.20l/l.h) were obtained at a HRT of 0.5h. Wu *et al.,*(2006) further developed such a granular- sludge with silicone- immobilized sludge at 40°C and pH 6.6±0.2, and reactor performance was examined at a HRT of 0.5-6h and an influent sucrose concentration of 10-40g-COD/l. Self- flocculated granular sludge occurred at a HRT of 0.5h, reached a concentration of up to 35.4g-VSS/l, and resulted in a significant increase in H_2 production rate (15l/l.h). A two- fold increase in specific H_2 production rate was found after formation of self-flocculated granular sludge due to transition in bacterial community structure.

Several other high- rate H_2 producing systems based on granular sludge techniques have been developed. Packing of small quantity of carrier matrices at a bottom of up flow reactor significantly stimulated sludge granulation that can be accomplished within 100h in a novel carrier- induced granular sludge bed (CIGSB) bioreactor (Lee *et al.,* 2004). CIGSB bioreactor, started up with a low HRT of 4-8h (corresponding to an OLR of 2.5-5gCOD/l.h), enabled stable operation at an extremely low HRTs (0.5h) without experiencing biomass washout. Granular sludge was rapidly formed in CIGSIB supported with activated carbon, reaching a maximum concentration of 26g/l at a HRT of 0.5h. Ability to maintain high biomass concentration at low HRTs corresponding to high OLRs highlights remarkable H_2 production efficiency of CIGSB processes. Reactor achieved an optimum volumetric H_2 production rate at 7.3l/l.h (7.15mol/l.d) and maximum H_2 yeild (3.03 mol H_2/ mol sucrose), when operated at a HRT of 0.5h on an influent sucrose concentration of 20g COD/l. Under optimum conditions, H_2 content and substrate conversion exceeded to 40 and 90 per cent respectively. H_2 production rate of CIGSB system further improved (Lee *et al.,* 2006) (9.31l/l.h) by optimizing reactor column height and diameter at a ratio of 12 and with agitation. After altering physical configuration of CIGSB bioreactor, concentration of granular sludge increased to 40g-VSS/l.

Chapter 12

Hydrogen and Fuel Cells

One of the biggest challenge we are facing today relates tothe problem of anthropogenic-driven climate change and its inextricable link to our global society's present and future energy needs. Hydrogen and fuel cells are now widely regarded as key energy solutions for the 21st century. The technologies are needed contribute significantly to a reduction in environmental impact, enhanced energy security, diversity and the creation of new energy industries. Hydrogen and fuel cells can be utilised in transportation, distributed heat and power generation and energy storage systems. However, the transition from a carbon-based (fossil fuel) energy system to a hydrogen-based economy involves significant scientific, technological and socioeconomic barriers to the implementation of hydrogen and fuel cells as the clean energy technologies of the future.

Why Hydrogen and Fuel Cells?

Global drivers for sustainable energy vision of our future center on the need to:

☆ Reduce CO_2 emissions and improve local (urban) air quality

☆ Ensure security of energy supply

☆ Create a new industrial and technological energy base, crucial for our economic prosperity.

Hydrogen is a very attractive alternative fuel. It can be obtained from diverse resources, both renewable (hydro, wind, solar, biomass, geothermal) and non-renewable (coal, natural gas, nuclear). Hydrogen can then be utilised in high-efficiency power-generation systems, including fuel cells for both vehicular transportation and distributed electricity generation. Fuel cells convert hydrogen or a hydrogen-rich fuel and an oxidant (usually pure oxygen or oxygen from the air) directly into electricity by an electrochemical process.

Fuel cells, operating on hydrogen or hydrogen-rich fuels, have the potential to become major factors in catalysing the transition to a future sustainable energy system with low-CO_2 emissions. The importance attached to such developments is rapidly increasing. Many countries are now compiling roadmaps, in many cases with specific numerical targets for the advancement of fuel-cell and hydrogen technologies. As just one potent example, Japan's Ministry of Economy, Trade and Industry has now set a target of 5 million hydrogen-fuel-cell vehicles and 10 million kW for the total power generation by stationary fuel cells by the year 2020!

At the present time, there are three major technological barriers that must be overcome for a transition from a carbon-based (fossil fuel) energy system to a hydrogen-based economy. First, the cost of efficient and sustainable hydrogen production and delivery must be significantly reduced. Second, new generations of hydrogen storage systems for both vehicular and stationary applications must be developed. Finally, the cost of fuel-cell and other hydrogen-based systems must be reduced.

The vision of such an integrated energy system of the future would combine large and small fuel cells for domestic and decentralised heat and electricity power generation with local (or more extended) hydrogen supply networks that would also be used to fuel conventional (internal combustion) or fuel-cell vehicles.

Unlike coal, gas or oil, hydrogen is not a primary energy source. Its role more closely mirrors that of electricity as an 'energy carrier', which first is produced using energy from another source and

Version 1.0 Foresight 2Hydrogen and fuel cells: towards a sustainable energy future then transported for future use, where its stored chemical energy can be utilised. Hydrogen can be stored as a fuel and utilised in transportation and distributed heat and power generation using fuel cells, internal combustion engines or turbines, and, importantly, a hydrogen fuel cell produces only water and no CO_2.

Hydrogen can also be used as a storage medium for electricity generated from intermittent, renewable resources such as solar, wind, wave and tidal power. It therefore provides the solution to one of the major issues of sustainable energy, namely the vexing problem of intermittency of supply. As long as the hydrogen is produced from non-fossil-fuel feed stock, it is a genuinely green fuel. Moreover, locally produced hydrogen allows the introduction of renewable energy to the transport sector, provides potentially large economic and energy security advantages and the benefits of an infrastructure based on distributed generation. It is this key element of the energy storage capacity of hydrogen that provides the potent link between sustainable energy technologies and a sustainable energy economy, generally placed under the umbrella term of 'hydrogen economy'.

Hydrogen Production, Distribution and Storage

Even though hydrogen is the third most abundant chemical element in the Earth's atmosphere, it is invariably bound up in chemical compounds with other elements. It is therefore produced from other hydrogen-containing sources using energy such as electricity or heat.

Hydrogen can be produced from natural gas, coal, hydrocarbons, biomass and even municipal waste using a variety of techniques as well as by splitting water. Such a diversity contributes significantly to the security of fuel supply (Figure 12.1).

Today, hydrogen is produced in large quantities by steam reforming of hydrocarbons, generally methane. This method yields CO_2 as a by-product but no more than from burning the same amount of methane. CO_2 emissions, the principal cause of global climate change, can be managed at large-scale facilities through CO_2 sequestration, which involves the capture and storage of CO_2 underground (*e.g.* in depleted natural gas and oil wells or geological formations). However, CO_2 sequestration is not yet technically and commercially proven. Another promising route would appear to be high-temperature pyrolysis (decomposition in the absence of oxygen) of hydrocarbons, biomass and municipal solid waste into hydrogen and (solid) carbon black, accompanied by its industrial use and/or easy sequestration. At present the cost of this process is higher than that of steam reforming of natural gas.

Hydrogen can be produced by splitting water through various processes including electrolysis, photo-electrolysis, high-temperature decomposition and photo-biological water splitting. The commercial production of hydrogen by electrolysis of water achieves an efficiency of 70–75 per cent. However, the cost of hydrogen is several times higher than that produced from fossil fuels (Dutton 2002; International Energy Agency 2006). Renewable sources of energy (*e.g.* wind, tidal, biomass) might provide local sources of hydrogen, but certainly will not match the volumes of hydrogen required globally for the new energy source. The use of nuclear energy (both fission and fusion) to supply future needs for hydrogen energy is also under consideration. A recent US Department of Energy report suggests that solar is most likely the only source of energy capable of producing enough hydrogen required to supply a hydrogen economy.

The present options for transporting hydrogen include compressed gas (200 bar) in tube cylinders, liquid hydrogen tanks and a few examples of local networks of hydrogen pipelines. All these options contribute significantly to the cost of hydrogen for end users and, in some cases, decentralised local hydrogen production using methane reforming or electrolysis of water will be economically feasible.

One of the crucial technological barriers to the widespread use of hydrogen as an effective energy carrier is the lack of a safe, low-weight and low-cost hydrogen storage method with a high energy density (Harris *et al.*, 2004; Crabtree *et al.*, 2004). Hydrogen contains more energy on a weight-for-weight basis than any other substance. Unfortunately, since it is the lightest chemical element, it also has a very low energy density per unit volume (Table 12.1).

One of the crucial technological barriers to the widespread use of hydrogen as an effective energy carrier is the lack of a safe, low-weight and low-cost hydrogen storage method with a high energy density (Harris *et al.*, 2004; Crabtree *et al.*, 2004). Hydrogen contains more energy on a weight-for-weight basis than any other substance. Unfortunately, since it is the lightest chemical element, it also has a very low energy density per unit volume (Table 12.1).

Fuel cells are classified according to the nature of their electrolyte, which also determines their operating temperature, the type of fuel and a range of applications. The electrolyte can be acid, base, salt or a solid ceramic or polymer that conducts ions. Table 12.2 summarises the characteristics of various fuel cell types.

Table 12.1: Gravimetric and Volumetric Energy Content of Fuels, Hydrogen Storage Options and Energy Sources (Container weight and volume are excluded)

Fuel	Specific Energy (kWh/kg)	Energy Density (kWh/dm³)
Liquid hydrogen	33.3	2.37
Hydrogen (200 bar)	33.3	0.53
Liquid natural gas	13.9	5.6
Natural gas (200 bar)	13.9	2.3
Petrol	12.8	9.5
Diesel	12.6	10.6
Coal	8.2	7.6
LiBH$_4$	6.16	4.0
Methanol	5.5	4.4
Wood	4.2	3.0
Electricity (Lithium-ion battery)	0.55	1.69

Unlike internal combustion engines or turbines, fuel cells demonstrate high efficiency across most of their output power range. This scalability makes fuel cells ideal for a variety of applications, from mobile phones to large-scale power generation. However, at present, fuel cells can't compete with conventional energy conversion technologies in terms of cost and reliability.

High-temperature solid oxide fuel cells (SOFCs) and molten carbonate fuel cells (MCFCs) are ideal for distributed energy supply operating today with natural gas, which enables the development and use of this technology independently from the establishment of a hydrogen infrastructure. Indeed, they offer an interesting transition to the hydrogen economy, with significant efficiency gains on today's commercially available hydrocarbon fuels, while operating effectively on renewable biofuels should these become cost-effective, and ultimately operating with high efficiencies on hydrogen when this becomes widely available. They are also being pursued for use as auxiliary power units (APUs) for vehicles, and in off-grid applications to replace small diesel generators. These types of fuel cells do not require an external reformer to convert hydrogen-rich fuels to hydrogen, which enables the use of a variety of fuels and reduces the cost associated with adding a reformer to the system. They are particularly well suited to CHP applications as they produce high-grade waste heat (or cooling) as well as electrical power. The technology has been already proven by several demonstration projects showing continuous operation over tens of thousands of hours.

Low-temperature proton exchange membrane (PEM) and alkaline fuel cells offer an order of magnitude higher power density than any other fuel cell systems.

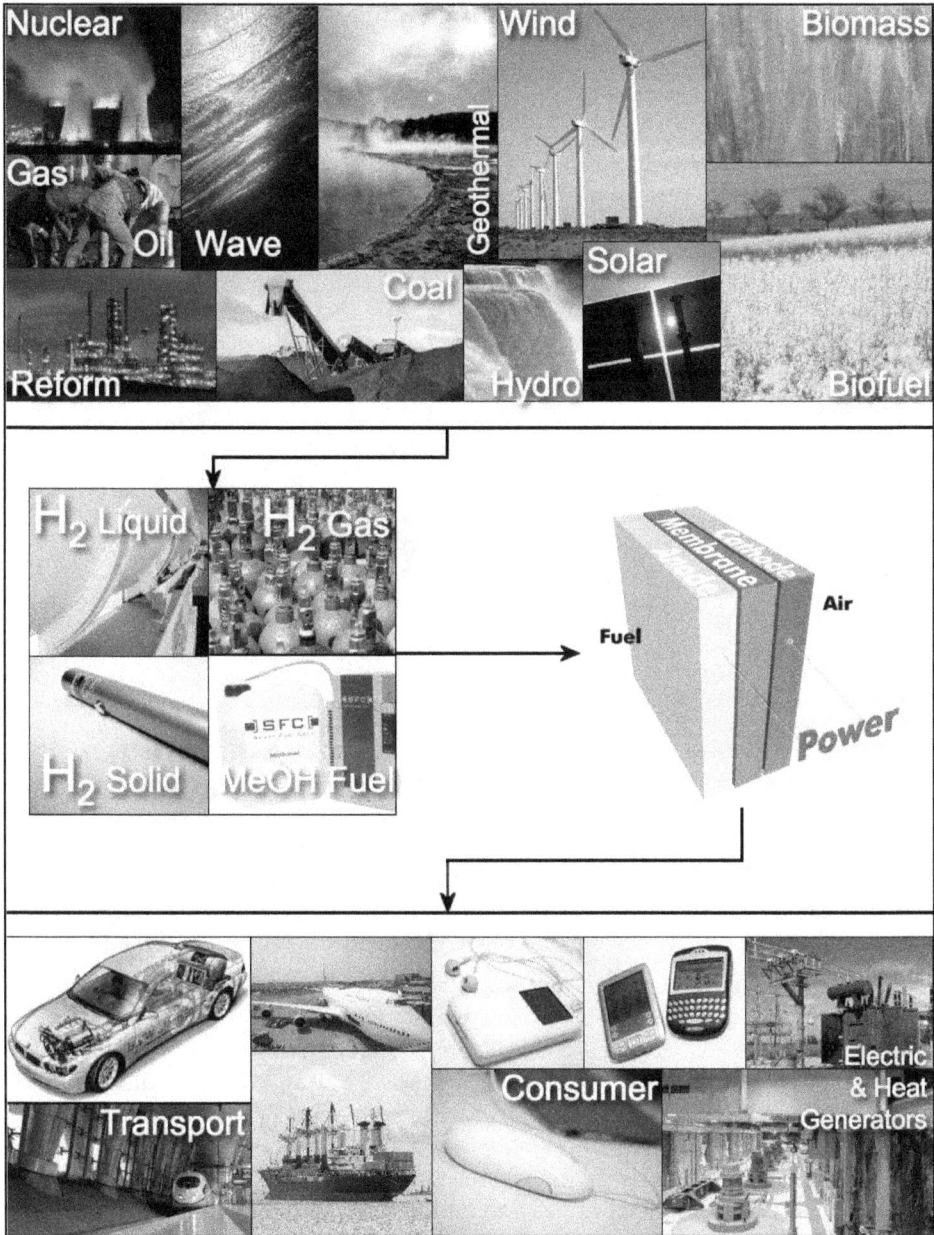

**Figure 12..1: Hydrogen as an Energy Carrier Linking Multiple
Production Methods and Sources to Various Fuel Cell Applications
(figure by Karl Harrison, University of Oxford).**

A major drawback, however, is that they require a costly platinum catalyst and need very pure hydrogen. PEM and alkaline fuel cells have been developed since the 1950s and used in the NASA space programme. PEM fuel cells are most favoured for

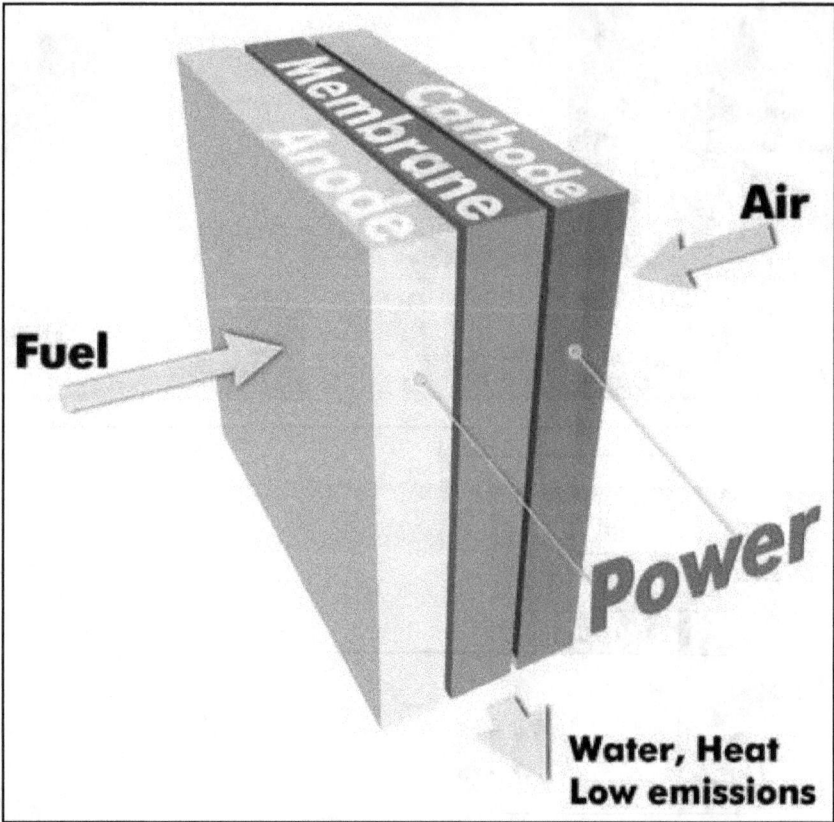

Figure 12.2: Diagram of a Fuel Cell Model
(Figure by Karl Harrison, University of Oxford).

mass-market automotive and small-scale CHP applications, and there is a massive global effort to develop commercial systems.

Phosphoric acid fuel cells (PAFCs) are more tolerant to impurities in hydrogen than PEMFCs or AFCs. PAFCs are typically used for stationary power generation but also to power large vehicles such as city buses. They are commercially available today, but their relatively high cost has restricted market uptake. Direct methanol fuel cells (DMFCs) are powered by methanol and are considered for a number of applications, particularly those based around replacing batteries in consumer applications such as mobile phones and laptop computers.

The Key Scientific and Technical Challenges

By 2050, the global energy demand could double or triple and oil and gas supply is unlikely to be able to meet this demand. Hydrogen and fuel cells are considered in many countries as an important alternative energy vector and a key technology for future sustainable energy systems in the stationary power, transportation, industrial and residential sectors (European Commission 2003; US Department

Table 12.2: Summary of Fuel Cell Types and their Present Characteristics
(Appleby and Foulkes 1993; Powell et al., 2002; US Department of Energy 2003)

Fuel Cell Type (Type of electrolyte)	Operating Temperature (°C)	Applications	Electrical Power Range (kW)	Electrical Efficiency (per cent)
Proton exchange membrane (PEMFC)	60–110	Mobile, portable, low power generation	0.01–250	40–55
Alkaline (AFC)	70–130	Space, military, mobile	0.1–50	50–70
Direct methanol (DMFC)	60–120	Portable, mobile	0.001–100	40
Phosphoric acid (PAFC)	175–210	Medium- to large-scale power and CHP	50–1,000	40–45
Molten carbonate (MCFC)	550–650	Large-scale power generation	200–100,000	50–60
Solid oxide (SOFC)	500–1,000	Medium- to large-scale power and CHP, vehicle APUs, off-grid power and micro-CHP	0.5–2,000	40–72
Solid oxide (SOFC)	500–1,000	Medium- to large-scale power and CHP, vehicle APUs, off-grid power and micro-CHP	0.5–2,000	40–72

of Energy 2004). However, as with any major changes in the energy industry, the transition to a hydrogen economy will require several decades.

The timescale and evolution of such a transition is the focus of many 'roadmaps' emanating from the USA, Japan, Canada and the EU (and many others). For example, the European Commission has endorsed the concept of a Hydrogen and Fuel Cell Technology Platform, with the expenditure of €2.8 billion over a period of 10 years. The introduction of hydrogen as an energy carrier has been identified as a possible strategy for moving the UK towards its voluntary adopted targets for CO_2 reduction of 60 per cent of current levels by 2050 (Department of Trade and Industry 2003).

Table 12.3 summarises the forecasts of several roadmaps for deployment status and targets for hydrogen technologies and fuel-cell applications that will use heat and a chemical process to dissociate water. Fusion power, if successfully developed, could be the ultimate source of a clean, abundant, and carbon-free resource for hydrogen production.

The components of a national hydrogen delivery and distribution network (including hydrogen pipelines) will need to be developed, providing a reliable supply of low-cost hydrogen to end users. If the hydrogen is produced from hydrocarbons, the hydrogen network will need to be coupled to the infrastructure necessary for carbon capture and storage. The use of hydrogen-fuelled vehicles will depend on the successful development of an affordable and widespread refuelling infrastructure. The basic components of a hydrogen delivery and dispensing infrastructure need to be developed, initially to supply local refuelling stations.

Table 12.3: Key Assumptions on Hydrogen and Fuel-Cell Applications (International Energy Agency 2006; European Hydrogen and Fuel Cell Technology Platform 2005)

Technology	Today	2020–2025	2050
Carbon capture and sequestration (CSS) (€/ton CO_2)	20–30	4–8	3–6
Hydrogen produced from coal with CCS (€/GJ)	8–10	7–9	3–5
Hydrogen transportation/storage cost (pipeline, 5,000 kg/h, 800 km) (€/GJ)	10–15	3	2
PEM fuel cells (€/kW)	6,000–8,000	400	40
High-temperature fuel cells (€/kW)	8,000–10,000	800	200
EU: portable fuel cells, sold per year	N/A	250 million	N/A
EU: fuel-cell vehicles, sold per year	N/A	0.4–1.8 million	N/A
EU: stationary fuel cells (CHP), sold per year	N/A	2–4 GW	N/A
Japan: fuel-cell vehicles, cumulative sale target	N/A	5 million	N/A
International Energy Agency forecast: global fleet of fuel-cell vehicles	N/A	N/A	700 million

For hydrogen to become a viable energy carrier, advanced hydrogen storage technologies will be required. For hydrogen fuel-cell transportation use – widely regarded as the first major inroad into the hydrogen economy – neither cryogenic nor high-pressure hydrogen storage options can meet the mid-term targets (US

Department of Energy 2004). More compact, low-weight, low-cost, safe and efficient storage systems operating at near-room temperatures and low pressures will need to be developed for automotive as well as for stationary applications. It is becoming increasingly accepted that solid-state hydrogen storage using hydrides of light elements is the only method that will enable a high weight percent and high volume density of stored hydrogen. At present, no known material meets these critical requirements.

Fuel cells have the potential to replace a very large proportion of current energy systems, from mobile phone batteries through vehicle applications to centralised or decentralised stationary power generation. Fuel cells offer a very attractive technology evolution path in that they can deliver significant efficiency gains on today's commercially available hydrocarbon fuels while also offering high efficiency in the future when hydrogen becomes widely available. The key scientific and technical challenges facing fuel cells are cost reduction and increased durability of materials and components..

Chapter 13

Hyvolution

1. Introduction

The novel approach in HYVOLUTION is based on a combined bioprocess employing thermophilic and phototrophicmicroorganism, to provide the highest hydrogen production efficiency in small-scale, cost effective industries.

The process starts with the conversion of biomass to make a suitable feedstock for the bioprocess. The subsequent bioprocess is optimized in terms of yield and rate of hydrogen production through integrating fundamental and technological approaches. Dedicated gas upgrading developed for efficiency at small-scale production units. Production costs will be reduced by system integration combining mass and energy balances.

This approach, work as Framework Programme project for BIOHYDROGEN, and enormous since it allows the greatest reduction in CO_2 emission and provides independence of fossil imports. Both topics are dominant in all global agreements on climate protection and because of the urgency in mitigating the greenhouse effect, it is of prime importance to start this research now.

1.1 Required Technology

The main objective of this project will be to development a technology for the cost-effective production of pure hydrogen from multiple biomass feedstocks. The bioprocess required starts with a thermophilic fermentation of feedstock to hydrogen, CO_2 and intermediates. So that through consecutive photo-heterotrophic fermentation, all intermediates will be converted to more hydrogen and CO_2, to achieve an efficiency of 75 per cent.

Some technical processes needed to adopt the sub-objectives contributing to the main scientific objective are:

1. Pretreatment technologies for optimal biodegradation of energy crops and bio-residues.

2. Maximum efficiency in conversion of fermentable biomass to hydrogen and CO_2.

3. Assessment of dedicated installations for optimal gas cleaning and gas quality protocols.

4. Minimal energy demand and maximal product output through innovative system integration.

5. Identification of market opportunities for a broad feedstock range spread over country as well as the whole world.

The technological prototype modules, form the basis of a blue print for the whole chain for converting biomass to pure hydrogen.

The sub-objectives are:

1. Equipment for mobilisation of fermentable feedstock

2. Reactors for thermophilic hydrogen production

3. Reactors for photo-heterotrophic hydrogen production

4. Devices for monitoring and control of the hydrogen production processes

5. Equipment for optimal gas cleaning procedures

Besides scientific and technological objectives, also socio-economic activities are included to increase public awareness and societal acceptance, and for identification of future opportunities, stakeholders and legal consequences of this specific bioprocess for decentral hydrogen production.

1.2 State of the Art

Distinct advanced strategies for the production of H_2 from biomass:

☆ Thermal processes like gasification or supercritical water gasification

☆ Non-thermal (biological or fermentative) processes, which are the issue in HYVOLUTION Non-thermal processes have a specific advantage for the efficient conversion of biomass with high moisture content to pure hydrogen. Secondly, fermentative processes do not require large installations for economy of scale. In this way, small-scale installations can be constructed for cost-effective conversion of the locally produced biomass on-site and loss of energy through transport is prevented.

The potential of biologically produced H_2 is recognized worldwide. The Netherlands is leading in research on application of bacteria for H_2 production from biomass, followed by Hungary, Turkey and the UK. Also in Asia and Canada and especially in the USA, biological H_2 production is seen as one of the options for renewable H_2 production on the longer term. The progress in research on hydrogenase enzymes and basic physiological parameters of hydrogen producing bacteria forms the basis for continuation and expanding the developments.

Most research in this topic has been performed with hydrogen producing bacteria, which have optimum growth and hydrogen production at ambient temperatures. The main drawback of these bacteria is that, besides hydrogen, they produce other reduced intermediates which compete with hydrogen production. As a result, the efficiency of biomass conversion to hydrogen is low.

The new approach focuses on employing thermophilic bacteria which grow at temperatures of 70 °C or above. These bacteria produce hydrogen together with acetic acid. In this process, the amount of hydrogen produced per unit of biomass is about twice as high in comparison to fermentation at ambient temperatures. Furthermore, the co-product acetic acid, is a prime substrate for H_2 production in a consecutive photofermentation for further increase of the final amount of H_2 produced per unit biomass. The combination of a thermophilic fermentation with a photofermentation enables the complete conversion of biomass to hydrogen with the highest efficiency theoretically possible.

The consortium for 'HYVOLUTION' was formed to exploit the acquired knowledge and make a breakthrough with a new taskforce aiming at the development of a hydrogen industry producing H_2 at a cost price of 10 Euro/GJ. The necessary advances will be established by a greater critical mass of specialists from different disciplines, representing academia, research organizations, SMEs and industries, in the European Research Area. As shown in Figure 13..1, the price target will be achieved by reducing costs in the biomass pretreatment, by optimizing the efficiency and rate of the fermentations enabling low cost thermo- and photo-bioreactors, by developing dedicated, low cost gas upgrading procedures and optimum system integration for making economic balances with respect to energy and heat utilization.

2. Relevance to the Priority Thematic Areas of the 6th EU Framework Programme

2.1 Relevance Scientific and Technical Objectives

The main scientific and technological objectives of HYVOLUTION are the development of a bioprocess for hydrogen production from biomass and the construction of prototypes of modules of the future industrial plant.

Shown in Figure 13..1 which fully covers one of the activities addressed in: New and advanced concepts in renewable energy technologies/BIOMAS S/Production of hydrogen-rich gas from multiple biomass feedstock. The composition of the produced raw gas leaving the bioreactors is estimated at 50-80 per cent H_2, the remainder being CO_2. Through dedicated gas cleaning the concentration of hydrogen will be increased to the requirements for use in power or bio-fuel production. Furthermore, the basis of the biomass to hydrogen chain in HYVOLUTION is the conversion of biomass, ranging from energy crops to bio-residues from agro-industries. The diversity of the biomass in HYVOLUTION will be warranted through the participation of different industrial stake-holders and specific patenity.

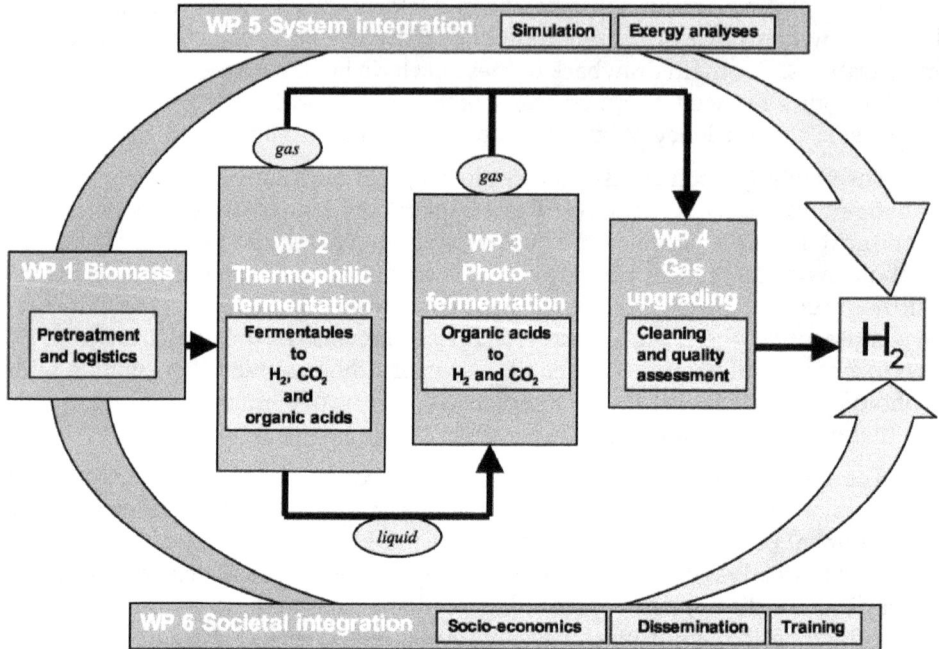

Figure 13.1: Scheme of HYVOLUTION: An integrated approach for non-thermal hydrogen production, which covers the whole chain from biomass to hydrogen, including societal integration for implementation in society.

IP will have strong crosscutting dimension because It will also relates to several elements in: New technologies for energy carriers. It is aim to collaborate closely with research activities undertaken in Fuel cells, including their applications, which can result in low cost fuel cells at the power range appropriate for electricity generation from the local H_2 production plant envisaged here. Sustainable produced H_2 will also be required in the implementation of alternative motor fuels. Finally, the socio-economic studies in HYVOLUTION, related to establishing decentral energy production units, integrate into the objectives of Large-scale integration of renewable energy sources into energy supplies.

HYVOLUTION required specialists to integrate the required research efforts in different disciplines, enabling eradication of the current fragmentation and increasing the coherence in research in non-thermal hydrogen production. In this way long-term collaborations for reinforcement of the Research Area are ensured and a global need in renewable hydrogen production will be achieved.

The vertical integration needed in HYVOLUTION, involvement in knowledge production through to technology development and knowledge transfer to end-users and principal stake-holders.The integration is achieved through the development of a coherent set of activities. The multidisciplinary and cross-disciplinary nature of IP, and the variety of the tasks involved, from laboratory-based research through to prototype development, knowledge dissemination and training, mean

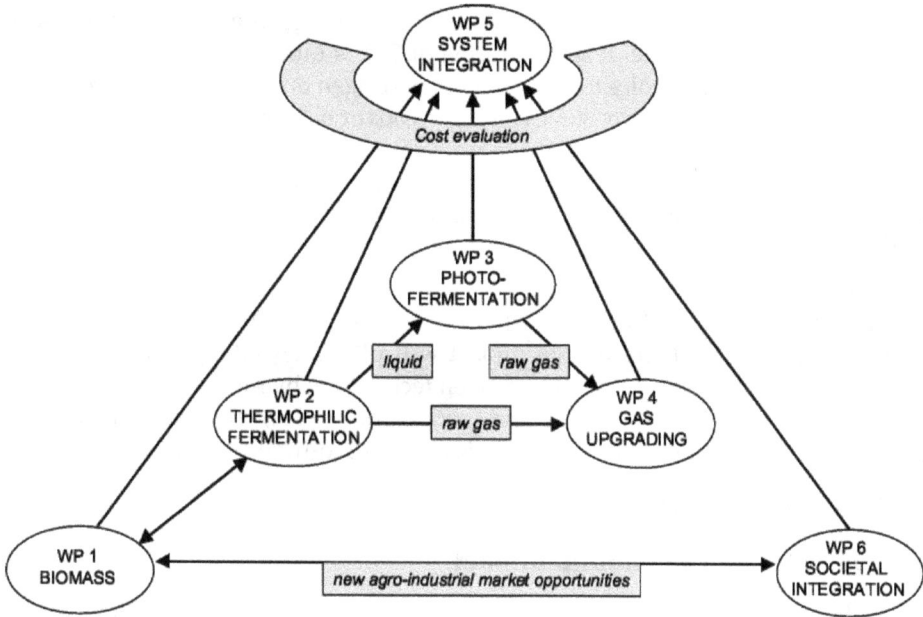

Figure 13.2: Coherence of Workpackages in HYVOLUTION.

that horizontal integration is also important. Intersectoral integration through partnerships between academia and industry including presence of SMEs for present and financial integration the project involves both public and private sector funding. It would be beneficial to develop the links of with international programmes or organisations as *e.g.* IENICA, EUREC, BFCNet, European Hydrogen and Fuel Cell Technology Platform, IEA, and HyNet are numerous and will act as accelerator through knowledge exchange in biomass and hydrogen R&D.

2.2 Relevance to the Wider Social Objectives

The major aim of HYVOLUTION would be establishment of decentral industries for hydrogen production at small-scale from locally produced biomass. The application of hydrogen for power or as bio-fuel will have a great beneficial effect on the quality of life. First of all, the benefits to the society are in common with other systems based on the application of hydrogen as an energy carrier. However, the hydrogen in HYVOLUTION offers the additional benefit of originating from renewable resources. As such, the benefits for the environment are enormous as the emissions of greenhouse gases and other pollutants associated with energy production are eliminated. The introduction of a decentral hydrogen industry, to enable a local or regional energy autonomy will provide sizable opportunities for increased industrial activities and employment.

2.3 Objectives

HYVOLUTION will make a very strong contribution to the agreements in the Kyoto protocol on the global reduction of greenhouse gases. The envisaged bioprocess is CO_2-neutral and even offers the advantage of easy capture of clean CO_2 at the production site. If sequestration of CO_2 is introduced, this process will be the opportunity by choice to start, linking hydrogen production to CO_2 removal. Furthermore Local energy resources (energy crops or bioresidues) providing power at small-scale sites close to the users will decrease the dependence on foreign energy supplies and also play an important role in delivering security of supply and freedom from geopolitical constraints. This approach minimizes the transport and infrastructure costs, associated with long-distance energy distribution and subsequently improves energy efficiency.

HYVOLUTION conforms entirely to the Environmental Technologies Action Plan (ETAP). All partners in this project will ultimately play an active role in rapidly spreading the developed hydrogen technology by means of commercially introducing small-scale equipment and process knowledge, providing the essential means of implementation of the newly developed eco-friendly technology.

3. Potential Impact

3.1 Societal and Economic Impact

The transition to the Hydrogen Economy will have a significant impact upon new employment opportunities and alternative markets for existing supply chains. The projected employment opportunities in the renewable energy sector by 2020 suggest that almost 1 million full time positions will arise if targets are achieved.It is obtained that H_2 obtained from biomass would have significant share to coverup. In tandem with significant labour gains, also increased labour incomes and increased profit as a result of increased demand in the supply chain and new enterprises can be expected. In monetary terms this will have a significant multiplier effect as a result of the re-spending of additional moneys as well as impacting upon the surrounding societies.

Besides reducing the to global energy prices the decentralised energy systems foreseen as a result of HYVOLUTION will evoke proliferation of new companies in rural areas for specialist services within the production, storage and distribution of hydrogen technologies.This will strengthen the competitive position between outside and inside based companies because being at the forefront of hydrogen system technology allows related companies to have first mover advantage.

The overall societal issues are perhaps more complex to determine, but the proliferation of decentralised specialist facilities is assumed to encourage inward investment into less favoured rural areas. The provision of good quality employment opportunities will, amongst others, mitigate rural migration.

3.2 Impact on Current and Future Policies

This project will significantly add to the number, diversity and choice of H_2 production routes available, giving greater security of H_2 supply at the local and regional level in countries. As such, this technology will highlight alternative and complementary strategies/routes to contribute to the increased share of renewable hydrogen as advocated. Furthermore, the industries developed through HYVOLUTION will contribute to the required interaction between centralised and decentralised energy supplies. This non-thermal approach, enabling the local conversion of the biomass, will initiate new agro-industrial production chains. As noted above, this will stimulate economic development for greater independency with respect to energy supply, and also with respect to hydrogen production in remote areas. More specifically, the output of the project will increase the economic diversity and prosperity of rural areas, through crop diversification, local industry creation and regional infrastructure development. Finally, HYVOLUTION will strongly contribute to the facilitation of the transition to mass hydrogen markets by realising the rapid erection of many small-scale production plants.

3.3 Impact on Business and Industry

HYVOLUTION provides improved scientific, technical and socio-economic understanding of the non-thermal production of hydrogen from biomass to give a reliable and cost-effective process by extending the findings in the preceding FP project, BIOHYDROGEN. Here a global lead in fermentative H_2 production from biomass has been established that will further materialise the scientific and technological advance to reinforce the strategic impact of being a leader in clean energy technology. The current fragmentation in non-thermal hydrogen production research is eradicated and a solid strategic position for new industrial ventures in sustainable energy systems will be the result.

3.4 Innovation

The main innovation in HYVOLUTION will be application of thermophilic bacteria which are superior in terms of efficiency in the first step in non-thermal hydrogen production from biomass.

The main innovation will be integrated approach of IP, where a multidisciplinary team of specialists in different sectors combine their efforts in a joint commitment to produce hydrogen from energy crops or bio-residues.

These innovation-related activities will be achieved by following workpackages:

☆ WP 1 relates to the introduction of crop-to-hydrogen chains in agricultural systems and the systematic utilisation of bio-residues in hydrogen generation. This approach will offer novel findings since detailed examination of the logistics has not previously been carried out for production of H_2 from biomass.

☆ WP 2 fully addresses the unique application of thermophilic bacteria using novel findings through an increased understanding of metabolism, genomics and proteomics. This would be aimed at the development of dedicated bioreactor prototypes with associated monitoring and control. This will underpin process optimisation and facilitate final industrial production.

☆ WP 3 addresses the coupling of the fermentative phase and the photofermentative H_2 production, which has not been performed before with feedstock derived from biomass. Similar as in WP 2, dedicated reactor prototypes with associated monitoring and control will be developed.

☆ WP 4 aims to develop dedicated, highly efficient gas upgrading systems. The units must be designed to handle small and frequently changing flow rates with different compositions. To control this integrated process, special gas sensor systems will also be developed.

☆ WP 5 involves the modelling and simulation of the unit processes needed to produce control strategies for the novel bioprocess.

☆ WP 6 evaluates the project from a socio-economic point of view. To identify the markets which will benefit from the envisaged local industry for hydrogen production from biomass?

3.5 Added Value in Carrying Out this Work at the Global Level

There are an increasing number of national, and international research activities focussing on H_2 production, storage and use. These are called participants and many of the participants are involved in various activities. A significant share of these investments in hydrogen R&D, reflected by the emphasis observed.

To date, world over the research in the EU with respect to non-thermal hydrogen production from biomass, have been under-funded and in lack of coordination, unlike that in Japan and the US. Besides the finalised FP project BIOHYDROGEN, there are no other international projects on non-thermal hydrogen production from biomass. In spite of this, several specialistic groups addressing biohydrogen production in the Netherlands, Hungary, and Turkey have developed. HYVOLUTION offers the eradication of this current fragmentation and fortifies the Research Area in biohydrogen research in itself, but also through the participation of Russia, adding to the project its highly qualified expertise in hydrogen research.

A strong contribution to fortification of the European Research Area also achieved through the integration of the different disciplines addressing solutions to complex scientific, technical and socio-economic issues. The composition, availability, logistics etc. of the biomass is of paramount importance and governs pretreatment and subsequent H_2 production routes. This consideration requires a pan-European approach involving the best specialists in the respective disciplines. Pooling of resources and expertise at a European level in HYVOLUTION gives the EU the opportunity to obtain leadership in fermentative H_2 production and to make a real breakthrough in non-thermal hydrogen production from biomass.

3.6 Account of Other National or International Research Activities

Many participants are represented in international organisations like IEA, the Hydrogen and Fuel Cell Platform, The European Hydrogen Association, HyWeb, HyNet, Hyways, Hysafe, biomass related networks like IENICA, BioMatNet and BCFNet, international standardization networks like ISO and IEC and smaller organisations like several national Hydrogen Associations, COST Action 841: Biological and Chemical Diversity of Hydrogen Metabolism, the Netherlands Biohydrogen Network, the Polish Scientific Network "Sustainable Energy Systems", to name just a few. Even though fermentative hydrogen production is in its infancy, these links show that HYVOLUTION is perfectly embedded in international organisations to take account of international hydrogen research activities.

Non-thermal hydrogen production from biomass as envisaged in HYVOLUTION will be complementary to the thermal processes SUPERHYDROGEN and AER-GAS and CHRISGAS. The latter two projects are complementary in the sense of biomass utilisation, as gasification requires dry biomass in contrast to a bioprocess where wet biomass is more suitable. In SUPERHYDROGEN higher moisture contents of the biomass are allowed but the extreme process conditions (as compared to a bioprocess) may require large-scale operation for making a cost-effective technology. According to Dutch co-ordinators of HYVOLUTION and SUPERHYDROGEN have already established their mutual interest to make an alliance of a bioprocess as proposed in HYVOLUTION for conversion of the wet fermentable fraction, with the technology of supercritical water gasification as in SUPERHYDROGEN for conversion of the drier non-fermentable fraction.

There are some other projects such as SOLAR-H, needed to explore sunlight directly for hydrogen production, mimicking biological hydrogen production with living organisms or isolated enzymes. In both approaches, the hydrogenase enzymes are central issues and thus, progress in both projects will have mutual benefits. Knowledge exchange between co-ordinators from HYVOLUTION as well as SOLAR-H provide greater benefit to the present and future world.

4. Implementation Plant

The expected aim of HYVOLUTION would be to deliver prototypes of process modules required to produce hydrogen of high quality in a bioprocess which is fed by multiple biomass feedstock.

In HYVOLUTION the approach may be based on the combination of a thermophilic fermentation (also called dark fermentation) with a photoheterotrophic fermentation. This approach required initiation for BIOHYDROGEN which is very challenging and promising. The novel issue is the application of thermophilic bacteria to start the bioprocess. This offers 2 important benefits in non-thermal hydrogen production. First of all, thermophilic fermentation at ~70 °C is superior in terms of hydrogen yield as compared to fermentations at moderate temperatures. In thermophilic fermentations, glucose is converted to, on the average, ~3 moles of hydrogen and ~2 moles of acetic acid as the main by-product. In other, mesophilic fermentations at ambient temperatures, the average yield is only 1 to 2 moles of hydrogen, at the most, per mole of glucose. This is due to the production of more reduced by-products like butyrate, propionate, ethanol or butanol under mesophilic growth conditions. The second advantage lies in the production of acetic acid as the by-product of the first fermentation. Acetic acid is a prime substrate for photoheterotrophic bacteria. Energy from light enables photoheterotrophic bacteria to overcome the thermodynamic barrier in the conversion of acetic acid to hydrogen. Through the combination of thermophilic fermentation with photoheterotrophic bacteria, complete conversion of the substrate to hydrogen and CO_2 can be established, resulting in 75 per cent conversion efficiency or 9 moles of hydrogen per mole of glucose.

WP 1 Biomass

The development of bioprocess for non-thermal production of hydrogen required acquisition of suitable feedstock on the one hand and gas upgrading on the other. Again, the application of thermophilic bacteria is outstanding.

WP 2 Thermophilic Fermentation and WP3 Photofermentation

To make this bioprocess technically and economically feasible, several issues are addressed in HYVOLUTION. Through extended research of the physiology, biochemistry and genomics of pure cultures of thermophilic as well as phototrophic bacteria, insight in metabolic pathways will be obtained. This is needed to model fermentations for optimal productivity and adjustment of the consecutive fermentations. This insight will be the basis for identifying and/or developing improved strains and creating mixed cultures which are generally known for robustness, with the elucidation of the fundamental microbiological properties, the development of dedicated bioreactors also resides in WP 2 and WP3 with the construction of prototype bioreactors being part of the technological objectives. The thermophilic bacteria are, as observed until now, inhibited by a partial concentration of 20 per cent hydrogen. This confers specific challenges for hydrogen removal in the thermo-bioreactor (WP 2). With photoheterotrophic bacteria, the technological challenge is the optimal penetration of light in the photo-bioreactor to establish high photochemical efficiency (WP 3). Both issues are extensively addressed in HYVOLUTION, because they are of paramount importance for overall efficiency.

WP 3 Gas Upgrading

Besides the upstream processing as described in WP 1, HYVOLUTION also addresses downstream processing. The raw gas produced in the bioreactors requires specific gas treatment. This is primarily because of the fairly high concentration of CO_2 in the raw gas and the relatively small and perhaps fluctuating quantity of hydrogen, and to a lesser extent because of the presence of contaminants, which are considered to contribute little to the final composition. It is known that fuel cells can be economic on small-scale. It is the area in between production and application that will be addressed to deliver technically and economically feasible gas cleaning devices, handling and safety procedures suitable to a small-scale hydrogen production plant.

WP 4 System Integration and WP 6 Societal Integration

System integration and societal integration form a basis to secure the scientific and technical objectives. These issues are fundamental to develop this new process for small-scale hydrogen production and to make it viable in terms of process-economics and socio-economics, including environmental impact. Both disciplines are integrated in HYVOLUTION to enable identified adjustments right from the start. This is necessary to avoid routes which will have no economic future or do not adhere to sustainability, and to make optimal use of the integrated approach

Chapter 14

Hydrogen as a Possible Alternative to Fossil Fuels: "State-of-the-Art"

Introduction

Already since the oil crisis of the mid-seventies an alternative to the production of energy from fossil materials most frequently imported, even from politically unstable countries, has been being sought. The application of biotechnologies to the energy sector makes it possible to use renewable raw materials, such as biomasses to produce energy carriers that integrate traditional oil-based carriers or other fossil fuels. Of these carriers the most important one is hydrogen. The main obstacle to that application arises when it comes to make these biotechnological processes competitive from an economic point of view. To achieve competitiveness would not only improve the reliability of energy supply but would also meet the European Union environmental objectives thus favouring the restriction of CO_2 emissions and the reduction of SO_2 and NOx emissions. The EU measures in the energy field also aim at combining the production of alternative energy sources with energy saving. The importance given to hydrogen as an alternative energy source is witnessed by the recent trend that has been characterizing the EU policy. In fact, after August/September 2002 Johannesburg's Conference, the EU has decided to appropriate more than 2 billion Euro to support researches in this sector, within the 6th Research and Technological Development Framework Programme. Two main aspects have directed the EU policy on that way: on one hand hydrogen is a *clean* and almost inexhaustible energy resource, and on the other it has all the characteristics that make it one of the fundamental energy systems for our future development, not to mention an environmental impact close to zero. In the U.S. the present situation is more or less similar to the European one, the only difference being that in the U.S., thanks to corn overproduction, more importance is given to power alcohol production destined to automotive fuel. It is important to point out that the origin of *Green Chemistry* or *Sustainable Chemistry* is attributed to the U.S. but, as the matter of fact, it came from the intuition of an Italian Professor born in 1857, Mr. Ciamician. In 1912 he stated his ideas and the results of his researches in a famous lecture entitled *The Photochemistry of the Future*, held in New York in 1912. His theory was based on the experiments he had been carrying out in Padua and in Bologna for 25 years. The main processes suitable for biological hydrogen production are taken into consideration herebelow.

H₂ Biological Processes Production

Hydrogen Production by Means of Photosynthetic and/or Anaerobic Microorganisms

In nature there exist several species of microorganisms, both prokaryotic and eukaryotic ones, capable of producing hydrogen. Hydrogen producing microorganisms belong to two big categories: photosynthetic ones and chemotrophic ones. Photosynthetic microorganisms, that include phototrophic bacteria, algae and cyanobacteria, can produce hydrogen thanks to their photosynthetic metabolism using solar energy. Chemotrophic organisms, on the contrary, produce hydrogen by means of fermentation processes. The main characteristics of the different groups of hydrogen producing microorganisms are listed in the following Table 14.1. The metabolism of phototrophic bacteria is both photolithotrophic and photoorganotrophic. In this group of bacteria, unlike cyanobacteria and green algae, the photosynthetic process is anaerobic. There are three families of phototrophic bacteria: sulphurless purple bacteria, purple thiobacteria and green thiobacteria. The first ones, through photosynthesis and under anaerobic conditions, use organic substances as carbon sources and hydrogen donors to reduce carbon dioxide, yet they do not oxidize the organic compounds containing sulphur. Some species, however, can anaerobically grow also in the darkness by getting energy through organic compound oxidation. Yet in the darkness hydrogen production is lower than in light conditions. Therefore photosynthetic bacteria, by exploiting the sun light, are able to transform several organic compounds, simultaneously producing hydrogen. The number of organic substances these bacteria can metabolize is quite high (carbohydrates, organic acids, fatty acids, etc.). The treatment of waste water polluted by organic material coming from several industry sectors such as feed, beverage and paper ones is a possible application for this type of microorganisms. In the U.S., some experiments were carried out for the treatment of pollutant streams coming from milk, whey, sugar or fruit-juice production industries. These experiments were successful both at laboratory conditions and scale and in real conditions using sun light. Purple thiobacteria are anaerobic or they grow in very low oxygen atmosphere. They use sulphur reduced compounds (in particular H_2S) as hydrogen donors to decrease carbon dioxide. Sulphur produced by this reaction accumulates inside the cell. Green thiobacteria oxidize sulphur inorganic compounds (*i.e.* H_2S) and carbon dioxide is reduced by means of a process similar to the one carried out by purple thiobacteria. Unlike what happens with the latter, sulphur does not settle inside the cell but only outside. Green sulphur bacteria oxidize the sulphur inorganic compounds (*i.e.* H_2S) and carbon dioxide is reduced by means of a process similar to the one typical of purple sulphur bacteria.

The removal of sulphur compounds, highly polluting for the environment and often present in industry and agriculture outputs, waste waters, rivers and pollutant sediments it is possible thanks to the application of this type of microorganisms. The optimum growth temperature of phototrophic bacteria ranges from 30 to 40°C. In general, like for most photosynthetic microorganisms, phototrophics are microorganisms living in an aquatic environment. The conversion efficiency of light energy into hydrogen when using purple bacteria is about 3 per cent. Phototrophic bacteria are able to remove polluting organic substances simultaneously producing hydrogen. However, since it is difficult to separate the bacteria after their use, for these applications a bacterial immobilized biomass is generally used. Bacteria immobilization also has the advantage of stabilizing hydrogen production, of reducing oxygen inhibiting effect and of improving production yields. The main difficulty when using these bacteria is their extreme sensitivity to the presence of oxygen in their growth environment. Cyanobacteria (or blue algae) are the biggest prokaryotic microorganisms capable of carrying out the photosynthesis and, unlike phototrophic bacteria, use the photosynthetic system to fix carbon dioxide, produce oxygen and use only water as electron donor. When exposed to the light some species of cyanobacteria produce hydrogen with a high intensity. Like for phototrophic bacteria the quantity of hydrogen produced depends on the quantity of solar energy and on its wave length. The conversion efficiency of solar energy achieves about 3 per cent. Unlike other photosynthetic microorganisms, cyanobacteria are able to produce hydrogen also in aerobic conditions and, for some species of cyanobacteria, the presence of oxygen does not inhibit the hydrogen production. Several species of eukaryotic microorganisms that can be classified as algae, specially green algae, can produce oxygen. These algae use the photosynthetic system of assimilating carbon dioxide with oxygen production. The hydrogen production rate is not very high and varies from 0.1 to 0.3 mmol H_2 h^{-1} g of dry biomass. To produce hydrogen the algae do not need an organic substratum but the presence of organic compounds can increase the production rate until achieving values of about 2 mmol H_2 h^{-1} g^{-1} of dry biomass. Unlike cyanobacteria, the presence of molecular nitrogen or ammonium ions does not inhibit H_2 production. With green algae a conversion efficiency of solar energy of about 10 per cent is achieved. Also for the algae oxygen concentration is a critical factor that may cause a rapid inhibition of the hydrogen photoproduction process. As far as a today use of algae to produce oxygen is concerned, results obtained up to now do not yet allow for immediate applications. The last group of hydrogen producing microorganisms is the chemotrophic one. These bacteria are characterized by the fact of producing hydrogen by means of a fermentation process, using as a substratum several organic substances including fatty acids and cellulose. The metabolism of these bacteria preferably occurs under anaerobic conditions, by oxidising organic substances. Under aerobic conditions the production of hydrogen is very low. The production yields for this kind of microorganisms vary from 1.1 to

25.7 μmoles of H_2 per μmoles of metabolized substratum depending on the species and the specific substratum. The main application of these microorganisms, which some groups of researchers are investigating, is to directly obtain hydrogen by means of fermentation. Up to now, if one consider the present efficiency of conversion into hydrogen of the chemical energy contained in the organic substance, the industrial production of H_2 through fermentation seems to be still far away.

Table 14.1 at the end of this paragraph summarizes the main characteristics of the microorganisms previously described in relation with hydrogen production. From a practical point of view, the hydrogen concentration in the gas biologically produced needs to be high since, in general, hydrogen production is associated with the simultaneous oxygen production. Therefore it is evident that too high or too low a concentration would lead to a low hydrogen gas or to the risk of explosion. Furthermore, the relatively high cost to separate hydrogen would suggest a direct combustion with oxygen by means of a thermal process. Up to now the different research programs concerning the present implementations of hydrogen bioproduction point to the following main possible sectors of use of this biological process:

1. To produce hydrogen through sun light by using green or blue algae in a water system;

2. To produce hydrogen from waste water polluted by organic and inorganic compounds using phototrophic bacteria and solar energy. In conclusion, as far as hydrogen for research and development is concerned, the different programs approved in the U.S., Europe and Japan during these last ten years have been aimed at defining and planning, on a pilot scale first and then on a large one, a system capable of producing hydrogen by using only water and sun light through microbial biomasses. Once optimized, this system should be compared from an economic point of view with hydrogen produced from other sources: fossil fuels, H_2O, biomasses.

Table 14.1: Main Characteristics of H_2 Producing Microorganisms

Microorganisms	Metabolism characteristics	Preferable Substratum	Sensitivity to Oxygen
Phototrophic bacteria:			
Purple bacteria	Photosynthetic and non-; anaerobe	Numerous organic substances	yes
Purple thiobacteria	Photosynthetic; anaerobe	Sulphur inorganic compounds	yes
Green thiobacteria	Mainly photosynthetic; anaerobe and aerobe	Sulphur inorganic compounds	yes
Cyanobacteria	Mainly photosynthetic; anaerobe	Inorganic	poor
Algae	photosynthetic and non-; anaerobe	Inorganic	yes
Chemotrophic	Non-photosynthetic; obligate and facultative anaerobe	numerous organic substances	yes, only for obligate aerobe

Other Hydrogen Production Processes

Hydrogen is mainly obtained from hydrocarbons through "r eforming" processes, gasification or electrolysis of water. These processes will be discussed in other papers on this specific field.

In any case, it is necessary to look for other possible hydrogen sources, to take a rentable way as independent as possible from oil. Nowadays the most interesting options seem to be production from biomasses and from renewable energy sources.

Use as fuel

Table 14.2: Comparison of Main Fuel Characteristics

Properties	Gasoline	Methane	Hydrogen
Formula		CH_4	H_4
Boiling point °C	30–190	−161	−253
Inflammability limits (air vol. per cent)	1.00–706	5.3–15	4–75
Combustion velocity in air m/s	0.50	0.40	2.9–3.5
Net ignition energy mJ	0.24	0.29	0.02
Minimum caloric value MJ/kg	40–45	50	120
Energy density (15°C, 100 kPa)			
MJ/m	33750	33.4	10.3

The strategic choice of an economy based on hydrogen finds its natural starting point in the production of this fuel from renewable sources or traditional fuels, to be implemented both in long-tested technologies (hydrogen engines) and in those being developed (fuel cells). Besides conversion technologies of natural gas, methanol, virgin naphtha into hydrogen, presently under development for future applications to fuel cells, researches are also focusing on the conversion of biomasses into hydrogen.

Chapter 15

Solution to the Climate Problem through Hydrogen Energy

Our position is clearly in favor of alternative energy and they can and must be the solution to fossil fuels, as well as a means of slowing climate change which we are witnessing. The world is doing many efforts to mitigate climate change, researching and developing various sources of the renewable energy such as wind, photovoltaic, solar, tidal, geothermal, aero thermal, bioenergy, undimotriz or hydropower, They will be overshadowed by the use of some countries like China or India of massive amounts of coal, which generate huge emissions of CO_2. This can lead to result in many developed countries resort to the use of coal to be competitive in this field.

As shown in multiple studies, alternative energy may be the weapon of the future in energy production, fossil fuels being relegated to a residual employment. Many countries are turning to so-called intelligent buildings for energy saving, with photovoltaic solar panels that provide electrical energy to the building or buildings with geothermal energy. This is still starting.

Other countries are designing models for efficient energy production from the oceans and seas, tidal energy or undimotriz energy. Although, like other renewable energies are in initial processes.

This leads us to consider for the moment, as it indicates, are alternative energy, namely, they not yet ready to be a substitute for fossil fuels and for the time being complementary energy from fossil fuels. At this time, these energies can not make the various societies to abandon the use of petroleum and its derivatives as energy.

We in this chapter we will try to address not a source of renewable primary energy, but an energy vector, the hydrogen. This, in the future may be the replacement of fossil fuels. However, despite the efforts of some countries, we are still far from turning the hydrogen energy future, and being the main energy source that solves our energy problems and climate change.

1. Hydrogen and Energy

Hydrogen is the most abundant chemical element in the universe and under normal conditions is in a gaseous state. The hydrogen is the most common and ubiquitous of the chemical elements. Hydrogen is an inexhaustible energy. Despite being the most abundant, but he is not pure in the nature, but is present as molecular or ionic.

As noted Gutierrez (2005) the hydrogen reservoir in pure state is found in Jupiter. It is inaccessible for the moment.

Therefore, hydrogen is not a primary source of energy but a vector, and therefore needed to obtain chemical decomposition of the element that he is associated. The most common way to find the hydrogen is in the water, to obtain it must be separated from oxygen by supplying an electric current (electrolysis) on a primary energy source.

A cost effective way to obtain hydrogen is to use renewable energy sources (wind, solar, geothermal, and so on.).

On consideration of hydrogen as an energy vector appropriate to highlight a number of characteristics: 1) low boiling points and proximity to the critical temperature. 2) Low densities of gas and liquid, 3) the content of deuterium, which may be one of the foundations of nuclear fusion (Gutiérrez, 2005).

The most attractive features to hydrogen as a transmitter of energy are in cleaning. In the absence of carbon in the transition process to generate energy that does not produce CO_2 emissions. Secondly, it is energy potential (Valero-Matas, 2010). Cleaning and energy potential opened up great expectations for climate change advocates, and some car companies saw a new kind of engine with hydrogen fuel. Science and technology have spent decades trying to find or design suitable components to make hybrid vehicles competitive with petrol cars. For now, hybrid cars are powered by so-called fuel cells or fuel cells.

2. The Hydrogen Economy and Social Impact

When we refer to the social effects produced by something, it should define what is going to be evaluated, then proceed to the analysis of the facts. This requires a life cycle assessment and impact of technology, becoming part a series of elements such as economic, environmental, social, health, risk, human needs, sensitivity, development objectives of society and political impacts decision making.

Therefore, taking these issues on the horizon, and emphasizing that the implementation of technology can never make the risks significantly outweigh the potential social, as any damage to the citizenship for their implementation, and undermining their quality of life instead of getting beneficial technologies reported. Reflect on the effects of the hydrogen economy.

Depending on what the source of production of hydrogen vary substantially the price. Given this fact we must ask two questions: If the question is to eliminate energy dependence on fossil fuels, or reduce the emission of CO_2 into the atmosphere. According to the election, the effects on the social impact vary substantially. If the intention is to reduce emissions of greenhouse gases as stated in previous sections, the current method of producing hydrogen via natural gas is the most profitable. Companies, universities and research centers continue to explore different components to reduce costs, making it competitive and produce fewer emissions of CO_2.

Another way is through acquiring hydrogen from biomass. The energetic use of the biomass is the gasification, which allows obtaining gas of synthesis (CO + H_2). The synthesis gas obtained can be used as direct fuel, as well as H_2 source or chemical feedstock to make other fuels. The production of hydrogen by gasification of biomass is an interesting option, it has the advantage over the conventional (steam methane reforming of water) to use a waste and not a chemical feedstock. It arises as a hope to the energy consumption, though it issues CO_2 there diminishes the dependence of the fossil fuels. It has an important disadvantage, his corrosive effect reducing considerably the life of the fuel cells and pipelines of transport.

On the other hand, if you wanted to eliminate the dependence on fossil fuels and emit no CO_2, the process is quite different. To achieve carbon-free hydrogen the most common is water, and electrolysis treatment. This method requires a lot of energy to break water molecule and split it into hydrogen and oxygen. The power supply either through renewable energy or nuclear energy is higher than that after the hydrogen is exploited. For example, a kilogram of hydrogen produced through natural gas comes to cost about 2 €, where 45 per cent of its production is due to the cost of natural gas. According to the current price of electricity, a kg of hydrogen produced by electrolysis would cost about 5€, where 85 per cent of product cost is the price of energy. In other ways, depending on a broad range of factors, an estimate would be around roughly between 3.5 € per kg to 8 € per kg of hydrogen.

To this must be added transportation costs, marketing, so on This issue is important to reference when analyzing the economic impact of a product over another. This will appreciate the difference in cost and economic impact on a society.

Another investigation that is under development is the production of hydrogen from green algae. More than 60 years ago it was discovered that a microscopic green algae, -whose scientific name is *Chlamydomonas reinhardtii* and we all know as the ponds scum -, can split water into hydrogen and oxygen in laboratory conditions. The possibility of using algae as microscopic power plants were the brainchild of Hans Gaffron, who observed in 1939, "for reasons unknown at this time, which stopped producing algae began producing oxygen and hydrogen for a short period. Green algae have a hydrogen-producing enzyme known as iron-iron hydrogenase which has evolved a structure that makes its particularly susceptible to attacking oxygen molecules. Green algae can produce hydrogen gas, H_2, in a process called "biophotolysis" or "photobiological hydrogen production. This process is carried out by photosynthetic enzymes, which split water to get electrons, photons excite electrons, and finally, use these electrons to reduce 2H + to H_2. The scientific challenge associated with this approach to hydrogen production is the enzyme that actually releases the hydrogen, called "reversible hydrogenase, is sensitive to oxygen. The process of photosynthesis, of course, produces oxygen and this normally stops the production of hydrogen very quickly in green algae. A team of biologists led by Raymond Surzycki Jean-David Rochaix and the University of Geneva, and Cournac Laurent and Gilles Peltier, both from the Atomic Energy Commission, National Center for Scientific Research and the University of the Mediterranean, the algae cell research *"Chlamydomonas reinhardtii"* that through the use of copper in the cells to block the generation of oxygen, it achieves a cycle of hydrogen production.

Algae to produce hydrogen, they must have sunlight and be in an anaerobic environment (without oxygen) to prevent oxygen toxicity to the top eventually responsible for hydrogen production. The alga synthesizes up "hydrogenase" that is ultimately responsible for producing hydrogen by combining with electrons derived from photosynthesis.

Another strategy is to modify the hydrogenase using genetic engineering to be more tolerant of oxygen

One might screen microorganisms in nature for the presence of oxygen-tolerant hydrogenases. The genes of these enzymes could then be introduced into algal cells and tested for hydrogen production under less stringent anaerobic conditions. This is the case of Dr. Melis of the University of Berkeley has managed to "design" a seaweed that contains less chlorophyll density, or that is, is more transparent. This means that sunlight can penetrate deeper into the mass of algae without being stopped by floating on the surface layer, leading to increased production of hydrogen in the amount of algae. Genetic modification of algal

Chlorophyll density decreases from 600 to 300 in the chloroplasts, the body of the cell where photosynthesis takes place. During normal photosynthesis, the algae use sunlight as energy to convert CO_2 in water and glucose, besides providing oxygen to the atmosphere and a small amount of hydrogen.

According to Melis acres of algae could produce about 200 kilograms of hydrogen for day To realize a balance sheet of the production of hydrogen not only attends to the final values of consumption and money-cost, also involved and the resources needed for production. In this case, tons of water needed to produce hydrogen fuel. Only in the United States was estimated in 2006 (Norskov and Christensen, 2006) about 150 million tons water per year to meet the transportation needs, if this is added, the demand for buildings, business and domestic use, consumption shoots to 500 million tons per year in USA. Does society and water resources and energy are prepared to meet this challenge, the hydrogen economy?

The fuel cells, they are the gadgets needed to transform chemical energy through a substance into electrical energy, and make it work external combustion engine. Currently, there are different types of fuel cells, suitable to various needs and demands. As the alkaline fuel cell (AFCs) that performs better. NASA uses this type of fuel cell in their space activities, but presents a major problem, its high price. Why is this price so high? Basically its components, the catalysts used are platinum/ruthenium, electrodes contain large amounts of noble metals, for example, the anode can be made with platinum and palladium, and gold and platinum cathode, if this is happening worldwide, decreasing costs is impossible, because the demand for gold, platinum, palladium and ruthenium are fired, increasing the value of these metals by the scarcity and the need to use. In the case of reaching the market, would be a matter of asking him, because if the oil runs out, the noble metals also. And on this there is a greater lack of durability. In the immediate future no one can see a drop in prices in the world there are millions upon millions of cars and their production is increasing.

In an attempt to make viable hydrogen-powered vehicles, automotive companies have developed a technology stack PEM (Proton Exchange Membrane) rather the product cheaper. The problem is not solved completely, because the platinum catalyst remains. Returning to the same matter, and if you choose to design all hybrid cars with this type of batteries, the fired platinum price, increasing the cost of the fuel cell. Enter the fray again the question of durability. Platinum is perishable and we are talking about millions of cars and piles of life between 10 and 15 years at best, and do not forget that the noble metal platinum is a finite.

At this point, several developments have occurred in lower prices and fuel cell performance without precious metals from catalysts. The research of Barnett and Zhan (2006) have developed a new fuel cell solid oxide, or SOFC, that converts iso-octane, high-purity compound similar to gasoline, hydrogen, which is used by the cell to energy. Also developed by the team of Michel Lefèvre *et al.* which have taken a giant step to reduce costs, using new catalysts that use iron instead of platinum without performance degradation. Iron-based catalysts for the oxygen-reduction reaction in polymer electrolyte membrane fuel cells have been poorly competitive with platinum catalysts, in part because they have a comparatively low number of active sites per unit volume. We produced microporous carbon–supported iron-based catalysts with active sites believed to contain iron cations coordinated by pyridinic nitrogen functionalities in the interstices of graphitic sheets within the micropores.

Store large amounts of hydrogen safely and cheaply, and enable its use (through fuel cells or direct combustion) is another challenge for this product. Hydrogen can be stored in different ways: aerated, high pressure compressed, liquid, by chemical or carbon nanostructures (Fakioglu *et al.*, 2004). Currently, hydrogen is the most used aerated, transported in cylinders and high pressure gas. This form of storage is not optimal if you are used as energy for a vehicle, due to the high volume of these cylinders. If the interest is in providing a group of houses (as they currently do propane tanks) will need a silo which will triple the amount currently used by the propane storage tanks. This is easy to observe, one must look at the main tank (where the liquid hydrogen) of the space shuttle, whose volume is greater than the aircraft itself[1]. Gasified hydrogen tanks must be made with special materials that maintain safety and avoid the risk (West, 2004). The chemical composition involves hydrogen use by many transition metals and alloys to store hydrogen in metal hydrides. This process presents a major problem, the heavy weight of the storage system as a result of low levels of hydrogen retention are achieved (Conte *et al.*, 2004). Finally, carbon nanostructures involve inserting inside a solid material at a temperature and pressure to later extract it with other values of pressure and temperature. This form of storage can accumulate a larger amount of hydrogen in volumes of the above dimensions. It is not the optimal model, although recent research has put a glimmer of hope to the nanoparticles such as hydrogen storage system (Ares, 2008).

The liquid hydrogen and hydrogen gas under pressure are the most widespread in the storage and transportation of hydrogen. In the case of hydrogen gas under pressure for transport, some suggest the network of natural gas pipelines. This is not possible because the natural gas will still be using for many years. In the course of doing so for the same natural gas pipelines, they do not serve as the fragility of the steel in the hydrogen makes the pipes require special insulation, carbon fiber for example. Therefore implies a high cost. The other option corresponds to submit to high pressure and hydrogen storage. For transporting liquid hydrogen needs to be exposed to hydrogen at a temperature of -252 degrees Celsius, that is to say, cryogenic fuel. In the United States, NASA and companies working with liquid hydrogen is transported in cryogenic tanks either truck, rail car or barge specially prepared. Bossel and Eliasson (2003) advise against this practice as domestic consumption for two reasons: 1) The power consumed by a tank of pressurized hydrogen becomes a significant fraction of the energy content of hydrogen consumed. For example, for a supply of 40 kilometers, the energy used in the route of supply is equal to 20 per cent of the hydrogen energy delivered. 2) You need a great transport fleet. As the ratio of 15 trucks of hydrogen for a fuel truck 25 tons. Select this possibility is irrelevant and irrational. First, the truck fleet overflows traffic; its implementation would entail a considerable increase in jobs. Secondly, the hydrogen fuel would shoot.

Safety is another important principle in assessing the impact of technology on society. As announced in its day Beck (1998) we face the risk society, it symbolizes not take more risks than necessary, and many are the result of side effects of technology. Regarding the safety of hydrogen there is no common approach. Some as Braun (2003) attest to the high security of hydrogen compared to other fossil fuels, and indicate that the numbers of accidents resulting from hydrogen are currently about 1 per cent. In contrast, other theorists believe hydrogen more dangerous than gasoline (Hordeski, 2005).

Hydrogen cars are the best opportunities in life, certain companies Ford, Toyota, Honda, Volkswagen, Chrysler, and so on. have developed hybrid cars combining petrol and electric power, reducing fuel consumption and therefore CO_2 emission. These cars have greatly two problems: 1) the need to introduce two motors inside, increasing the size and reduce the space for passengers or trunk, 2) the price of a hybrid between 12,000 € and 18,000 € more than a petrol or diesel, depending on model and automobile company.

Another idea lies in the use of clean hydrogen, *i.e.* 100 per cent hydrogen car, whose impact will be of another magnitude. As happens with vehicles powered by gasoline or diesel fuel will need storage, but in the case of hydrogen as the volume is larger deposit is needed a larger and heavier, added to the engine problem, determine the size of the vehicle. For benefits equal to their gasoline counterparts need more power, resulting in higher consumption and cost. A liquid, the evaporation capacity of hydrogen is very high. A NASA study (Los Alamos) showed that in ten days had evaporated hydrogen from a tank car.

The environmental impact to a hydrogen economy would be the ideal of sustainability, but only when conditions were appropriate. And this is not the case. To stop issuing between 70 per cent and 80 per cent of CO_2 into the atmosphere, it would mean the possibility of regeneration in a relatively small space of time, as well as, improve air quality in cities, especially the most polluted as Mexico. Even with only operate worldwide with hydrogen cars, only the issuer would reduce greenhouse gases by 20 per cent. These values refer to hydrogen fuel with no carbon. Not taken into account, the hydrogen obtained from natural gas. If your application is given in optimal conditions, the hydrogen would be a sustainable energy vector.

Political action of this magnitude, in a globalized world where oil is currency, and the big multinationals of oil and oil producing countries are strangling the world economies, it would happen to be a history. The decision would not prevent the various problems announced by Rifkin (2002). Some of them even become worse over. Rifkin declares that with the economy of the hydrogen it was coming near to a social justice, prosperity and equality.

This will not happen; the hydrogen economy requires a very high technological level which lacks the most disadvantaged countries and oil producing countries. The formation of the citizens of these countries is very limited. Therefore continue to rely on the industrialized countries (first world). Second, if the technological and production levels depend on first world countries, how will they be able to produce such materials, wind turbines, photovoltaic systems, water electrolysis, fuel cells, and so on? Especially, when some living on the threshold of subsistence, and even below. How can promise the power of freedom? The incursion of the hydrogen economy in society will produce an unprecedented social change and energy; we might even attend a social revolution. Your application will not only transform the concept of energy also ways of life.

3. Hydrogen Economy may be the Solution to Dependence on Fossil Fuels?

There is every reason to not consider the economics of hydrogen energy as XXI century, for example

a. The inability to produce hydrogen for all cars in the world. And not just the millions of existing vehicles, but the millions of cars to the growing demand from China and India (remember that these countries are the main cause of increased oil).

b. The necessary water to produce this quantity of hydrogen. We are before a problematic substance since it is the water. His shortage has led to the War to several countries. With the time they will be sharpening before the major shortage of this liquid.

c. The cost to address the need to produce hydrogen through electrolysis. This is not resolved in the same terms as the oil where a plant can refine million tons of crude oil. The hydrogen production plants have less capacity, so it will require thousands.

d. The number of hydro and other tools necessary for citizens to fill up your fuel tank, regardless of the proposal (more media than real), Rifkin (2002) that all citizens have a hydrogen charger at home or with a proximity 10 meters.

e. The costs in their development, production, storage, distribution networks, product modification and furnishings of everyday life (adjustment of housing supply hydrogen, cooker, washing machine, and so on).

Today, most hydrogen is produced from natural gas through steam reforming of methane, and although this can be understood as a first foray into the hydrogen economy, represents only a modest reduction in emissions from automobiles hybrid vehicles. Along with this model also can use the electrolysis of water, heat, wind, geothermal, solar and biomass processing (using a variety of technologies ranging from reforming to fermentation).

The Biomass processing techniques can benefit greatly from the wealth of research carried out over the years in refining and fuel conversion of liquid and gaseous fossil. Biomass can be easily converted into liquid fuels, including methanol, ethanol, biodiesel and pyrolysis oil can be transported easily and generate hydrogen on site. Although biomass is clearly (and necessarily) sustainable, it can be transformed into the hydrogen supply taking into account the amounts required for global hydrogen supply. Fundamentally, the limitation of food as it has been observed with ethanol driving up grain prices. Still, even if they can pay the money, would leave the neediest people without food. On the other hand, continues to emit CO_2 to the atmosphere, to a lesser extent. It is clearly not sustainable. Hydrogen energy may not be a substitute for fossil fuels by the huge amount of problems and impossibilities that make it unfeasible

Hydrogen energy may not be a substitute for fossil fuels by the huge amount of problems and impossibilities that make it unfeasible. The hydrogen economy was again a false alarm, before the world need to address energy issues have taken place in the world. This may be another case like cold fusion. Where they had placed many hopes, it did not produce environmental impact of nuclear waste were not hazardous and generated large amounts of energy.

In 1989, Pons and Fleischman scientists announced they had achieved cold fusion production with the corresponding energy release. It was all a hoax. Why?, it is quite clear, when in the 70's of last century opened the door to cold fusion research, scientists, politicians, enterprises, and so on, saw it, the energy solution and began to demand

research, funds and materials to address the issue. After much money spent, time and human capital, it has not seen results. To avoid abandonment of the project, the researchers invented the fraud. Hopes were cooled cold fusion. Although there is research but are being developed with less enthusiasm and expectations of previous years. Pass the same to the hydrogen economy, this will not be a hoax but it will be a succession of isolated research and she will not be motivated by a project of general interest. At the bottom of this issue, there is another political reality commercial interests, and do not forget the politics and economics, is that the potential importance of hydrogen as an energy carrier may seem exaggerated, but very significant. The energy value of hydrogen as a substitute for oil is the main objective of the policies of OECD countries, occupying a secondary interest environmental benefit (Andrews, 2005). Like there is no economic profitability, the project will suffer a drop, though, this energy carrier is important to keep clean our planet from greenhouse gases.

environment is good news, but in order to achieve this there must be a common policy designed to achieve sustainable development using renewable energy, and for this it is necessary to provide a large global project, the style of the Manhattan Project, United States. If countries get to work on a large project, possibly within ten years would have a commercially viable hydrogen economy. But today, this seems more like a siren song that a reality.

Not only is the hydrogen energy that we can draw from this dependence on oil, and consequently to reduce CO_2 emissions, there are many other alternative energy sources, which were developed in the past, and as of today, research and employment is very low. In 1870, geothermal energy, took a quantum leap, as scientists saw it as a good source of energy, then increased research to study the ground thermal regime.It was not until the twentieth century, and the discovery of radiogenic heat (heat balance). In 1967 he was inaugurated in the Race, France, and the first floor of harnessing the tides (tidal energy).

O wave energy, which knowledge dates from the French Revolution, however, does not begin to be studied in depth until the late twentieth century.

Concerns about clean energy are creating a collective imagination, demand for countries to seek energy alternatives, and we do not stop to think about the economic and political interests of big oil companies. Automation's companies discussed the many efforts towards a hydrogen economy, and what we observe, is that this companies go years in it, and we still have a viable production car of its kind.

Chapter 16

Sustainable Development and Hydrogen Production

Introduction

Hydrogen was first discovered in 1766 by Henry Cavendish (1731-1810) in London, England when he collected it over a metal and described it as "inflammable air". Hydrogen was named in 1783 by Antoine Lavoisier (1743-1794), with the origin of name from the Greek words "hydro" and "genes" meaning "water" and "generator", because when hydrogen burns, water is is the basis of the so-called hydrogen energy, as summarized in Table 16.1.

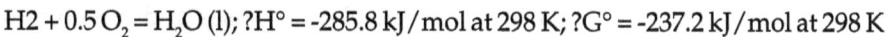

H2 + 0.5 O_2 = H_2O (l); ?H° = -285.8 kJ/mol at 298 K; ?G° = -237.2 kJ/mol at 298 K

Hydrogen energy has the potential to solve many of the major environmental problems that the US and the world are encountering as a result of fossil fuel combustion. However, there remain major challenges in, and thus major opportunities for, hydrogen energy research. There are some general reviews and numerous research articles on various aspects of hydrogen energy (Momirlan and Veziroglu, 2002; Ogden, 1999; Winter and Nitsch, 1988). The US Department of Energy has been a strong proponent for research and development related to hydrogen energy development in the US. In his State of the Union Address on January 28, 2003, the US President George W. Bush announced a new hydrogen initiative that is dubbed Freedom Fuel (related to the fuel cell-based Freedom Car) (DOE EERE, 2003). The Freedom Car and Freedom Fuel initiatives have certainly stimulated great public interest and more attention in the research community on hydrogen energy. The development of H_2-based energy system requires multifaceted studies on hydrogen sources, hydrogen production, hydrogen separation, hydrogen storage, H_2 utilization and fuel cells, H_2 sensor and safety aspects, as well as infrastructure and technical standardization. This article discusses the technical processing options for hydrogen production in conjunction with hydrogen utilization, fuel cells, and mitigation of CO_2 emissions, and offers some personal perspective.

Table 16.1: Principle Thermodynamics of Hydrogen Energy

Thermodynamic Property	Values	State of Water
Reaction	$H_2 + 0.5\ O_2 = H_2O$	
Enthalpy of reaction at 25°C	?H° = -285.8 kJ/mole (HHV)	H_2O as liquid
Enthalpy of reaction at 25°C a	?H° = -241.8 kJ/mole (LHV)a	H_2O as vapora
Entropy of reaction at 25°C	?S° = - 163.3 J/mole•K	H_2O as liquid
Free energy of reaction at 25°C	?G° = -237.2 kJ/mole	H_2O as liquid
Free energy of reaction at 80°C	?G° = -228.1 kJ/mole	H_2O as liquid
Free energy of reaction at 80 °C	?G° = -226.1 kJ/mole	H_2O as vapor

a) The heat of vaporization of H_2O is 44 kJ/mole. When the reaction heat involving water as a vapor product is used for calculation, the value is referred to as lower heating value (LHV).

H_2 Production Options

It should be emphasized that molecular hydrogen is an energy carrier but not an energy resource, and thus hydrogen must be produced first. Table 16.2 summarizes the possible options for H_2 production. By energy and atomic hydrogen sources, hydrogen can be produced from coal (gasification, carbonization), natural gas and propane gas (steam reforming, partial oxidation, autothermal reforming, plasma reforming), petroleum fractions (dehydrocyclization and aromatization, oxidative steam reforming, pyrolytic decomposition), biomass (gasification, steam reforming, biological conversion), and water (electrolysis, photocatalytic conversion, chemical and catalytic conversion).

Figure 16.1 shows the current commercial processes and possible future options for H_2 production and related research issues. Figure 16.2 compares the current commercial process technologies for H_2 production versus the scale of H_2 production based on studies by industrial gas producers as reported in literature (Rostrup-Nielsen, *et al.*, 2003; Gunardson, 1998; HP-Gas, 2002). Excellent reviews have been published on H_2 production technologies by Rostrup-Nielsen (2002) and on catalysis involved in H_2 production by Armor (1999). The relative competitiveness of different options depends on scale of production, H_2 purity requirement, catalytic processing methods and energy sources available. Current commercial processes for H_2 production largely depends on fossil fuels both as the source of hydrogen and as the source of energy for the production processing. Fossil fuels are non-renewable energy resources, but they provide a more economical path to hydrogen production in the near term (next 5 to 20 years) and perhaps they will continue to play an important role in the mid-term (20 years to 50 years from now). Alternative processes need to be developed that do not depend on fossil hydrocarbon resources for either the hydrogen source or the energy source, and such alternative processes

Table 16.2: Options of Hydrogen Production Processing Regarding Atomic Hydrogen Source, Energy Source for Molecular Hydrogen Production and Chemical Reaction Processes

Hydrogen Source	Energy Source	Reaction Processes
1. Fossil Hydrocarbons	1. Primary	1. Commercialized Process
Natural gas[a1]	Fossil energy[b1]	Steam reforming[c1]
Petroleum[a2]	Biomass	Autothermal reforming[c1]
Coal[a1,a2]	Organic Waste	Partial oxidation[c1]
Tar Sands, Oil Shale,	Nuclear energy	Catalytic dehydrogenation[c2]
Natural Gas Hydrate	Solar-thermal	G a s i
		carbonization[c1]
2. Biomass	Photovoltaic	Electrolysis[c3]
3. Water (H_2O)	Hydropower	2. Emerging Approaches
4. Organic/Animal Waste	Wind, Wave, Geothermal	Membrane reactors
5. Synthetic Fuels	2. Secondary	Plasma Reforming
MeOH, FTS liquid, etc.		
6. Specialty Areas	Electricity[b2]	Photocatalytic
Organic Compound	H_2, MeOH, etc.	Solar-thermal chemical
		Solar-thermal catalytic
Metal hydride, ch	3. Special Cases	Biological
complex hydride		
Ammonia, Hydrazine	Metal bonding energy	Thermochemical cycling
Hydrogen sulfide	Chemical bonding energy	Electrocatalytic
7. Others	4. Others	3. Others

a1: Currently used hydrogen sources for hydrogen production.

a2: Currently used in chemical processing that produces H_2 as a byproduct or main product.

b1: Currently used as main energy source.

c1: Currently used for syngas production in conjunction with catalytic water-gas-shift reaction for H_2 production.

c2: As a part of industrial naphtha reforming over Pt-based catalyst that produces aromatics.

c3: Electrolysis is currently used in a much smaller scale compared to steam reforming.

need to be economical, environmentally friendly, and competitive. An example of an alternative, environmentally-friendly process is biological H_2 production from biomass (Logan *et al.*, 2003).

Some organic substances can be used for hydrogen production, and they includes methanol synthesized via synthesis gas from natural gas and coal (steam reforming, decomposition), ethanol made from biological fermentation of crops and biomass (steam reforming), and sugars or carbohydrates (steam reforming, gasification). For some special applications, hydrogen-containing inorganic compounds such as ammonia and hydrogen sulfide have also been considered as source compounds for hydrogen production (catalytic decomposition).

Regardless of the hydrogen source and energy source, in order for hydrogen energy to penetrate widely into transportation and stationary applications, the costs

H-Source	Current H₂-Prod Process	New H₂-Prod Proc Options	Research Issues
Natural Gas	Steam Reforming & Water Gas Shift	Advanced Reforming &Membrane Processes	R&D Topic Areas
Petroleum	Dehydrocyclization & Dehydrogenation	On-board/On-site Reforming & Adsorption Desulfurization	Catalysis & Materials
Coal	Gasification,Cleanup & Water Gas Shift	Adavanced Integrated Process & O₂/H₂ Membranes	Chemical Reactions
Biomass	Gasification/Reforming & Water Gas Shift	Catalytic Production/Separation Biological H2 Production	Reactor Configuration
Water	Electrolysis Using Electrical Energy	Thermochem Cycle & Membrane Photo-catalytic/Photo-electrochem	Processing Scheme
			Product Separation
			Membrane Separation
			Computer Modeling
			System Optimization
			Integrated Systems
			Innovative Concepts

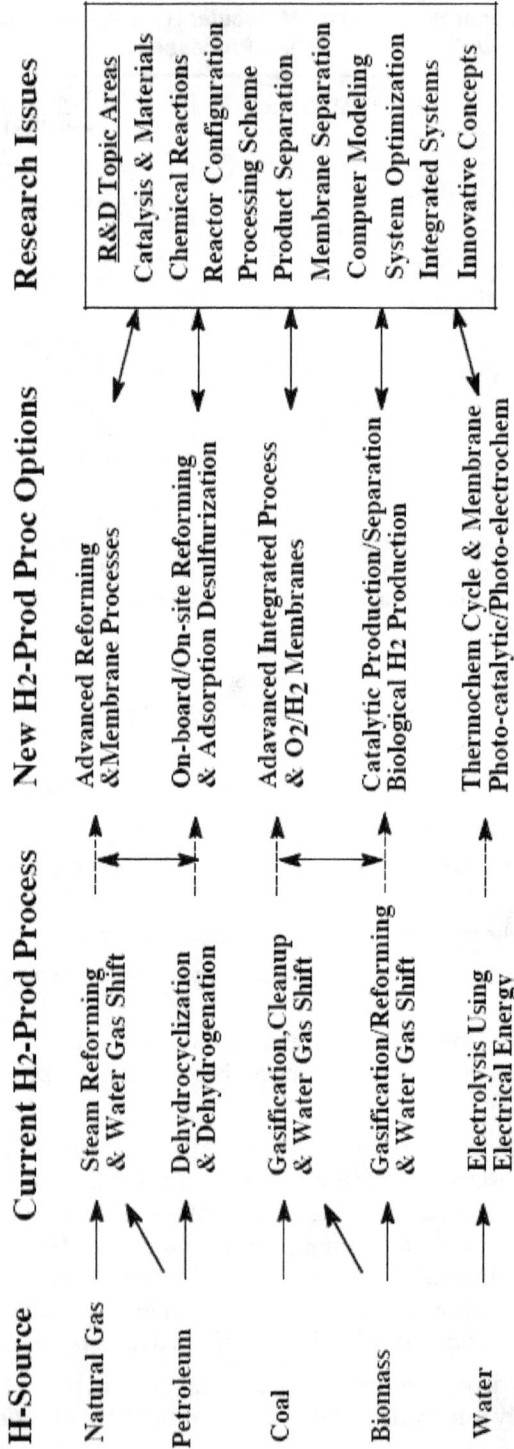

Figure 16.1: Current Processes and Possible Future Options for H₂ Production and Related Research Issues.

of H_2 production and separation need to be reduced significantly, *e.g.*, probably by a factor of 2 or more from the current technology.

H2 separation is also a major issue as H_2 coexists with other gaseous products from most industrial processes, such as CO_2 from chemical reforming or gasification processes. Pressure swing adsorption (PSA) is used in current industrial practice. Several types of membranes are being developed that would enable more efficient gas separation.

H2 production and separation can be integrated in novel membrane processes that incorporate reaction and separation in the same unit. There are several types of membranes, gasdiffusion membrane, ion-conducting membrane, and catalytic membranes.

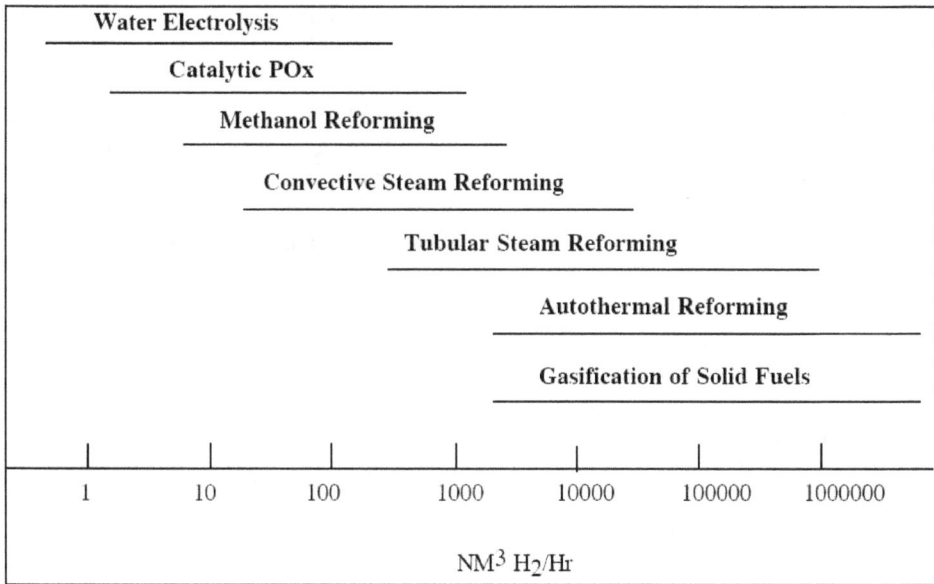

Water Electrolysis

Catalytic POx

Methanol Reforming

Convective Steam Reforming

Tubular Steam Reforming

Autothermal Reforming

Gasification of Solid Fuels

| 1 | 10 | 100 | 1000 | 10000 | 100000 | 1000000 |

NM^3 H_2/Hr

Figure 16.2: Technology of Choices Based on the Scale of Hydrogen Demand (Sources: (1) J. R. Rostrup-Nielsen, J. Sehested, N. Udengaard, Paper presented at Am. Chem. Soc. Symp. H_2 Energy for the 21st Century, March 23-27, 2003, New Orleans, LA; (2) H. Gunardson, Industrial Gases in Petrochemical Processing. Marcel Dekker, New York, 1998, 283 pp; (3) HP-Gas. Gas Processes 2002. Hydrocarbon Processing, 2002, (81 (5), 61-121.). Note: 1 Nm3/h = 847.44 standard cubic feet per day (scfd); 1 million standard cubic feet (MMscfd) = 1180 Nm3/h

A Proposed Concept on CO-Enriched Gasification for H_2 Production from Coal

Because coal resource is much more abundant than natural gas and petroleum in US and many other countries, production of H_2 from coal via gasification is an important path to H_2 energy development in the foreseeable future. Statistically, coal is also the most abundant fossil energy resource in the world (Song and Schobert, 1996). An excellent review of coal gasification technologies has been published by

Stiegel and Maxwell (2001). There are major technical challenges for developing more economical system for coal-based hydrogen and synthesis gas production coupled with CO_2 capture and sequestration, including the following aspects: more efficient gasification for H_2 production purpose, sulfur-tolerant catalysts for water-gas-shift reaction; efficient removal of hydrogen sulfide from hot gas; effective separation of CO_2 from H_2 in the presence of steam (Xu *et al.*, 2003), and permanent sequestration of CO_2 as solid mineral (Maroto-Valer *et al.*, 2002).

A concept called CO-enriched gasification (COEG) is proposed here, as illustrated in Figure 16.3, for coal and biomass gasification that may be studied further. Coal gasification typically occurs at very high temperatures and the heat utilization (heat transfer to make steam) at very high temperature is not very efficient, thus incorporation of an endothermic chemical reaction may be helpful. CO-enriched coal gasification involves CO-enriching reaction (eq.(2)) using recycled CO_2 and exothermic oxidation reactions (eqs. 3 and 4). CO-enriched gasification of coal can be achieved by integrating reactions of 3 co-reactants-steam, CO_2, and oxygen with carbon (coal) (eqs. 1-4). It is similar in principle to the tri-reforming concept recently proposed by the author for natural gas (Song, 2003). CO_2 can be recovered by either conventional method or by using the CO_2 "molecular basket" (Xu *et al.*, 2002, 2003) and recycled for the gasification. The COEG could be a more effective way compared with H_2 production via steam gasification (eq. 1). Compared to conventional processes, CO-enriched gasification is expected to be superior for H_2 production purpose because each CO molecule gives an additional H_2 molecule upon water gas shift reaction (eq. 5). The proposed CO-enriched gasification is based in part on a recently reported concept of tri-reforming of natural gas which involves 3 reactions simultaneously-CO_2 reforming, steam reforming and partial oxidation of methane (Song, 2001; Song *et al.*, 2002). This reaction produces CO-rich gas due to addition of CO_2 as a co-reactant (eq. 2), and such CO-rich gas will produce more H_2 since CO will produce more H_2 molecule upon water-gas-shift reaction (eq. 5). Such gasification can also be conducted in the presence of biomass. Previous studies have shown that coal gasification reactivity is higher in steam than in carbon dioxide (Messenbock *et al.*, 1999). Certain catalysts can change the reactivity or reaction rate in CO_2 gasification, and it is possible that an integrated oxidative CO_2-steam gasification can proceed to the extents suitable for enhanced H_2 production.

$$C + H_2O = CO + H_2 \qquad (1)$$

$$C + CO_2 = 2\,CO \qquad (2)$$

$$C + 0.5\,O_2 = CO \qquad (3)$$

$$C + O_2 = CO_2 \qquad (4)$$

$$CO + H_2O = CO_2 + H_2 \qquad (5)$$

Sulfur-Tolerant Catalyst Needed for Water Gas Shift for H₂ Production from Coal

For conversion of coal-derived CO to H_2 and CO_2 by water-gas-shift reaction (eq. 6), sulfur-tolerant catalyst would be preferred. Because gas from coal gasification contain H_2S in high concentrations (*e.g.*, 20000 ppm), conventional approach is to

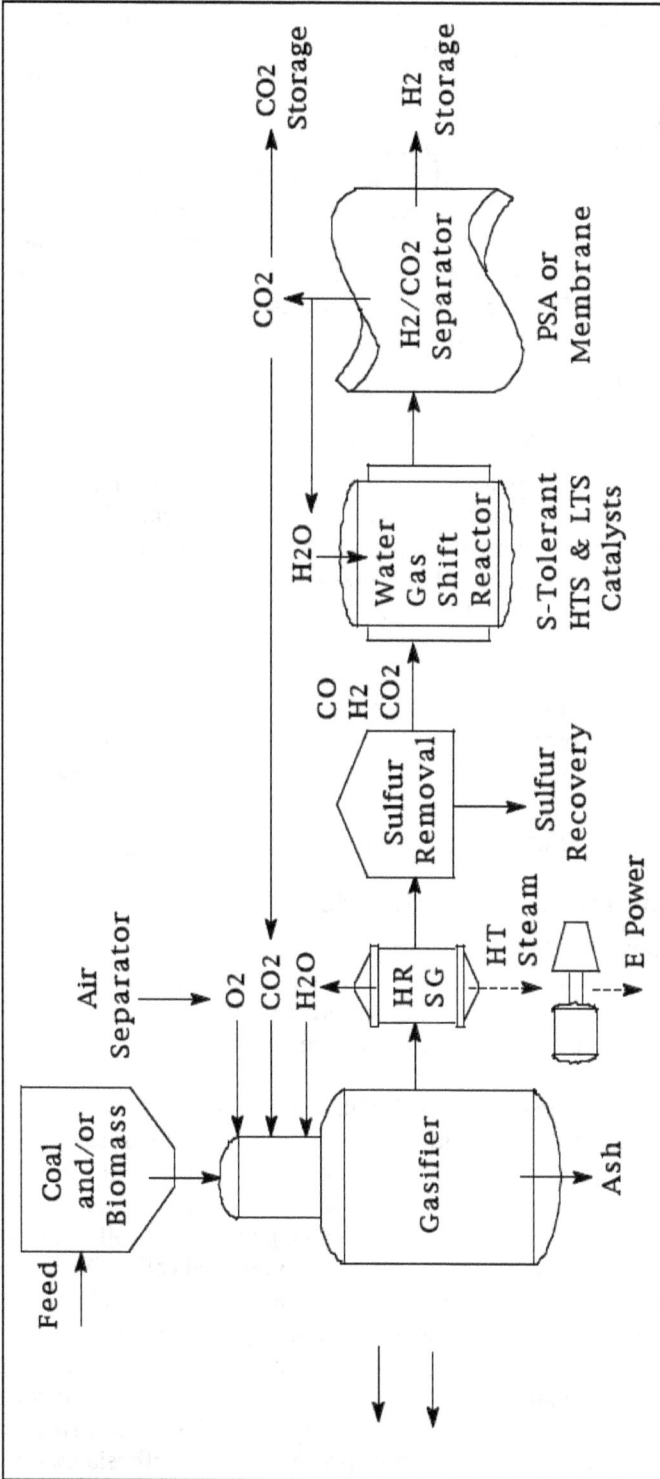

Figure 16.3: Conceptual Scheme of the Proposed CO-Enriched Gasification (COEG) Process for H$_2$ Production from Coal and/or Biomass.

remove H_2S before water gas shift reaction because the current industrial catalysts are extremely sensitive to sulfur of any form. While sulfur removal is necessary, it is inevitable for gas to contain a trace amount of H_2S. Current commercial Fe-Cr2O3-Al2O3 catalyst (HTS) and Cu-ZnO-Al2O3 (LTS) catalyst are sensitive to poisoning by sulfur (H2S), even at just a few ppm level. It would be desirable if sulfur-tolerant and active catalyst can be developed for water-gas-shift reaction. It may be worthy to explore more transition-metal sulfide based catalysts for developing sulfur-tolerant water-gas-shift catalysts. Based on our prior work on catalytic coal liquefaction in the presence of water (Song *et al.*, 2000) and sulfur-tolerant catalyst for hydrogenation (Song and Schmitz, 1997), we believe certain transition metal sulfide based catalyst can be effective WGS catalyst in the presence of hydrogen sulfide.

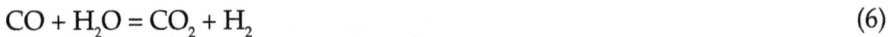

$$CO + H_2O = CO_2 + H_2 \qquad\qquad\qquad (6)$$

Separation of CO_2 from Gas Mixture following LT-WGS

For the proposed concept to work, we need effective separation of CO_2. We have proposed a CO_2 "molecular basket" concept for effective separation of CO_2 from H_2 in the presence of steam. Our recent results with mesoporous molecular sieve of MCM-41 modified with a branched polymer polyethyleneimide show that the CO_2 "molecular basket" of MCM-41- PEI type is very effective for separation of CO_2 from simulated flue gas, and the presence of steam further enhanced the capacity of the CO_2 separation (Xu *et al.*, 2002, 2003). This concept could be applied in principle for separation of CO_2 and H_2O from H_2, as in the mixture from lowtemperature water-gas-shift reaction (represented by eq. 7), thus enabling the CO_2 recovery and use for gasification by the proposed COEG process. The CO_2 can also be stored for other uses and for permanent sequestration such as mineral sequestration.

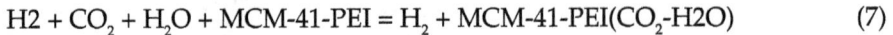

$$H2 + CO_2 + H_2O + MCM\text{-}41\text{-}PEI = H_2 + MCM\text{-}41\text{-}PEI(CO_2\text{-}H2O) \qquad (7)$$

On-board/On-site H_2 Production for Fuel Cells

Fuel cell converts chemical energy directly to electricity. It is intrinsically much more efficient than conventional combustion/heat-based energy conversion systems and is an important new path for efficient, clean and sustainable energy development., for which The original device for fuel cell was invented by Sir William Grove in UK in 1839, has emerged as a very promising energy device for the 21st century (Larminie and Dicks, 2000). A major reason for the greater interest in hydrogen energy now worldwide is that the technologies of fuel cells using H_2 as a fuel have advanced to the extent where many people begin to see its major commercial application potentials. There are five types of fuel cells including polymer electrolyte membrane fuel cell (PEMFC), alkali fuel cell, phosphoric acid fuel cell, molten carbonate fuel cell, and solid oxide fuel cell (SOFC). Among the five types, SOFC and PEMFC are the two most promising fuel cells (Larminie and Dicks, 2000). Figure 16.4 outlines the fuel processing steps and options for different fuel cell applications (Song, 2002). There are major challenges in the development of (1) fuel processor for on-site or on-board production of H_2 that meets the stringent requirement of CO (<10 ppmv) and H_2S (<20 ppbv) for H_2-based proton-exchange membrane fuel cell system, and (2) fuel processor for synthesis gas production

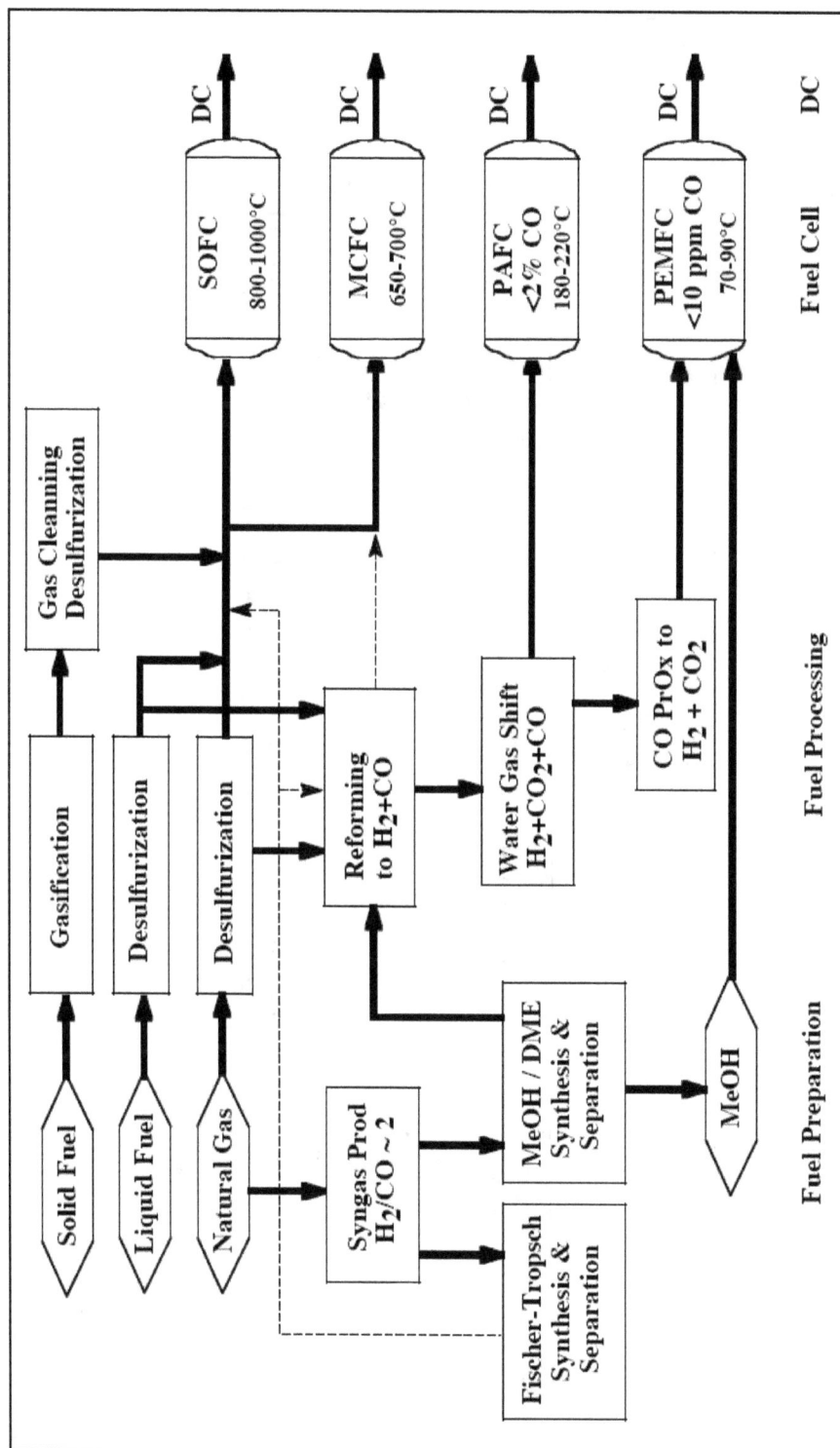

Figure 16.4: The Concepts for Fuel Processing of Gaseous, Liquid and Solid Fuels for High Temperature and Low-Temperature Fuel Cell Applications [Song, 2002].

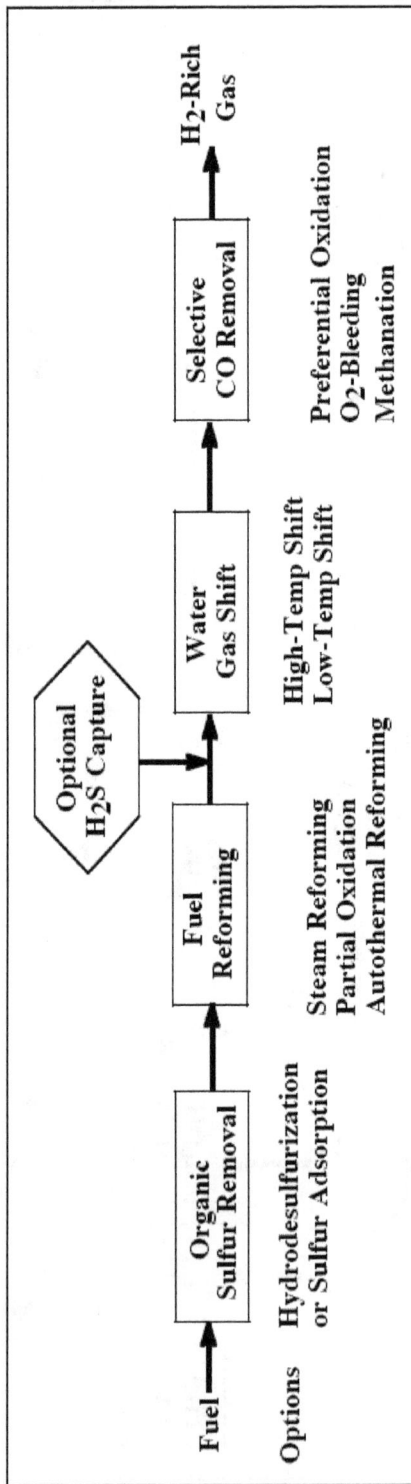

Figure 16.5: Steps and Current Options for On-site and On-board Processing Liquid and Gaseous Hydrocarbon Fuels and Alcohol Fuels to Produce H$_2$-rich Gas for Low-Temperature Fuel Cells (PEMFC) [Song, 2002].

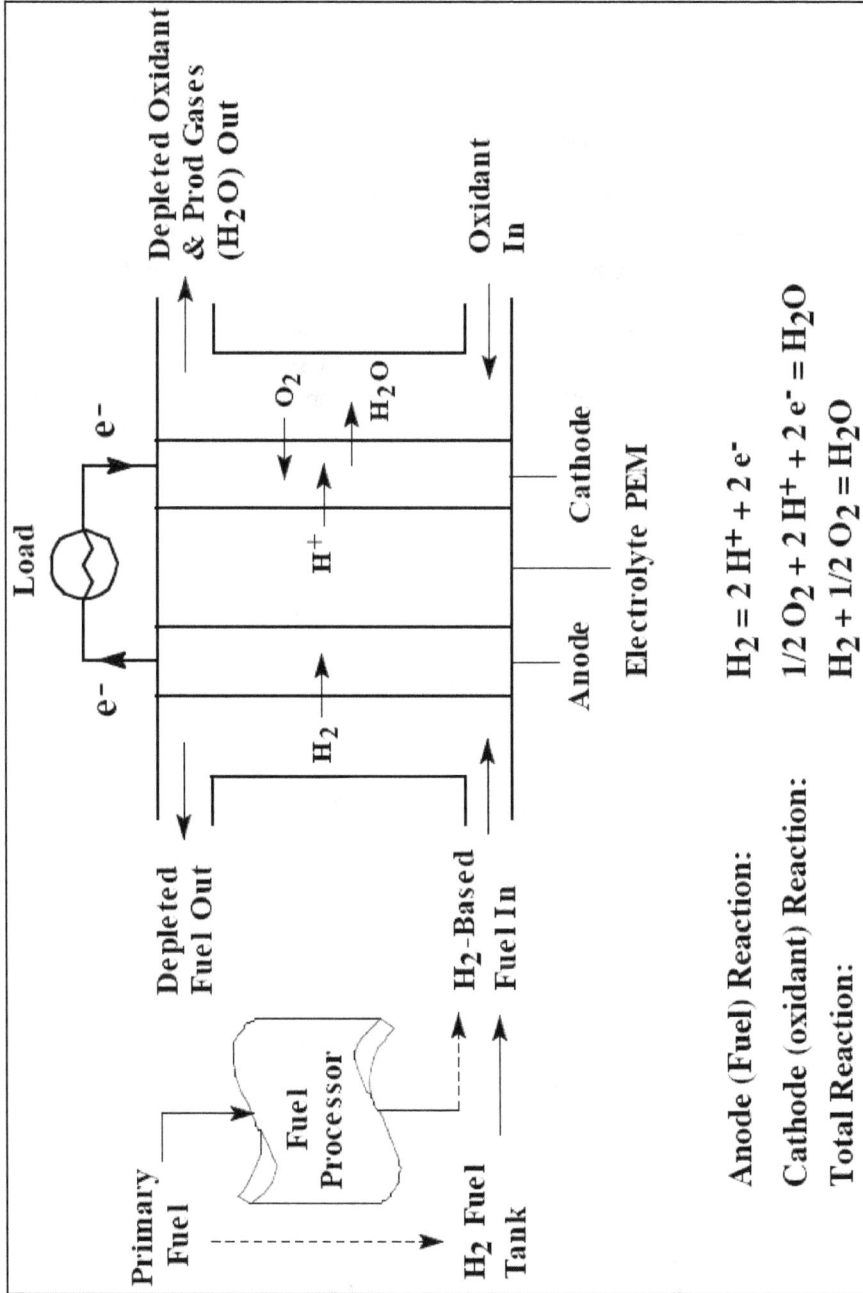

Figure 16.6: Concept of Proton-Exchange Membrane Fuel Cell (PEMFC) System using On-board or On-site Fuel Processor, or On-board H₂ Fuel Tank [Song, 2002].

Anode (Fuel) Reaction: $H_2 = 2 H^+ + 2 e^-$

Cathode (oxidant) Reaction: $1/2 O_2 + 2 H^+ + 2 e^- = H_2O$

Total Reaction: $H_2 + 1/2 O_2 = H_2O$

for solid-oxide fuel cells that use hydrocarbon fuels. The sulfur contents of most hydrocarbon fuels are too high for use in fuel cell reformer and in anode chamber, if when such fuels meet EPA sulfur requirements in 2006-2010 for automotive vehicles. Removal of organic sulfur before reforming and cleaning inorganic sulfur after reforming would be important for H_2 and syngas production for fuel cells, but conventional desulfurization methods are not suitable for fuel cell applications (Song, 2002; Song and Ma, 2003). Figure 16.5 and Figure 16.6 show the concepts and processing steps in an integrated fuel cell systems (Song, 2002).

CO_2 Emission Control Related to H_2 Production

Hydrogen energy and fuel cell development are closely related to the mitigation of CO_2 emissions. Fuel cell using hydrogen allows much more efficient electricity generation, thus can decrease CO_2 emission per unit amount of primary energy consumed or per kilowatt hour of electrical energy generated. For fossil fuel-based hydrogen production such as coal gasification, it would be desirable to develop new approaches that produce hydrogen in a more environmentally-friendly process that also includes effective CO_2 capture or CO_2 utilization as an integral part of the system.

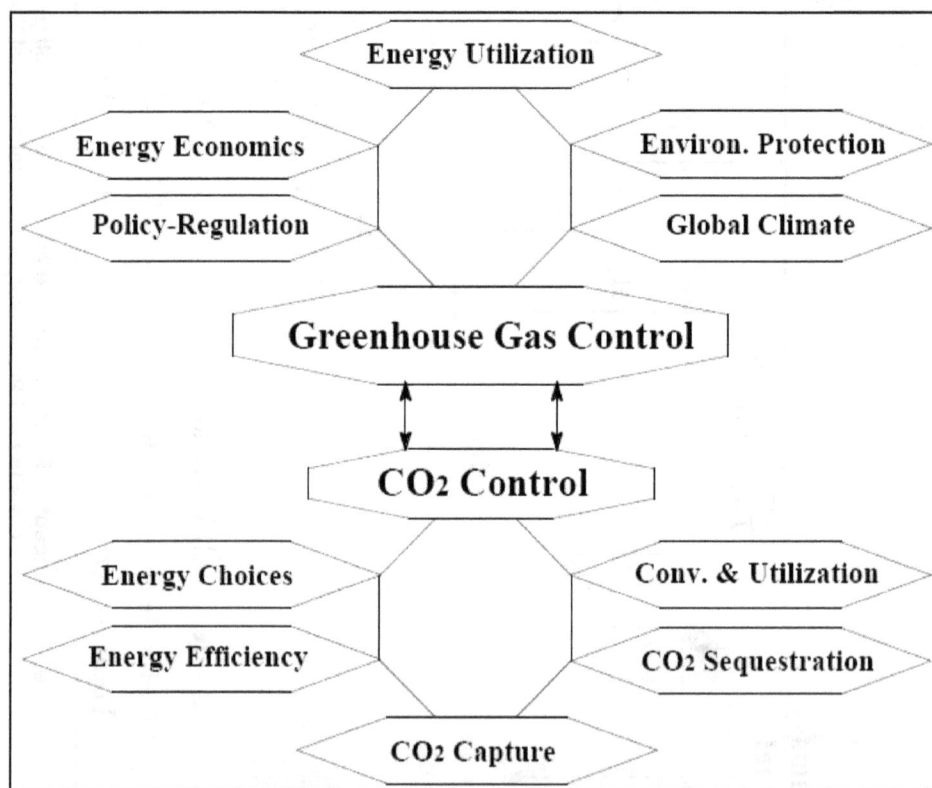

**Figure 16.7: Key Issues for Control of Greenhouse Gas
and Related Technical Areas for CO_2 Control (Song, 2002b).**

CO$_2$ capture involves chemical or physical separation of CO$_2$ from gas mixtures. Common methods include absorption using an agent such as monoethanol amine, physical adsorption using solid adsorbent, chemical adsorption using the above-mentioned CO$_2$ "molecular basket", and cryogenic separation at low temperatures, and membrane separation (see below). Depending on further study and verification, the proposed COEG process concept shown in Figure 16.3 could possibly facilitate the CO$_2$ recycling for high-temperature heat utilization and H$_2$ production on the one hand, and produce concentrated CO$_2$ that facilitates CO$_2$ storage/sequestration on the other hand. CO$_2$ sequestration refers to long-term storage of CO$_2$ in various reservoir locations with large capacity, such as mineral carbonation, geologic formations, ocean, aquifers, and forest. Permanent storage of CO$_2$ as carbonate has some intrinsic merit when compared to ocean sequestration (Maroto-Valer *et al.*, 2002). As can be seen from the thermodynamic analysis shown in Figure 16.8 (Song, 2002b), carbonate would be far more stable among most carbon-containing compounds, and reaction of CO$_2$ with calcium salt to form calcium carbonate is thermodynamically feasible.

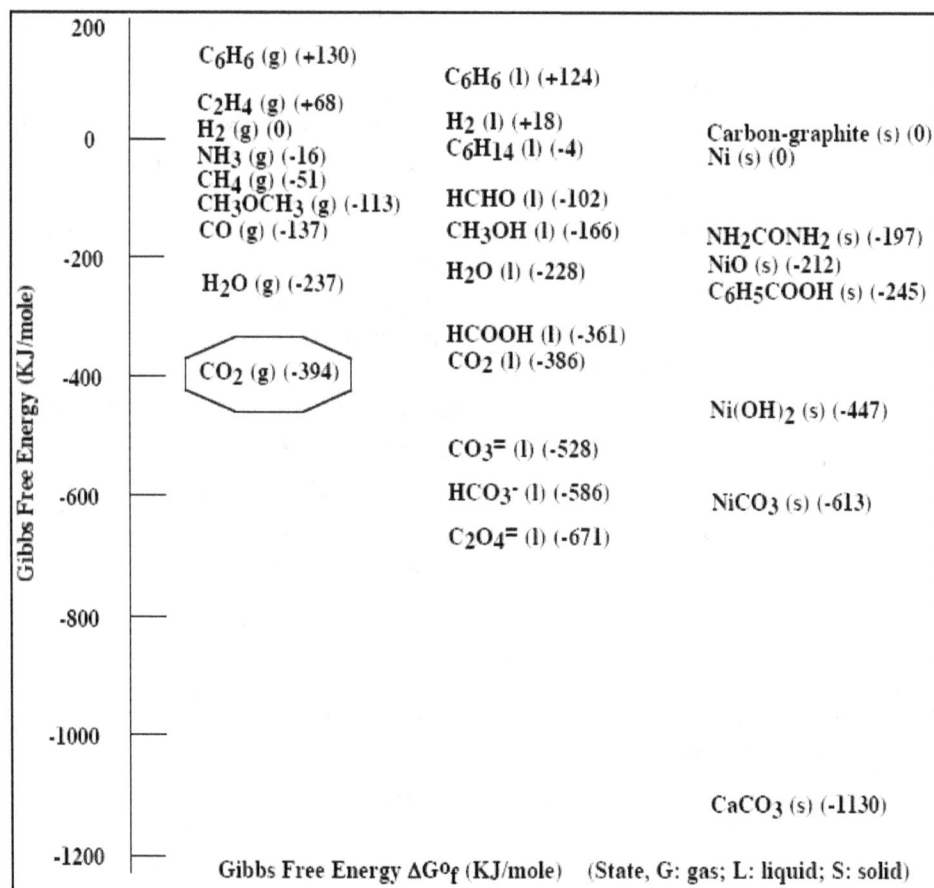

Figure 16.8: Gibbs Free Energy of Formation for CO$_2$ and Related Molecules (Song, 2002b).

Future Perspectives

Development of H_2-based energy system require multi-faceted studies on hydrogen sources, hydrogen production, hydrogen separation, hydrogen storage, H_2 utilization and fuel cells, H_2 sensor and safety aspects, as well as infrastructure and technical standardization. On the other hand, hydrogen energy is one form of energy but unlike the primary energy sources, hydrogen energy is a form that must be produced from chemical transformation of other substances first before being used. The production and utilization is also associated with various energy resources, fuel cells, CO_2 emissions, H_2 emission, safety and infrastructure issues. The technical communities of researchers also need to explore the following global questions that I believe are important, and more studies are needed to answer these questions:

☆ Should we use hydrogen in the future as energy carrier for transportation? Does the hydrogen energy solve the potential global warming problem by reducing CO_2?

☆ Should we rely on hydrogen for residential and stationary electrical power generation?

☆ Would distributed on-board and/or on-site H_2 production using conventional fuels be more desirable if the technology can be developed?

☆ Would transportation and storage of H_2 produced from centralized plants through various means be more desirable in the future?

☆ What are the key technical, environmental and socio-economical factors that must be clarified for global hydrogen energy development ?

☆ What should we use as hydrogen source for molecular H_2 production in the future, and what energy source should we use for the production processing in the future?

☆ If the currently dominant fossil fuels ever become so scarce to the extent that H_2 becomes a dominant energy carrier at some point in the long-term future, what will the world use as resources for materials and chemical feedstocks (that are currently derived largely from fossil fuels)? Should the societies reduce consumption of fossil hydrocarbon resources, regardless the location of such resources, and save them for future generations ?

Figure 16.9 presents a personal vision for research towards comprehensive and effective utilization of hydrocarbon resources in the 21st century. No single energy source will satisfy the needs of societies in the long term. The best-possible scenario for future energy system is a balanced mix of various forms of renewable energies and resources used in combination.

It is important to think about the above questions in the context of sustainable energy development. To begin with end in mind, I believe the world could be better served if the energy research community explores the future energy issues in a comprehensive and integrated fashion (Song, 2001, 2002c, 2002d). For comprehensive utilization research, we should look at different uses of hydrocarbon resources, for

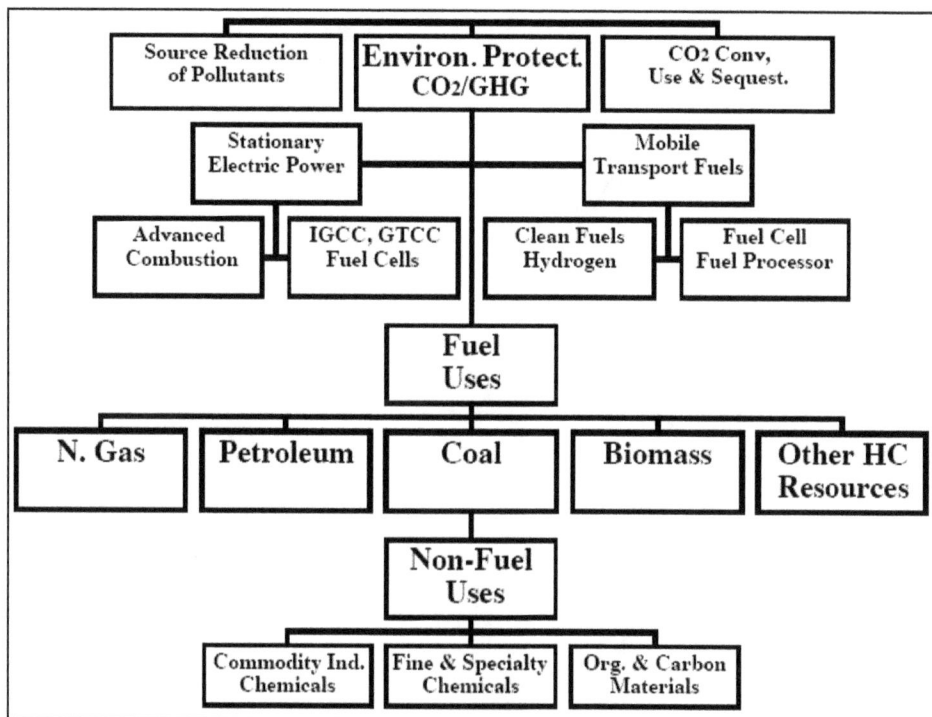

Figure 16.9: A Personal Vision for Research Towards Comprehensive and Effective Utilization of Hydrocarbon Resources in the 21st Century.

fuel uses and non-fuel uses, that are equally important although different in volume (Song and Schobert, 1993, 1996; Schobert and Song, 2000; Song, 2001, 2002c, 2002d).

Finally, it should be mentioned that a recent computational study at CalTech points out possible/potential environmental impacts of hydrogen energy in the future (hydrogen leak, estimated at up to 10-20 per cent of total H_2 produced in the future, could cool the Earth's stratosphere due to additional water formed from hydrogen at high altitudes and disturb the ozone chemistry) (NaTrompme *et al.*, 2003). Another study at MIT indicates limitations of hydrogen energy development for transportation by fuel cell cars with respect to its effect on greenhouse gas reduction (Weiss *et al.*, 2003). More experimental and theoretical studies regarding possible long-term effects are needed to clarify the related issues.

Chapter 17

Status of Biological Hydrogen Production

Integrated Process for Hydrogen Production

During dark fermentation, desired product is H_2, collected in gas phase. But there are some liquid by- products, which contain mainly organic acids (butyric acid, acetic acid, lactic acid, formic acid etc). Major disadvantage of this fermentive broth is disposal problem. On the other hand, fermentive broth can be treated as carbon source for photofermentation process. Many reports (Shi and Yu, 2006; Tao et al., 2007; Asada et al., 2006) showing in Table 17.1 total H_2 yield increase by a considerable amount; maximum yield is 7.1 mol H_2/mol of glucose (Asada et al., 2006).

Kim et al. (2006) reported a three step integrated process, in which biomass is produced by Chlamydomonas reinhardtii in the first step, then the algal mass (equivalent to starch- glucose) is used as a feedstock for dark fermentation by Clostridium butyricum that produces H_2 and organic acids.In the last step, fermentative broth is used for photo fermentation by Rhodobacter sphaeroids to produce H_2 Shi et al. (2006) shows steps in the conversion of glucose to H_2 by a combination of dark and photo fermentation.

Green algae were found to absorb visible light for solar radiation whereas photosynthetic bacterium absorbs some of infrared portion of solar radiation. So combining these two can increase overall light conversion efficiency. Melis et al. (2006) successfully combined Chlamydomonas reinhardtii and Rhodospirillium rubrum. Similar integrated processes have also been reported (Miura et al., 1992; Ohmiya et al., 2003). Such an integrated system would constitute a high yield, and more economically viable H_2 production.

Analysis Techniques

Metabolic Flux Analysis

Rate of H_2 production from these processes is very slow except dark fermentation. But the major problem associated with dark fermentation is that yield of H_2 production is low. Metabolic flux analysis (Edward *et al.*, 2002) (MFA) technique can be used to improve yields. Using this method, intracellular fluxes are calculated from extra cellular fluxes. If measured fluxes are not sufficient, optimization techniques can be used to determine unknown fluxes. Manish *et al.* (2007) used MFA to improve H_2 yield for *E. coli* with 31compounds and 27 pathway reactions, giving maximum yield possible at 2 mol/mol of glucose consumed in a particular pathway with maximum production of acetate and ethanol. The production based on strains lacking Idh indicated a possible methodology to increase H_2 production. Similar result was obtained by Oh *et al.* (2008) with *Citrobacter amalonaticus* Y 19 giving maximum H_2 yield at 8.7 mol/mol of glucose consumed when metabolism is directed to suggested pathway. So, MFA technique can define the upper limit of H_2 production in microbial fermentation and suggest the most favorable path. This technique can provide apriori directions for genetic modifications towards preferred pathways.

Net Energy Analysis

For new processes, sustainability is of prime importance. Manish *et al.* (2008) suggested that criteria should be net energy ratio (NER), energy efficiency and greenhouse gas (GHG) EMISSIONS. In first step, material and energy balance are computed. Total energy consumption is then classified into renewable method only when energy output of the process is more than nonrenewable energy input. So NER is defined as

NER = Hydrogen output (MJ)/Non-renewable energy input (MJ)

An NER value, more than one indicates renewable nature of processes. Using emission factors, GHG emission is calculated. Energy efficiency is defined as

Energy efficiency= Energy output/Energy input

Where, energy input indicates energy content in feedstock (chemical energy).

Manish *et al.* (2008) found that BHP process is viable from net energy and GHG reduction point of view. Integrated process of dark and photo fermentation has maximum efficiency. The result obtained in the same study shows that CO_2 reduction is 7.31-9.37 Kg/Kg H_2 production (57-73 per cent) and non renewable energy used is 123.2- 148.7 MJ/Kg H_2 yield (65-79 per cent) as compared to steam methane reforming (SMR) process. However, when by products from these bihydrogen processes are not utilized, efficiencies of bihydrogen processes are significantly lower than SMR. Hence by product removal and utilization is an important factor for BHP process. Net energy analysis is a useful tool to determine viability of process.

Hydrogen and Fuel Cell Application

H_2 can be utilized for electricity generation and transportation using fuel cells.

Levin *et al.* (2004) estimated amount of H_2 required running a proton exchange membrane fuel cell and also estimated bioreactor size required to run a 5KW PEM fuel cell. The methodology of Levin *et al.* (2004) was use to estimate the size of bioreactor using the best performance data. To power a 5KW fuel cell required volumetric flow rate was 119.7 mol/h. It shows that required size of bioreactor for four processes reduced significantly in the last four years. In few studies (Nakada *et al.*, 1999; Lin *et al.*, 2007) bihydrogen processes have been combined with fuel cells (Table 17.2).

Table 17.1: Recalculated Size of Bioreactor

System	HPR (l/l/h)	Temperature of Effluent Gas °C	HPR mmol/h	Required Size of Bioreactor, l
Direct photolysis[12]	0.0029	28	0.117	1.0×106
Indirect photolysis[13]	0.025	30	1.006	1.2×105
Photo fermentation[32]	0.118	30	4.75	2.5×104
Dark fermentation[58]	9.31	25	380.78	3.2×102

Table 17.2: Experimental Results for Bioreactor-Fuel Cell Combination

Culture	Conditions Mentioned	Observation
Photosynthetic bacteria[26] (*Rhodobacter sphaeroides*)	☆ Reactor volume, 1l1 ☆ Light illumination, 107W/m² for 100h ☆ Gas input to fuel cell, 0.6-2.01 l/h	Gas produced 35.7 litre with purity 88.1% ☆ Output power, 1W ☆ Conversion efficiency- ● Light to hydrogen 4.3% ● Fuel cell, 44% ● Overall, 1.9%
Anaerobic fermentative bacteria [82] (*Clostridium pasteurianum*)	☆ Eactor volume, 3l ☆ Sucrose as substrate ☆ Fuel cell fed, 1.72l/h ☆ Input H_2 purity, 99% ☆ Fuel cell efficiency, 50%	☆ HPR, 1.5L/H ☆ H_2 in biogas, 40.6% ☆ Voltage output, 2.28V ☆ Current output, 0.38A ☆ Power output, 0.87W

Cost Analysis of Bihydrogen Production Processes

Yield is directly proportional to the operating cost and rate is directly proportional to the reactor cost or installation cost. H_2 storage cost that is almost same for all kinds of process related to H_2 production. In case of photosynthesis water is used a substrate so operating cost is very low and mainly consists of the maintenance cost. As production rate of this process is very low, required size of reactor would be high and hence installation cost is high. Zaborsky (1999) reported that reactors (10 per cent light conversion efficiency) should cost about $100/m² for two- stage system, and suggested that near- horizontal tubular reactor (NHTR) might cost $50/m² and thus projected cost of H_2 will be $15/GJ. As reported in NREL annual report (Amos, 2008), an 11000 m² pond type bioreactor (at Phoenix, Arizona) using green alga had a reactor cost of $10/m². Total capital investment is $5,168,000, whereas operating cost including storage cost is $1480000 and thus H_2 production cost is \$13.5Kg of H$_2$ (US Department of Energy, DOE set a goal for renewable H$_2$ production at a cost of \$2.60/kg H$_2$). Madamwar *et al.* (2000) reported that cost of H$_2$ from cyanobacterial photosynthesis I s\$25/m^3 that is far less than reported cost of H$_2$ from photovoltaic splitting of water (\$170/m^3) (Block and Melody, 1992). For viability the costs of these processes have to be reduced to less than 20 per cent of its present cost.

Chapter 18
Future Lines of Investigation

Considerable progress has been made during the last thirty years in the study of physiology, biochemistry and genetics of microorganisms. The results have show that hydrogen is metabolized in various organisms by three enzymatic reactions : (1) nitrogenase which catalyses unidirectional, ATP-dependent hydrogen evolution, and can function either in the light or in the dark under aerobic or microaerobic conditions; (2) uptake hydrogenase, which is membrane bound and, although capable of both hydrogen evolution and uptake, functions physiologically in the direction of hydrogen oxidation; and (3) 'classical' or membrane bound and functions mainly during dark aerobic fermentation. However it may be a long time in future when biological hydrogen production can become an economic alternative to fossil fuels. Considering the literature, advantages and available technology, it is evident that photosynthetic bacteria and cyanobacteria, the two groups of microorganisms, are most competitive candidates for future researches on hydrogen production. The following lines summarize their advantages, problems and possible solutions.

1. Hydrogen Production-Photosynthetic by Bacteria

Advantage

1. Hydrogen production from organic substrates (or inorganic substrates) is easily demonstrable.
2. System is stable either as cell suspension or after cell immobilization.
3. Uses infra-red spectral light range not used in conventional agriculture.
4. Can be successfully utilized in sewage clean-up programme or in scavenging the otherwise toxic trace elements from polluted water bodies.

Problems

1. Cost (and availability) of hydrogen donor.
2. Need to assure absence of other contaminating organisms which may consume hydrogen produced in the ambient atmosphere.

Approaches

It can be taken as the most ideal system for hydrogen production provided alternate, cheap and abundant hydrogen donors can be identified.

Strain Screening and Selection Programme

1. Marine strains : No attempt has been made in our country to utilize the marine resources for hydrogen production.

    A search in this direction is desirable.
2. Isolation of thermostable strains.
3. Isolation of strains with high nitrogenase content and activity.
4. Isolation of uptake hydrogenase deficient strains.

Economical Substrates

1. Production of hydrogen at the expense of agricultural and domestic wastes.
2. Waste depollution.
3. Cell stabilization by immobilization.

2. Hydrogen Production by Cyanobacteria

Advantages

1. Water is cheap available substrate.
2. Grow luxuriantly, often better than photosynthetic bacteria, in almost all possible habitats (rice fields, polluted ponds, in hot springs, usar soils etc.).
3. Hydrogen production is easily demonstrable.
4. Uses visible light.
5. System is stable either as cell suspension or after immobilization (more efficient, convenient and stable than chloroplasts).

Problems

1. Hydrogen production rates are lower than photosynthetic bacteria.
2. Simultaneous oxygen production in light.
3. Need to assure absence of other contaminating organisms which may use the hydrogen produced.

Possible Approaches

1. Screening programme for organisms (both fresh water as well as marine) which produce hydrogen at high rates for (a) future studies, (b) maximization of hydrogen evolution.
2. Focus on conditions which could eliminate oxygen evolution through photosystem II activity, attempts to modify physiologically photocapture system.
3. Genetic approaches to provide more (oxygen) stable hydrogenase/higher nitrogenase content and photocapture system.
4. Large scale cultivation of cyanobacteria (liquid cultures/immobilized system) and studies on hydrogen production based on existing knowledge.

3. Hydrogen Production by Algae

Advantages

1. Water is cheap available substrate.
2. Grows luxuriantly in almost all possible habitats, often to cyanobacteriaq.
3. Uses visible light.
4. Hydrogen production demonstrable even in dark.
5. System is stable either as cell suspension or after immobilization, convenient and stable than chloroplasts.

Problems

1. The mechanism of hydrogen production varies considerably depending on the species of algae and experimental conditions.
2. Simultaneous production of hydrogen and oxygen.
3. A period of adaptation is required for anaerobic conditions.

Possible Approaches

1. Screening programme for organisms (fresh water as well as marine) which produces hydrogen at high rates.
2. Focus on conditions which could eliminate oxygen evolution through photosystem II activity.
3. Genetic approaches for obtaining mutants which exhibit not lag for anaerobic adaptation.

4. Hydrogen Evolution by Chloroplasts

Advantages

1. Isolation and purification of chloroplasts from various sources is a well established technique.
2. Uses visible light.
3. Hydrogen production is demonstrable.

The Problems

1. It is least stable hydrogen production system.
2. Oxygen and its reduced forms, superoxide and hydrogen peroxide, if allowed to accumulate, cause oxidation of thylakoid membrane and inactivates hydrogenase.
3. Light inhibition of photosynthetic electron transport.
4. Auto oxidation of electron mediator/catalysts.
5. Expensive organic molecules serve as electron donor.

Possible Approaches

1. Focus on conditions which could stabilize the chloroplast preparations.
2. Search for a cheaper source of electron donor and conditions which regenerates the electron donor.
3. Improvement of immobilizing conditions.

References

Abdel–Basset, R. and Bader, K.P., (1998). Physiological analysis of the hydrogen gas exchange in cyanobacteria. *J. Photochem. Photobiol. B: Biol.*, 43: 146–151.

Abeles, F.B., (1964). Cell free hydrogenase from *Chlamydomonas*. *Plant Physiol.*, 39: 164–176.

Adams, D.G., (2000). Symbiotic interactions. In: *The Ecology of Cyanobacteria*, (Eds.) Whitton BA and Potts M, pp. 523–561. Kluver Academic Publishers, Dordrecht.

Adams, M.W. and Hall, D.O., (1979). Publication of the memberance bound hydrogenase of *Esherichia coli*. *Biochemic. J.* 183: 11–22.

Adams, M.W.W. and Mortenson, L.E., (1984). The physical and catalytic properties of hydrogenase II of *Clostridium pasteurianum*. *J. Biol. Chem.*, 259: 7045–7055.

Adams, M.W.W. and Stiefel, E.I., (1998). Biological hydrogen production, not so elementary. *Science*, 282: 1842–1843.

Adams, M.W.W., Mortenson, L.E. and Chen, J.S., (1980). Hydrogenase. *Biochim, Biophys. Acta* 594: 105–176.

Adams, M.W.W., Mortenson, L.E. and Chen, J.S., (1981). Hydrogenase. *Biochimica et Biophysica Acta*, 594: 105–176.

Aggag, M. and Schelgel, H.G., (1974). Studies on a gram positive hydrogen bacterium *Nocardia opaca* 1b. III. Purification, stability and some properties of the soluble hydrogen dehydrogenase. *Arch. Microbial.* 100: 25-39.

Aguey Zinsou, K.F. y J.R Ares Fernández, (2008). Synthesis of Colloidal Magnesium: A Near Room Temperature Store for Hydrogen. *Chemistry of Materials*, 20(2): 376–378.

Akhmanova, A. *et al.*, (1998). A hydrogenosome with a genome. *Nature*, 396: 527–528.

Akkerman, I., Janssen, M., Rocha, J., Wijffels, R.H., (2002). Photobiological hydrogen production: photochemical efficiency and bioreactor design. *Int. J. Hydrogen Energy*, 27: 195–208.

Albracht, S.P.J., Graf, E.G. and Thauer, R.K., (1982). The EPR properties of nickel in hydrogenase from *Methanobacterium thermoautotrophicum*. *FEBS Lett.*, 140: 311–313.

Albracht, S.P.J., Kalkman, M.L. and Slater, E. C., (1983). Magnetic interaction of nickel (III) and the iron-sulphur cluster of hydrogenase from *Chromatium vinosum*. *Biochim. Biophys. Acta*, 724: 309–316.

Albrecht, S.L., Mairer, R.J., Hanus, F.J., Russell, S.A., Emerich, D.W., Evans, H.J., (1979). Hydrogenase in *Rhizobium japonicum* increase nitrogen fixation by nodulated soybeans. *Science*, 230: 1255–1257.

Almon, H. and Boger, P., (1984). Nickel-dependent uptake hydrogenase activity in the blue-green alga *Anabaena variabilis*. *Z. Naturforsch*. 39: 90–92.

Amos, W.A., (2008). Updated cost analysis of photobiological hydrogen production from *Chlamydomonas reinhardtii* green algae, *Milestone Completion Report* (NREL US Department of Energy, USA): http: //www. Osti.gov/bridge, visited on 3rd July 3.

Anderson, K., Tait, R.C. and King, W., (1981). Plasmids required for utilization of molecular hydrogen by *Alkaligenes eutrophus*. *Arch. Microbiol*. 129: 384–390.

Andrews, C.J. (2005). Energy Security as a Rationale for Governmental Action. *IEEE Technology and Society Magazine*. Summer, pp. 16–25.

Aoyama, K., I. Uemura, J. Miyake, and Y. Asada, (1997). Fermentative Metabolism to Produce Hydrogen Gas and Organic Compounds in a *Cyanobacterium, Spirulina platensis*. *J. Fermentation and Bioengineering*, 83: 17–20.

Appel, J. and Schulz, R., (1996). Sequence analysis of an operon of a NAD(P)-reducing nickel hydrogenase from the Cyanobacterium *Synechocystis* sp. PCC 6803 gives additional evidence for direct coupling of the enzyme to NAD(P) H–dehydrogenase (complex 1). *Biochim Biophys Acta* 1298: 141– 147.

Appel, J. and Schulz, R., (1998). Hydrogen metabolism with oxygenic photosynthesis: hydrogenases as important regulatory device for a proper redox poising? *J. Photochem. Photobiol B: Biol.*, 47: 1–11.

Appel, J., Phupruch, S., Steinmuller, K. and Schulz, R., (2000). The bi-directional hydrogenase of *Synechocystis* sp. works as an electron valve during photosynthesis. *Arch. Microbiol.*, 173: 333–338.

Appleby, A.J. and Foulkes, F.R. (1993). *Fuel Cell Handbook*. New York: Van Nostrand Reinhold.

Aragno, M. and Schlegel, H.G., (1981). The hydrogen oxidizing bacteria. In: *The Prokaryotes: A Handbook on Habitats, Isolation and Identification of Bacteria* (Eds.) Starr. M.P., Steep, H. Truper, H.G., Ballows, A and Schlegel, H.G.) Vol. I. Springer–Verlag, Berlin, Heidelberg New York, pp. 865–893.

Aristarkhow, A.T., Nikandrov, V.V. and Krasnovakii, A.A., (1986). Conditions of hydrogen photoproduction by subcholroplast fragments containing photosystem I. *Mol. Biol.* (Moscow), 20: 1344–1355.

Arkhipov, V.N., Boichenko, V.A. and Litvin, F.E., (1980). Connection between chlorophyll fluorescence induction and hydrogen photoevolution in *Chlorella vulgaris*. *Biofizika*, 25: 246–249.

Armor, J.N., (1999). The Multiple Roles for Catalysis in the Production of H₂. *Applied Catalysis A: General*, 176(2): 159–176.

Arnon, D.I., Losada, M., Nozaki, M. and Tagawa, K., (1961a). Protoproduction of hydrogen, photofixation of nitrogen and a unified concept of photosynthesis. *Nature* 190: 601–606.

Arnon, D.I., Mixtsui, A and Paneque, A., (1961b). Photoproduction of hydrogen gas coupld with photosythetic photophosphorylation. *Science*, 134: 1425.

Arp, D.J. and Burris, R.H., (1979). Purification and properties of the particulate hydrogenase (EC 1.12.2.1) from the bacteroids of soybean root nodules. *Biochim. Biophys. Acta*, 570: 221–230.

Arp, D.J. and Burris, R.H., (1981). Kinetic mechanisms of hydrogen oxidizing hydrogenase from soybean nodule bacteroils. *Biochemistry*, 2: 2234–2240.

Asada, Y. and Kawamura, S., (1985). Hydrogen evolving activity among the genus *Microcystis*, under dark and anaerobic conditions. Report of the Fermentation Research Institute, Japan, No. 63.

Asada, Y. and Kawamura, S., (1986). Aerobic hydrogen accumulation by a nitrogen fixing cyanobacterium, *Anabaena* sp. *Appl. Environ. Microbiol.* 51: 1063–1066.

Asada, Y. and Miyake, J., (1999). Photobiological hydrogen production. *J Biosci Biotechnol*, 88: 1–6.

Asada, Y., Kuwamura, S. and Ho, K.K., (1987). Hydrogenase from the unicellular cyanobacterium *Microcystis aeruginosa*. *J. Phytochem.* 26: 637–640.

Asada, Y., Tmizuka, N. and Kawamura, S., (1985). Prolonged hydrogen evolution by a cyanobacterium (Blue-green alga), *Anabaena* sp. *J. Ferment. Technol.*, 63: 85–90.

Asada, Y., Tokumoto, M., Aihara, Y., Okum, Ishimi, K., Wakayama, T., Miyake, J., Tomiyama, M. and Khono, H., (2006). Hydrogen production by co-cultures of *Lactobacillus* and a photosynthetic bacterium, *Rhodobacter sphaeroides* RV. *Int J Hydrogen Energy*, 31: 1509–1513. (73).

Asada, Y., Tonomura, K. and Nakayama, O., (1979). Hydrogen evolution by an isolated strain of *Anabaena*. *J. Ferment. Technol.*, 57: 280–286.

Ashley, S., (2005). On the road to fuel cell cars. *Scientific American*, 292: 52–55.

Avtges, P., Scolnik, P.A. and Haselkorn, R., (1983). Genetic and physical map of the structural genes (nif HDK) coding for the nitrogenase complex of *Rhodopsudomonas capsulata*. *J. Bacteriol*, 156: 251–256.

Axelsson, R., Oxelfelt, F. and Lindblad, P., (1999). Transcriptional regulation of Nostoc uptake hydrogenase. *FEMS Microbiology Letters*, 170: 77–81.

Bagramyan, K., Mnatsakanyan, N., Poladian, A., Vassilian, A. and Trchounian, A., (2002). The roles of hydrogenases 3 and 4, and the F0F1–ATPase, in H_2 production by Balban International Sciences Services, Philadelppia, Pennsylvania, pp. 639–654.

Bálint, B., Bagi, Z., Tóth, A., Rákhely, G., Perei, K., Kovács, K.L., (2005). Utilization of keratin-containing biowaste to produce biohydrogen. *Appl Microbiol Biotechnol,* 69(4): 404–410.

Ballentine, S.B. and Boxer, D.H., (1985). Properties of two hydrogenases from *Escherichia coli.* In: *Microbial Gas Metabolism: Mechanistic, Metabolic and Biotechnological Aspects* (Eds.) Poole, R.K. and Dow, C.S.. Academic Press, New York, pp. 103–107.

Banerjee, M., Kumar, A. and Kumar, H.D., (1989). Factors regulating nitrogenase activity and hydrogen evolution in *Azolla Anabaena* symbiosis. *Int. J. Hydrogen Energy* 14: 871–879.

Barbosa, M.J., Rocha, J.M.S., Tramper, J. and Wijffels, R.H., (2001). Acetate as a carbon source for hydrogen production by photosynthetic bacteria. *J. Biotechnol.,* 85: 25–33.

Barron, R.M. and Worman, H.J., (1999). Prenylated prelamin A interacts with Narf, a novel nuclear protein. *J. Biol. Chem.,* 274L 30008–30018.

Beck, U., (1998). *La sociedad del riesgo: hacia una nueva modernidad.* Barcelona: Paidos.

Becking, J.H., (1970) Frankiaceae Fam. Nov. (Actinomycetales) with one new combination and six new species of the genus *Frankia Brunchorst* 1866. 174. *Int. J. Syst. Bacteriol.,* 20, 201–220.

Becking, J.H., (1983). The *Parasponia parviflora* and *Rhizobium symbiosis,* isotopic nitrogen fixation, hydrogen evolution and nitrogen fixation efficiency and oxygen relation. *Plant and Soil,* 75: 343–360.

Beent, R., Rigopoulos, N. and Fuller, R.C., (1964). The pyruvate phosphoroclastic reaction and light dependent nitrogen fixation in bacterial photosynthesis. *Proc. Natl. Acad. Sci., U.S.* 52: 762–768.

Behar, M., (2005). Warning: the hydrogen economy may be more distant than it appears. *Popular Science* 266: 65–68.

Belkin, S. and Padan, E., (1978). Hydrogen metabolism in the facultative anoxygenic cyanobacteria (blue-green algae) *Oscillatoria limnetica* and *Aphanothece halphytica.* *Arch. Microbiol.,* 116: 109–111.

Ben-Amotz, A., Erbes, D.L., Riederer-Henderson, M.A., Peavey, D.G. and Gibbs, M., (1975). H_2 metabolism in photosynthetic organism. Dark H_2 evolution and uptake by algae and mosses. *Plant Physiol.* 56: 72–77.

Benemann, J., (1977). Hydrogen and methane production through microbial photosynthesis. In: *Living systems as Energy Converters* (Eds.) Buvet, R., Allen, M.J. and Massue, J.P., pp. 285. Elsevier/North Holland Biomedical Press, Amsterdam.

Benemann, J.R. and Hallenbeck, P.C., (1978) Basic and applied studies of hydrogenase in cyanobacteria. In: *Hydrogenases: Their Catalytic Activity, Structure and Functions* (Eds.) Schlegel, H.G. and Schneider, K., p. 395. Goltze, Cottingen.

Benemann, J.R. and Hallenbeck, P.C., (2002). Biological hydrogen production, fundamentals and limiting processes. *Int. J. Hydro Energy*, 27: 1185–1193.

Benemann, J.R. and San Pietro, A., (2001). Workshop on Biohydrogen R&D. Behtesda, MD, September 29–30, 2000 (Final Report Submitted to the Dept. of Energy Hydrogen Program, May).

Benemann, J.R. and Weare, N.M., (1974). Hydrogen evolution by nitrogen-fixing *Anabaena cylindrica* cultures. *Science*, 184: 174– 175.

Benemann, J.R. and Weare, N.M., (1974a). Hydrogen evolution by nitrogen-fixing *Anabaena cylindrica* cultures. *Science* 184: 174–175.

Benemann, J.R. and Weare, N.M., (1974b). Nitrogen fixation by *Anabaena cylindrica*. III. Hydrogen-supported nitrogenase activity. *Arch. Microbial.*, 101: 401–408.

Benemann, J.R., (1990). The Future of Microalgae Biotechnology. In *Algal Biotechnology*, (Eds.) R.C. Cresswell, T.A.V. Rees, and N. Shah. Longman, London pp. 317–337.

Benemann, J.R., (1995). "Feasibility Analysis of Photobiological Hydrogen Production" in D.L. Block and T.N. Versiroglu, eds., *Hydrogen Energy Progress X, Proc. 10th World Hydrogen Energy Conf.*, Cocoa Beach, Florida, June 20–24, 1994, pp. 931–940.

Benemann, J.R., (1996). Hydrogen biotechnology, progress and prospects. *Nature Biotechnol.*, 14: 1101–1103.

Benemann, J.R., (1996). Hydrogen Biotechnology: Progress and Prospects. *Nature Biotechnology*, 14: 1101 –1103.

Benemann, J.R., (1996). *Photobiological Hydrogen Production*. Final Report to the Pittsburgh Energy Technology Center, U.S. Dept. of Energy. pp. 225.

Benemann, J.R., (1998a). *Processes Analysis and Economics of Biophotolysis of Water. A Preliminary Assessment*. Report to the International Energy Agency Hydrogen Program, Subtask B, Annex 10, Photoproduction of Hydrogen.

Benemann, J.R., (1998b). In: *Defense of the Utility of Nitrogenase-based Systems for Biohydrogen, Specifically Biophotolysis, R&D*. Report Submitted to the DOE/NSF Biohydrogen Workshop, 4/26–27, 1998, Under Contract TXE–7–17011–01 with the National Renewable Energy Laboratory.

Benemann, J.R., (1998c). The Technology of Biohydrogen. In: *BioHydrogen*, (Eds.) O. Zaborsky *et al.* Plenum Press, New York, pp. 19–30.

Benemann, J.R., (2000). Report to the US Department of Energy Hydrogen Program. Biohydrogen Production. Hawaii Natural Energy Institute, University of Hawaii, p. 7.

Benemann, J.R., and O. Zaborsky, (1996). Biohydrogen: Market Potential. Proc. 7th Annual U.S. Hydrogen Meeting, April 2–4, Alexandria, VA., pp. 369–379, National Hydrogen Association, Washington D.C.

Benemann, J.R., Berenson, J.A., Kaplan, N.O. and Kamen, M.D., (1973). Hydrogen evolution by a chloroplast-ferredoxin-hydrogenase system. *Proc. Natl. Acad. Sci. USA*, 70: 2317–2320.

Benemann, J.R., Miyamoto, K. and Hallenbeck, P.C., (1980). Bioengineering aspects of biophotolysis. *Enzyme Microbiol. Technol.* 2: 103–111.

Benemann, J.R., Miyamoto, K., Hallenbeck, P.C. and Murray, M.A., (1982). Hydrogenase activity in the thermophile *Mastigocladus laminosus*. *Biochem. Biophys. Res. Commun.* 106: 1196–1202.

Bennemann, J.R., Berenson, J.A., Kaplan, N.O. and Kamen, M.D., (1973). Hydrogen evolution by a chloroplast-ferredoxin-hydrogenase system. *Proc. Natl. Acad. Sci.*, USA, 8: 2317–2320.

Bennet, M. and Weetall, H.H., (1976). Production of hydrogen using immobilized *Rhodospirillium rubrum*. *J. Solid–Phase Biochem.* 1: 137–142.

Bennett, B., Lemon, B.J. and Peters, J.W., (2000). Reversible carbon monoxide binding and inhibition at the active site of the Fe-only hydrogenase (CpI). *Biochemistry*, 39: 7455–7460.

Bennt, R. and Fuller, R.C., (1964). The pyruvate phosphoroclastic reaction in Chromatium: A probable role for ferredoxin in a photosynthetic bacterium. *Biochem. Biophys. Res. Commun.*, 16: 300–307.

Benson, D., Daniel, R., Arp, J. and Burris, R.H., (1980). Hydrogenase in actinorhizal root nodules and root nodule homogenates. *J. Bacteriol.*, 142: 138–144.

Beral, E. and Zapan, M., (1977). Inorganic Chemistry, E.D.P. Bucharest.

Berezin. I.V. and Varfolomeev, S.D., (1979). Energy related applications of immobilized enzymes. *Appl. Biochem. Bioeng.*, 2: 259–290.

Beringer, J.E., (1984). The application of molecular genetics in agriculture. In: *Current Perspectives in Nitrogen Fixation* (Eds.) Gibson, A.H. and Newton, W.E., pp 3–6. The Hague and Wageningen, nijhoff, Junk, pudoc.

Berk, R.S. and Canfield, J.H., (1964). Bioelectrochemical energy conversion. *Appl. Microbiol.* 12: 10–12.

Bernhard, T.H. and Gottschalk, G., (1978). Cellyields of *Esherichia coli* during anaerobic growth on fumarate and molecular hydrogen. *Arch. Microbiol.* 116: 235–238.

Bernstein, J.D. and Olson, J.M., (1981). Biochemical factors affecting the quantum efficiency of hydrogen of hydrogen production by membrances of green photosythetic bacteria. *Photochemi. Photobiol.*, 34: 725–731.

Billinger, R., Zurrer, H. and Bachofen, R., (1985). Photoproduction of molecular hydrogen from wastewater of a sugar refinery by photosynthetic bacteria. *Appl. Microbiol. Biotechnol.*, 23: 147–151.

Bishop, N.I. and Gaffron, H., (1963). On the interrelation of the mechanism for oxygen and hydrogen evolution in adopted algae. In: *Photosynthetic Mechanisms of Green Plants* (Eds.) Kok, B. and Jagendorf, A., pp. 441–454. Publication 1145, Natl. Acad. Sci., National Research Council, Washington, D.C. (U.S.A.).

Bishop, N.I., Frick, M. and Jones, L.W., (1977). Photohydrogen production in green algae: water serves as the primary substrates for hydrogen and oxygen production. In: *Biological Solar Energy Conversion*, (Eds.) Mitsui, A., Miyachi, S., San Pietro, A. and Tamura, S., pp. 3–22. Academic Press, New York.

Block, D.L. and Melody, I., (1992). Efficiency and cost goals for photoenhanced hydrogen production processes. *Int. J. Hydrogen Energy*, 17: 853–861.

Blokesch, M. and Bock, A., (2002). Maturation of (NiFe)–hydrogenases in *Escherichia coli*: The HypC cycle. *J. Mol. Biol.*, 324: 287–296.

Bohme, H., 1998. Regulation of nitrogen fixation in heterocyst-forming cyanobacteria. *Trends Plant Sci.*, 9: 346–351.

Boichenko, V.A., (1980). Effect of anoxia on the activity of photosystem II in *Chlorella vulgaris*: Role of hydrogenase system. *Fiziol. Rast. (Mosc.)* 27: 42–51.

Boichenko, V.A., Arkhipov, V.N. and Litvin, F.E., (1983). Simultaneous measurements of fluorescence induction and hydrogen photoproduction in *Chlorella vulgaris* under anaerobic conditions. *J. Biofizika* 22: 976–979.

Boison, G., Bothe, H. and Schmitz, O., (2000). Transcriptional analysis of hydrogenase genes in the cyanobacteria *Anacystis nidulans* and *Anabaena variabilis* monitored by RT–PCR. *Curr. Microbiol.*, 40: 315–321.

Boison, G., Schmitz, O., Mikheeva, L., Shestakov, S. and Bothe, H., (1996). Cloning, molecular analysis and insertional mutagenesis of the bi-directional hydrogenase genes from the cyanobacterium *Anacystis nidulans*. *FEBS Letters*, 394: 153–158.

Boison, G., Schmitz, O., Schmitz, B. and Bothe, H., (1998). Unusual gene arrangement of the bidirectional hydrogenase and functional analysis of its diaphorase subunit HoxU in respiration of the unicellular cyanobacterium *Anacystis nidulans*. *Curr. Microbiol.*, 36: 253–258.

Bone, D.H., (1960). Locatization of hydrogen activating enzyme in *Psedomonas saccharophila*. *Biochemi. Biophys. Res. Commun.*, 3: 211–214.

Borronibira, C.H., (2001). "Fuel Cells". En John Zumerchik. *Macmillan Encyclopedia of Energy*. New York: Macmillan.

Bossel, U. and Eliasson, B., (2003). Energy and the Hydrogen Economy *www. methanol.org/pdfFrame.cfm?pdf=HydrogenEconomyReport2003.pdf*.

Bossel, U., (2006). "Does a Hydrogen Economy Make Sense?" *Proceedings of the IEEE*, 94: 34–48.

Bossel, U., (2007). *Why a hydrogen economy doesn't make sense, http: //www.physorg. com/* news85074285.html – 37k.

Bothe, H., (1982). Nitrogen fixation. In: *The Biology of Cyanobacteria* (Eds.) Carr, N.G. and Whitton, B.A., pp. 87–104. The University of California Press, Berkeley, Los Angeles.

Bothe, H., Distler, E. and Eisbrenner, G., (1978). Hydrogen metabolism in blue-green algae. *Biochimie*, 60: 277–289.

Bothe, H., Neuer, G. Kalbe, I. and Eisbrenner, G., (1980). Electron donors and hydrogenase in nitrogen-fixing microorganisms. In: *Nitrogen Fixation* (Eds.) Stewart W.D.P. and Gallon, J.R., pp. 83–112. Academic Press, New York.

Bothe, H., Tennigkeit, J. and Eisbrenner, G., (1977a). The utilization of molecular hydrogen by the blue-green alga *Anabaena cylindrica. Arch. Microbiol.*, 114: 43–49.

Bothe, H., Tennigkeit, J. and Eisbrenner, G., (1977b). The hydrogenase-nitrogenase relationship in the blue-green alga *Anabaena cylinidrica. Planta*, 133: 237–242.

Bott, M., Eikmanns, B., and Thauer, R.K., (1986). Coupling of carbon monoxide oxidation to CO2 and H2 with the phosphorylation of ADP in acetate-grown *Methanosarcini barkeri. European Journal of Biochemistry* 159: 393–398.

Bradley, M.J., (2000). Future Wheels Interviews with 44 Global Experts on the Future of Fuel Cells for Transportation and Fuel Cell Infrastructure and a Fuel Cell Primer, Northeast Advanced Vehicle Consortium, Boston, MA, p. 89.

Braun, H., (2003). Safety of Hydrogen. En Middleton, P. R. Larson. M. Nicklas. y B Collins, eds., *Renewable Hydrogen Forum*, Washington. D.C: American Solar Energy Society.

Bregoff, H.M. and Kamen, M.D., (1952). Studies on the metabolism of photosynthetic bacteria, XIV. Quantitative relation between malate dissimilation. Photoproduction of hydrogen and nitrogen metabolism in *Rhodospirlllum rubrum. Arch. Biochem. Biophys.*, 36: 202–220.

Brodelius, P., (1985). The potential role of immobilization in plant cell biotechnology. *Trends Biotechnol.*, 3: 280–285.

Brouers, M. and Hall, D.O., (1986). Ammonia and hydrogen production by immobilized cyanobacteria. *J. Biotechnol.*, 3: 307–322.

Bui, E.T. and Johnson, P.J., (1996). Identification and characterization of Fe-hydrogenase in the hydrogenosome of *T. vaginalis. Mol. Biochem. Parasitol.*, 76: 305–310.

Bulen, W.A., Lecomte, J.R., Burns, R.C. and Hinkson, J., (1965). Nitrogen fixation studies with aerobic and photosynthetic bacteria. In: *Non-heme Iron Proteins: Role in Energy Conversation* (Ed.) San Pietro, A. Yellow Springs: Antioch Press, pp. 261–274.

Buranakarl, L., Ito, K., Izaki, K. and Takahashi, H., (1968). Purification and characterization of a raw starch digestive amylase from non-sulfur purple photosynthetic bacetrium. *Enzyme Microb. Technol.*, 10: 173–179.

Buranakearl, L., Yoneyama, H., Iwata, T., Kim, J.S., Okuda, S., Kampee, T., Izaki, K. and Takahashi, H., (1987). Growth of non-sulphur purple photosynthetic bacteria under high temperature. *Bull. Japanese Soc. Microbial Ecol.*, 2: 13–19.

Burman, G., Shima, S. and Thauer, R.K., (2000). The metal-free hydro-genase from methanogenic archaea, evidence for a bound cofactor. *FEBS Lett.*, 24: 200–204.

Callaham, D., del Tredici, P. and Torrey, J.G., (1978). Isolation and cultivation *in vitro* of the actinomycete causing root nodulation in *Comptonia. Science* (Washington), 199: 899–902.

Calvin, M. and Taylor, S.E., (1989). Fuels from algae. In: *Algal and Cyanobacterial Biotechnology,* (Eds.) Cresswell, R.C., Rees, T.A.V. and Shah, N. Longman Group, UK Ltd.

Cammack, R. M. Frey. R. Robson, eds. (2001). *Hydrogen as a fuel. Learning for nature.* London: Taylor and Francis.

Cammack, R., Hall, D.O. and Rao, K.K., (1985). Hydrogenase: Structure and application in hydrogen production. In: *Microbial Gas Metabolism: Mechanistic, Metabolic and Biotechnological Aspects,* (Eds.) Poole, R.K. and Dow, C.S. Academic Press, London, New York, pp. 75–102.

Cammack, R., Patil, D.S., Aquirre, R. and Hatchikian, E.C., (1982). Redox properties of the ESR–detectable nickel in hydrogenase from *Desulfovibrio gigas. FEBS Lett.,* 142: 289–292.

Cao, H., Zang, L. and Melis, A., (2001). Bioenergetic and metabolic processes for survival of sulfur deprived *Dunaliella salina* (Chlorophyta). *J. Appl. Phycol.,* 13: 25–34.

Carnahan, J.E., Mortenson, K.E., Mower, H. and Castle, J.C., (1960). Nitrogenase in cell free extracts of *Clostridium pasteurianum. Biochim. Biophys. Acta,* 64: 530–535.

Carrasco, C.D., Buettner, J.A. and Golden, J.W., (1995). Programed DNA re-arrangement of a cyanobacterial hup L gene in heterocysts. *Proc. Natl. Acad. Sci.,* 92: 791–795.

Carter, K.R., Jennings, N.T., Hanns, J. and Evans, H.J., (1978). Hydrogen evolution and uptake by nodules of soybeans inoculated with different strains of *Rhizobium japonicum. Can. J. Microbiol.,* 35: 405–452.

Centrell, M.A., Haugland, R.A. and vans, H.J., (1983). Construction of a *Rhizobium japonicum* gene bank and use in isolation of a hydrogen uptake gene. *Proc. Natl. Acad. Sci., USA* 80: 181–185.

Chae, K.J., Choi, M., Ajayi, F.F., Park, W., Chang, S. and Kim, S., (2008). Mass transport through a proton exchange membrane (Nafion) in microbial fuel cells. *Energy and Fuels,* 22: 169–176.

Chain, Y.K., Nelson, L.M. and Knowler, R., (1980). Hydrogen metabolism of *Azospirillium brasiliensis* in nitrogen free medium. *Can. J. Microbial.,* 26: 1126–1131.

Chan Van, N., Hai, C. and Gogotov, I.N., (1983). Photosynthesis and hydrogen metabolism in *Anabaena azollae. Mikrobiologiya,* 52: 896–901.

Chang, F.Y. and Lin, C.Y., (2004). Bihydrogen production using an up-flow anaerobic sludge blanket reactor. *Int. J. Hydrogen Energy,* 29: 33–39.

Chang, J.S., Lee, K.S. and Lin, P.J., (2002). Bihydrogen production with fixed-bed bioreactors. *Int. J. Hydrogen Energy*, 27: 1167–1174.

Chen, C.C. and Lin, C.Y., (2000). Using sewage sludge as seed in an anaerobic hydrogen producing reactor. In: *Proceedings 25th Wastewater Treatment Technology Conference*, December 1–2, p. 368–72.

Chen, C.C. and Lin, C.Y., (2003). Using sucrose as a substrate in an anaerobic hydrogen producing reactor. *Adv. Environ. Res.*, 7: 695–699.

Chen, C.C. and Lin, Cy., (2003). Using sucrose as a substrate in an anaerobic hydrogen-producing reactor. *Adv. Environ. Res.*, 7: 695–699.

Chen, C.C., Lin, C.Y. and Chang, J.S., (2001). Kinetics of hydrogen production with continuous anaerobic cultures utilizing sucrose as the limiting substrate. *Appl. Microbiol. Biotechnol.*, 57: 56–64.

Chen, C.P., Almon, H. and Boger, P., (1989). Physiological factors determining hydrogenase activity in nitrogen fixing *Heterocystous cyanobacteria*. *Plant Physiol.* 89: 1035–1038.

Chen, J.S. and Mortenson, L.E., (1974). Purification and properties of hydrogenase from *Clostridium pasteurianum* W5. *Biophys. Biochem. Acta*, 37: 283–298.

Chen, J.S., (1978). Structure and function of two hydrogenases from the dinitrogen fixing bacterium *Clostridium pasteurianum* W5. In: *Hydrogenases: Their Catalytic Activity, Structure and Function* (Eds.) Schlegel, H.G. and Schneider, K. Goltze, K.G., Gotingen, pp. 57–81.

Chen, W.M., Tseng, Z.J., Lee, K.S. and Chang, J.S., (2005). Fermentative hydrogen producing with *Clostridium butyricum* CGS5 isolated from anaerobic sewage sludge. *Int. J. Hydrogen Energy*, 30: 1063–1070.

Cheng, S. and Logan, B. E., (2007). Sustainable and efficient biohydrogen production via electrohydrogenesis. *Proceedings of the National Academy of Science*, Vol. 104, pp. 18871–18873.

Chin, H.L., Chen, Z.S. and Chou, C.P., (2003). Fedbatch operation using *Clostridium acetobutylicum* suspension culture as biocatalyst for enhancing hydrogen production. *Biotechnology Progress*, 19: 383–388.

Chippaux, M.F. and Pascal, M.C., (1972). Isolation and phenotype of mutants from *S. typhimurium* Defectie in formate-hydrogen lyase activity. *J. Bacteriol.* 110: 766–768.

Christopher, G. Arges, Ramani, Vijay and Pintauro, Peter N., (2010). Anion Exchange membrane Fuel cells, The Electrochemical Society Interface: Summer.

Clayton, R.K. and Sistrom, W.R., (1978). *The Photosynthetic Bacteria*. Plenum Press, New York.

Cocquempot, M.F., Aguirre, R., Lissolo, T., Monson, P., Hatchikian, K.C. and Thomas, O., (1982). Co-immobilization effects on hydrogen production by a chloroplasts-membranes hydrogen system. *J. Biotechnol. Lett.*, 4: 313–318.

Cocquempot, M.F., Thommaset, B., Barbotin, J.N. Gelff, G. and Thomas, D., (1981). Conparative stabilization of biological of biological photosystems by several immobilization procedures. *Eur. J. Appl. Microbial, Biotechnol.*, 11: 193–198.

Cocquempot, M.P., Thgomas, D., Champigny, N.L. and Moyse, A., (1979). Immobilization of thylakoids in porous and stabilization of the photochemical process by glutaraldehyde action at sub-zero temperature. *Eur. J. Appl. Microbial, Biotechnol.*, 8: 39–41.

Cole, J.A., (1976). Microbial gas metabolism. *Adv. Microb. Physiol.* 14: 1–92.

Colebeau, A. and Vignais, P.M., (1983). The membrane bound hydrogenase of *Rhodopseudomonas capsulata* is inducible and contains nickel. *Biochim. Biophys. Acta* 748: 128–138.

Collet, C., Adler, N., Schwitzgu´ebel, J.P., P´eringer, P., (2004). Hydrogen production by *Clostridium thermolacticum* during continuous fermentation of lactose. *Int J Hydrogen Energy*, 29: 1479–85.

Conte, M.A., Iacobazzi, M., Ronchetti, Y.R. Vellone, (2002). "Hydrogen economy for a sustainable development: state of the art and technological perspectives". *Journal of Power Sources*, 100: 171–187.

Conte, M.P.P., Prosini, Y.S. Passerini, (2004). "Overview of energy/hydrogen storage: state of the art of the technologies and prospects for nanomaterials" *Materials science and Engineering*, 108: 2–8.

Cothran, H., (2003). *Global Resources: Opposing Viewpoints*. San Diego: Greenhaven Press.

Crabtree, G.W., Dresselhaus, M.S. and Buchanan, M.V., (2004). The Hydrogen Economy. *Physics Today*, 57(12): 39–44.

Cuendet, P. and Gratzel, M., (1982). New Photosystem I electron acceptors: Improvement of hydrogen photoproduction by chloroplast. *Photochem. Photobiol.*, 36: 203–210.

Cunningham, S.D., Kapulnik, Y., Kagan, S.A., Brewin, N.J. and Phillips, D.A., (1985). Hup activity determined by pRL6JI does not increase N_2–fixation. In: *Nitrogen Fixation Research Progress* (Eds.) Evans, H.J., Bottomley, P.J. and Newton, W.E., p. 227. Dordrecht: Nijhoff.

Dabrock, B., Bahl, H. and Gottschalk, G., (1992). Parameters affecting solvent production by *Clostridium pasteurium*. *Appl. Environ. Microbiol.*, 58: 1233–1239.

Daday, A. and Smith, G.D., (1983). The effect of nickel on the hydrogen metabolism of the cyanobacterium *Anabaena cylindrica FEMS Microbiol. Lett.*, 20: 327–330.

Daday, A. and Smith, G.D., (1987). The hydrogenase-nitrogenase relationship in a symbiotic cyanobacterium isolated from *Macrozamia communis* L. Johnson. *Aust. J. Plant Physiol.*, 14: 319–324.

Daday, A., Lambert, G.R. and Smith, G.D., (1979). Measurement *in vivo* of hydrogenase catalyzed hydrogen evolution in the presence of nitrogenase enzyme in cyanobacteria. *Biochem. J.* 177: 139–144.

Daday, A., Mackerras, A.H. and Smith, G.D., (1985). The effect of nickel on hydrogen metabolism and nitrogen fixation in the cyanobacterium *Anabaena cylindrical*. *J. Gen. Microbiol.*, 131: 231–238.

Daday, A., Mackerras, A.H. and SMith, G.D., (1985). The effect of nickel on hydrogen metabolism and nitrogen fixation in the cyanobacterium *Anabaena cylindrica. J. Gen. Microbiol.* 131: 231–238.

Daday, A., Platz, R.A. and Smith, G.D., (1977). Anaerobic and aerobic hydrogen gas formation by the blue-green alga *Anabaena cylindrica. Appl. Environ. Microbiol.*, 34: 478–483.

Darensbonrg, M.Y., Lyon, E.J., Zhao, X. and Georgakaki, I.P. (2003). The organometallic site of Fe-hydrogenase, model and entatic states. *Chem. Proc. Natl. Acad. Sci.*, 100: 3683–3688.

Das, D. and Veziroglu, T.N., (2001). Hydrogen production by biological processes: A survey of literature. *Int. J. Hydrogen Energy*, 26: 13–28.

Davis, D.D. and Stevenson, K.L., (1977). A recording gas microvilumeter. *J. Chem. Edu.*, 56: 394–395.

De Lacey, A.L., Sradler, C., Cavazza, C., Harchikian, E.C. and Fernandez, V.M., (2000). FTIR characterization of the active site of the Fe-hydrogenase from *Desulfovibrio desulfuricans. J. Am. Chem. Soc.*, 122: 11232–11233.

De Vrije, T., De Haas, G.G., Tan, G.B., Keijsers, E.R.P. and Claassen, P.A.M., (2002). Pretreatment of *Miscanthus* for hydrogen production by *Thermotoga elfii. Int. J. Hydrogen Energy*, 27: 1381–1390.

Degen, J., Uebele, A., Retze, A., Schmid-Staiger, U. and Trosch, W., (2001). A novel air-lift photobioreactor with baffles for improved light utilization through the flashing light effects. *J. Biotechnol.*, 92: 89–94.

Demirbas, A., (2007). Progress and recent trends in biofuels. *Prog. Energy Combust. Sci.*, 33: 1–18.

Department of Trade and Industry, (2003). *Energy White Paper: Our Energy Future: Creating a Low–Carbon Economy*: http: //www.dti.gov.uk/energy/whitepaper/ourenergyfuture.pdf (accessed 2006).

Der, A., Bagyinka, C., Pali, T. and Kovacs, K.L., (1985). Effect of enzyme concentration on apparent specific activity of hydrogenase. *Ann. Biochem.*, 150: 481–486.

Dickson, E.J. and Ruyan M. Smulyan, (1977). *Hydrogen Energy Economy: A Realist Appraisal of Prospects and Impacts*. New York: Praeger.

Diner, B. and Mauzerau, D., (1973). The turnover times of photosynthesis and redox properties of the pool of electron camers between the photosystem. *Biochim. Biophys. Acta*, 305: 353–363.

Ditzig, J., Liu, H. and Logan, B.E., (2007). Production of hydrogen from domestic wastewater using a bioelectrochemically assisted microbial reactor (BEAMR). *International Journal of Hydrogen Energy*, 32: 2296–2304.

Dixon, R.O.D., (1978). Nitrogenase-hydrogenase interrelationships in rhizobia. *Biochemie*, 60: 233–236.

DOE EERE. (2003). Freedom Car and Fuel Initiative.
http: //www.eere.energy.gov/hydrogenfuel/, Energy Efficiency and Renewable Energy, US Department of Energy, Viewed June 20.

Dommergues, Y.R., Diem, H.H., Hauthier, D.L., Dreyfus, B.L. and Cornet, F., (1984). Nitrogen fixing trees in the tropics: potentialities and limitation. In: *Advances in Nitrogen Fixation Research*, (Eds.) Veeger. C. and Newton, W.E. The Hague and Wageningen: Nijhoff, Junk, Pudoc, pp 7–13.

Drake, D., Leonard, J.T. and Hirsch, A.M., (1985). Symbiotic genes in Frankia. In: *Nitrogen Fixation Research Progress*, (Eds.) Evans, H.J., Bottomley, P.J. and Newton, W.E. Dordrecht, Nijhoff, p 147.

Dunn, S., (2001). *Hydrogen Futures: Towards a Sustainable Energy System*. Washington, D.C.: Worldwach Institute.

Dupuis, A., Peinnequin, A., Darouzet, E. and Lunardi, J., (1997). Genetic disruption of the respiratory NADH–ubiquinone reductase of *Rhodobacter capsulatus* leads to an unexpected photosynthesis negative phenotype. *FEMS Microbiol. Lett.*, 148: 107–114.

Dutton, A.G., (2002). *Hydrogen Energy Technology*. Tyndall Working Paper TWP 17. Tyndall Centre for Climate Change: *http: //www.tyndall.ac.uk/publications/ working_papers/wp17.pdf* (accessed October 2006).

Eady, R.R. and Postgate, J.R., (1974). Nitrogenase. *Nature*, 249: 805–810.

Edie, S.A., 1983. Acetylene reduction and hydrogen evolution by nitrogenase in a *Rhizobium* and legume symbiosis. *Can. J. Bot.*, 61: 780–185.

Edward, J.S., Convert, M. and Palsson, B.O., (2002). Metabolic modeling of microbes: The flux balance approach. *Environ. Microbiol.*, 4: 133–140. .

Eiringer, T. and Friedrich, B., (1991). Cloning, nucleotide sequence, and heterologous expression of a high-affinity' nickel transport gene from *Alcaligenes eutrophus*. *J. Biological Chemistry*, 266: 3222–3227.

Eiringer, T., and Mandrand-Berthelot, M.A., (2000). Nickel transport in microorganisms. *Arch. Microbiol.*, 173: 1–9.

Eisbrenner, G. and Bothe, H., (1979). Modes of electron transfer from molecular hydrogen in *Anabaena cylindrica*. *Arch. Microbiol.*, 123: 37–48.

Eisbrenner, G. and Evans, H.J., (1983). Aspects of hydrogen metabolism in nitrogen-fixing legumes and other plant microbe associations. *Ann. Rev. Plant Physiol.*, 34: 105–136.

Eisbrenner, G., Distler, E., Floener, L. and Bothe, H., (1978). The occurrence of the hydrogenase in some blue-green algae. *Arch. Microbiol.*, 118: 177–184.

Eisbrenner, G., Distler, E., Floener, L. and Bothe, H., (1978). The occurrence of hydrogenase in some blue-green algae. *Archives of Microbiology*, 118: 177–184

Eisbrenner, G., Roos, P. and Bothe, H., (1981). The number of hydrogenases in cyanobacteria. *J. Gen. Microbiol.*, 125: 383–390.

El-Shishtawy, R.M.A., Kawasaki, S., Morimoto, M., (1997). Biological H_2 production using a novel light-induced and diffused photobioreactor. *Biotechnol. Tech.*, 11: 403–407.

Emerich, D.W., Ruiz-Argueso, T., Ching, T.M. and Evans, H.J., (1979). Hydrogen dependent nitrogenase activity and ATP formation in *Rhizobium japonicum* bacteroids. *J. Bacteriol.*, 137: 153–160.

Emerson, R. and Arnold, W., (1932a). A separation of the reaction in photosynthesis by means of intermittent light. *J. Gen. Physiol.*, 5: 391–421.

Emerson, R. and Arnold, W., (1932b). The photochemical reaction in the photosynthesis. *J. Gen. Physiol.*, 16: 191–205.

Environmental Technologies Action Plan (ETAP) Communication from the Commission to the Council and the European Parliament Com, (2004). 38 final. Stimulating Technologies for Sustainable Development: An Environmental Technologies Action Plan for the European Union Brussels (January 2004).

Erbes, D.L., Burris, R.H. and Orme-Johnson, W.H., (1975). On the iron sulfur clusters in the hydrogenase from *Clostridium pasteurianum* W5. *Proc. Natl. Acad. Sci., USA*, 72: 4795–4799.

Erhardt, U., (1966). Uber das Wassenrstoff activierende system von *Hydrogenomonas* H16. Abnahme des activitat bie heterotrophen. Wachstum. *Arch. Mikrobiol.*, 54: 115–124.

Ernst, A., Kerfin, W., Spiller, H. and Boger, P., (1979). External factors influencing light-induced hydrogen evolution by the blue-green alga *Nostoc muscorum* Z. *Naturforsch*, 34: 820–825.

European Commission (2003). *Hydrogen Energy and Fuel Cells: A Vision of Our Future.* http://www.europa.eu.int/comm/research/energy/pdf/hydrogen-report_en.pdf (accessed October 2006).

European Hydrogen and Fuel Cell Technology Platform (2005). *Deployment Strategy*: https://www.hfpeurope.org/hfp/keydocs (accessed October 2006).

Evans, C.H., Hanus, F.J., Russel, S.A., Harker, A.R., Lambert, G.R. and Dalton, D.A., (1985). Biochemical characterization, evolution and genetics of H_2 recycling in *Rhizobium.* In: *Nitrogen Fixation and CO_2 Metabolism* (Eds.) Ludden, R.W. and Burris, L.E. Elsevier, New York, pp. 3–11.

Evans, H.J. and Barber, L.E., (1977). Biological nitrogen fixation for food and fiber production. *Science*, 197: 332–339.

Evans, H.J. Emerich, D.W., Lepo, J.E., Mairer, R.J. and Carter, K.R., (1980). The role of hydrogenase in nodule bacteroids and free-living rhizobia. In: *Nitrogen Fixation*, (Eds.) Stewart, W.D.P. and Gallon, J.R. Dordrecht, Nijhoff, pp 209–215.

Evans, H.J., Eisbrenner, G., Cantrell, M.A., Russel, S.A. and Hanus, F.J., (1982). The present status of hydrogen recycling in legumes. *Israel J. Bot.*, 31: 72–88.

Evvyernie, D., Morimoto, K., Karita, S., Kimura, T., Sakka, K. and Ohmiya, K., (2001). Conversion of chitinous wastes of hydrogen gas by *Clostridium paraputrificum* M–21. *J. Biosci. Bioeng.*, 91: 339–343.

Ewart, G.D. and Smith, G.D., (1989a). Immunochemical analysis of the soluble hydrogenase from the cyanobacterium *Anabaena cylindrica. Biochim. Biophys. Acta,* 997: 83–89.

Ewart, G.D. and Smith, G.D., (1989c). Purification and properties of soluble hydrogenase from the cyanobcterium *Anabaena cylindrica. Arch. Biochem. Biophys.,* 268: 327–337.

Fabiano, B. and Perego, P., (2002). Thermodynamic study and optimization of hydrogen production by *Enterobacter aerogenes. Int. J. Hydrogen Energy,* 27: 149–56.

Fakioglu, E.V., Yürüm, Y. and Veziroglu, T.N., (2004). A review of hydrogen storage systems based on boson and its compounds. *International Journal of Hydrogen Energy,* 29: 1371–1376.

Fang, H.H.P. and Liu, H., (2002). Effect of pH on hydrogen production from glucose by mixed culture. *Bioresour. Technol.,* 82: 87–93.

Fang, H.H.P., Liu, H. and Zhang, T., (2002). Characterization of a hydrogen producing granular sludge. *Biotechnol. Bioeng.,* 78: 44–52.

Fang, H.H.P., Liu, H. and Zhang, T., (2005). Phototrophic hydrogen production from acetate and butyrate in wastewater. *Int. J. Hydrogen Energy,* 30: 785–93.

Fang, H.H.P., Zhang, T. and Liu, H., (2002). Microbial diversity of a mesophilic hydrogen- producing sludge. *Appl. Microbiol. Biotechnol.,* 58: 112–118.

Fascetti, E. and Todini, O., (1995). *Rhodobacter sphaeroids* RV cultivation and hydrogen production in a one and two stage chemostat. *Appl. Microbial. Biotechnol.,* 44: 300–305.

Fascetti, E., D'Addario, E., Todini, O. and Robertiello, A., (1998). Photosynthetic hydrogen evolution with volatile organic acid derived from the fermentation of source selected municipal wastes. *Int. J. Hydrogen Energy,* 23: 753–760.

Fay, P. and Van Baalen, C., (1987). *The Cyanobacteria: Current Researches.* Elsevier Science Publishers, B.V., Amsterdam.

Federov, A.S., Tsygankov, A.A., Rao, K.K. and Hall, D.O., (1998). Hydrogen photo-production by *Rhodobacter sphaeroides* immobilized on polyurethane foam. *Biotechnol. Lett.,* 20: 1007–1009.

Felten, P. von, Zurrer, H. and Bachofen, R., (1985). Production of molecular hydrogen with immobilized cells of *Rhodopirillum rubrum. Appl. Micribiol. Biotechnol.* 23: 15–20.

Ferchichi, M., Crabbe, E., Hintz, W., Gil, G.H. and Almadidy, A. (2005). Influence of culture parameters on biological hydrogen production by *Clostridium saccharoperbutylacteonicum* ATCC 27021. *World J. Microbiol. Biotechnol.,* 21: 855–862.

Firzgerald, M.P., Rogers, L.J., Rao, K.K. and Hall, D.O., (1980). Efficiency of ferredoxins and flavodoxins as mediators in systems for hydrogen evolution. *Biochem. J.*, 192: 665–672.

Florin, L., Tsokoglou, A. and Happe, T., (2001). A novel type of iron hydrogenase in the green alga *Scenedesmus obliquus* is linked to the photosynthetic electron transport chain. *J. Biol. Chem.*, 276: 6125–6132.

Fork, D.C., (1980). Oxygen electrode. *Methods Enzymol.*, 69: 113–122.

Fornori, C.S. and Kaplan, S., (1982). Genetic transformatiuon of *Rhodopseudomonas sphaeroides* by plasmid DNA. *J. Bacteriol.*, 152: 89–97.

Forsberg, C.W., (2007). Future hydrogen markets for large-scale hydrogen production systems. *Int. J. Hydrogen Energy*, 32: 431–439.

Foyer, C.H. and Hall, D.O., (1980). Oxygen metabolism in active chloroplasts. *Trends in Biochemical Sciences*, 5: 188–191.

Francov, N. and Vignais, P.M., (1984). Hydrogen production of *Rhodopseudomonas capsulata* cell entrapped in carregeenan beads. *Biotechnnol. Lett.*, 6: 639–644.

Freidrich, B., Heine, E., Fing, A. and Friedrich, C.G., (1981). Nickel requirement for active hydrogenase formation in *Alkaligenes eutrophus*. *J. Bacteriol.*, 145: 1144–1149.

Frenkel, A., (1949). A study of the hydrogenase systems of green and blue green algae. *Bio. Bull.*, 97: 261–262.

Frenkel, A.H., Gaffron, H. and Battley, E.H., (1950). Photosynthesis and photoreduction in the blue-green alga, *Synechococcus elongatus* Nag. *Bio. Bulletin. Marine Biological Laboratory Woods Hole, Mass.*, 99: 157–162.

Frenkel, A.W. and Rieger, C., (1951). Photoreduction in algae. *Nature (London)*, 167: 1030.

Frias, J.E., Merida, A., Herrero, A., Martin-Nieto, J. and Flores, E., (1993). General distribution of the nitrogen control gene *ntcA* in cyanobacteria. *J. Bacteriol.*, 175: 5710–5713.

Friedrich, B. and Schwartz, E., (1993). Molecular biology of hydrogen utilization in aerobic chemolithotrophs. *Ann. Rev. Microbiol.*, 47: 351–383.

Fry, I., Robinson, A.E., Spath, S. and Packer, L., (1984). The role of sodium sulfide in anoxygenic photosynthesis and hydrogen production in the cyanobacterium *Nostoc muscorum*. *Biochem. Biophys. Res. Commun.*, 123: 1138–1143.

Fujii, T., Tarusawa, M., Miyanage, M., Kiyoto, S., Watanabe, T. and Yabaki, M., (1987). Hydrogen evolution from alcohols, malate and mixed electron donors by *Rhodopseudomonas* sp. 7. *Agric. Biol. Chem.*, 51: 1–7.

Gaffron, H. and Rubin, J., (1942). Fermentative and photochemical production of hydrogen in algae. *J. Gen. Physiol.*, 26: 219–240.

Gaffron, H., (1940). Oxyhydrogen reaction in green algae and the reduction of carbon dioxide in the dark. *Science* 91: 529.

Gallon, J.R. and Chaplin, A.E., (1987). *An Introduction to Nitrogen Fixation.* Cassel Educational Limited, London.

Garcin, E., Vernede, X., Hatchikian, E. C., Volbeda, A., Frey, M. and Fontecilla-Camps, J.C., (1999). The crystal structure of a reduced [Ni–Fe–Se] hydrogenase provides an image of the activated catalytic centre. *Struct. Fold. Des.*, p. 557–566.

Genty, B. and Harbinson, J., (1996). Regulation of light utilization for photosynthetic electron transport. In: *Photosynthesis and the Environment*, (Ed.) N.R. Baker. Kluwer, Dordrecht, pp. 67–99.

Gerasimenko, L.M. and Zavarzin, G.A., (1981). Anabolic hydrogen uptake by cyanobacteria. *Mikrobiologiya*, 50: 955–959.

Gest, H. and Kamen, M.D., (1949). studies on the metabolism of photosynthetic bacteria. IV. Photochemical production of molecular hydrogen by growing cultures of photosynthetic bacteria. *J. Bacteriol.*, 28: 238–245.

Gest, H. and Kamen, M.D., (1949a). Studies on the metabolism of photosynthetic bacterial per cent. Photochemical production of molecular hydrogen by growing cultures of photosynthetic bacteria. *J. Bacteriol.*, 58: 238–245.

Gest, H., (1954). Oxidation and evolution of molecular hydrogen by microorganisms. *Journal of Bacteriology*, 18: 43–73.

Gest, H., (1972). Energy conversation and generation of reduced power in bacterial photosynthesis. *Adv. Microbiol. Physicol.*, 7: 243–283.

Gest, H., Kamen, M.D. and Bregoff, H.M., (1950). Studies on the metabolism of photosythetic bacteria. V. Photoproduction of hydrogen and nitrogen fixation by *Rhodopspirillum rubrum. J. Biol. Chem.*, 182: 153–170.

Gest, H., Omerod, J.G. and Omorod, K.S., (1962). Photometabolism of *Rhodopspirillum rubrum:* Light dependent dissimulation of organic compounds to carbon dioxide and molecular hydrogen by and anaerobic citric acid cycle. *Arch. Biochem.*, 97: 21–23.

Gfeller, R.P. and Gibbs, M., (1984). Fermentative metabolism of *Chlamydomonas reinhardtii* I, analysis of fermentative products from starch in dark and light. *Plant Physiol.*, 75: 212–218.

Ghirardi, M.L., Zhang, L., Lee, J.W., Flynn, T., Seibert, M. and Greenbaum, E. *et al.*, (2000). Microalgae: A green source of renewable H2. *Tibtech*, 18: 506–11.

Gibbons, N.E. and Murray, R.G.E., (1978). Proposals concerning the higher taxa of bacterial. *Int. J. Systematic Bacteriol.*, 28: 1–6.

Gibbs, M., Gfeller, R.P. and Chen, C., (1986). Fermentative metabolism of *Chlamydomonas reinhardtii* II, photoassimilation of acetate. *Plant Physiol.*, 82: 160–166.

Ginkel, S.V., Oh, S.E. and Logan, B.E., (2005). Biohydrogen production from food processing and domestic wastewaters. *Int. J. Hydrogen Energy*, 30: 1535–42.

Gisby, P.E. and Hall, D.O., (1980). Bophotolytic hydrogen production using alginate-immobilized chloroplasts, enzymes and synthetic catalysts. *Nature (Lond.)*, 287: 251–253.

Gisby, P.E., Rao, K.K. and Hall, D.O., (1982). Hydrogen production from water using chloroplasts, enzymes and synthetic catalysts. In: *'Solar world Forum'*, Vol. 3 (Eds.) Hah, D.O. and Morton, J., Pergamon Press, Oxford, pp. 2242–2247.

Glick, B.R., Martin, W.G. and Martin, S.M., (1980). Purification and properties of perplasmic hydrogenase from *Desulfovibrio desulfulricans*. *Can. J. Microbiol.*, 26: 1214–1223.

Gogotov, I.N. and Kondratieva, E.N., (1969). Conditions of hydrogen formation by *Rhodopseudomonas* sp. *Izv. Akad. Nauk. SSSR Ser Biol.*, 1: 161–165.

Gogotov, I.N. and Kosyak, A.V., (1976). Hydrogen metabolism by *Anabaena variabilis* in the dark. *Mikrobiologiya*, 45: 586–591.

Gogotov, I.N. and Laurinavichene, T.V., (1975). The role of ferredoxin in hydrogen metabolism of *Rhodospirillium rubrum*. *Mikrobiologiya*, 44: 581–590.

Gogotov, I.N., (1978). Relationship in hydrogen metabolism between hydrogenase and nitrogenase in phototrophic bacteria. *Biochimle*, 60: 267–275.

Gogotov, I.N., Zorin, N.A. and Bogorov, L.N., (1974). Hydrogen metabolism and the ability for nitrogen fixation in *Thiocapsa roseopersicina*. *Micribiologiya*, 43: 1–6.

Gogotov, I.N., Zorin, N.A., Serebryakova, L. and Kondratieva, E.N., (1978). The property of hydrogenasee from *Thiocapsa roseopersicina*. *Biochim. Biophys. Acta*, 523: 335–343.

Gogotov, I.V., Kosyak, A.V. and Krupenko, A.N., (1976). Hydrogen production by the cyanobacterium *Anabaena variabilis* in light. *Mikrobiologiya*, 45: 941–945.

Golden, J.W., Robinson, S.J. and Haselkorn, R., (1985). Rearrangement of nitrogen fixation genes during, heterocyst differentiation in the cyanobacterium *Anabaena*. *Nature*, 314: 419–423.

Golden, S.S., Brusslan, J. and Haselkorn, R., (1987). Genetic engineering of the cyanobacterial chromosome. *Methods Enzymol.*, 153: 228– 229.

Gómez Romero, P., 2002. "Pilas de combustible: energia sin humos." *Mundo científico*, 233: 66–71.

Gordon, J.K., (1981). Introduction to nitrogen fixing prokaryotes. In: *The Prokaryotes: A Handbook on Habitats, Isolation and Identification of Bacteria*, (Eds.) M.P. Starr, H. Stolp, H.G. Truper, A. Balour and H.G. Schlegel. Vol. 1. Springer-Verlag, Berlin, Heidelberg, New York, pp. 781–794.

Gorell, T.E. and Uffen, R.L., (1978). Light dependent and light independent production of hydrogen gas by photosynthesizing *Rhodospirillum rubrum* mutant C. *Photochem. Photobiol.*, 27: 351–358.

Gorrell, T.E. and Uffen, R.L., (1977). Fermentive metabolism of pyruvate by *Rhodospirillum rubrum* after anaerobic growth in darkness. *J. Bacteriol.*, 131: 533–543.

Gotogov, I.N., Mitkina, T.V. and Glinskii, V.P., (1974). Effect of ammonium on hydrogen evoluation and nitrogen fixation in *Rodopseudomonas palustris*. *Mikrobiologiya*, 43: 586–591.

Graf, E.G. and Thauer, R.K., (1981). Hydrogenase *Methanobacterium thermoautotrophicum*: A nickel-containing enzyme. *FEBS Lett.*, 136: 165–169.

Graham, A., Boxer, D.H., Haddock, B.A., Mandrand-Berthelot, M.A. and Jones, R.W., (1980). Immunological analysis of the membrane bound hydrogenase of *Escherichia coli*. *FEBS Lett.* 113: 167–172.

Grande, H.J., Dunham, W.R., Averill, B., Dijk, C.V. and Sands, R.H., (1983). Electron paramagnetic resonance and other properties of hydrogenases isolated from *Desulfovibrio vulgaris* (strain Hilden borough) and *Megasphaera elsdenii*. *Eur. J. Biochem.*, 136: 201–207.

Gray, G.T. and Gest, H., (1965). Biological formation of molecular hydrogen. *Science* 148: 186–192.

Greenbaum, E., (1977). The molecular mechanisms of photosynthetic hydrogen and oxygen production. In: *Biological Solar Energy Conversion*, (Eds.) A. Mitsui, S. Miyachi, A. San Pietro and S. Tamura. Academic Press, New York, pp. 101–107.

Greenbaum, E., (1977). The photosynthetic unit of hydrogen evolution. *Science* 196: 879–880.

Greenbaum, E., (1980). Simultaneous photoproduction of hydrogen and oxygen by photosynthesis. *Biotech. Bioeng. Symp.*, 10: 1–13.

Greenbaum, E., (1982). Photosynthetic hydrogen and oxygen production, kinetic studies. *Science*, 196: 879–880.

Greenbaum, E., (1988). Energetic efficiency of hydrogen photoevolution by algal water splitting. *Biophys. J.*, 54: 365–368.

Greenbaum, E., Guillard, R.R.L. and Sunda, W.G., (1983). Hydrogen and oxygen photoproduction by marine algae. *Photochem. Photobiol.*, 37: 649–655.

Grigoriev, S.A., Porembsky, V.I. and Fateev, V.N., (2006). Pure hydrogen production by PEM electrolysis for hydrogen energy. *International Journal of Hydrogen Energy*, 31: 171–175.

Gsaffron, H. and Rubin, J., (1942). Fermentative and photochemical production of hydrogen in algae. *J. Gen. Physiol.* 265: 219–240.

Gst, H. and Kamen, M.D., (1949b). Photoproduction of molecular hydrogen by *Rhodospirillum rubrum*. *Science*, 109: 558–559.

Guan, Y., Deng, M., Yu, X. and Zang, W., (2004). Two stage photoproduction of hydrogen by marine green algae *Platymonas subcordiformis*. *Biochem. Eng. J.*, 19: 69–73.

Gubuli, J. and Borthakur, D., (1996). The use of a PCR cloning and screening strategy to identify lambda clones containing the hupB gene of Anabaena spa. strain PCC 7120, *J. Microbiological Methods*. 27: 175 – 182 preprint.

Gutiérrez Jodra, L. (2005). "El hidrógeno, combustible del futuro". *Rev.R. Acad. Cienc. Exact. Fís. Nat.*, 99: 49–67.

Hageman, R.V., Orme-Johnson, W.H. and Burris, R.H., (1980). Rol of magnesium ATP in the hydrogen evolution reaction catalyzed by nitrogenase from *Azotobacter vinelandii*. *Biochemistry* 19: 2333–2342.

Hagen, W.R., van Berkel-Arts, A., Kruse-Wolkers, K.M., Voordouw, G. and Veeger, C., (1986). The iron-sulfur composition of the active site of hydrogenase from *Desulfovibrio vulgaris* (Hildenborough) deduced from its subunit structure and total iron-sulfur content. *FEBS Lett.*, 203: 59–63.

Halbleib, C.M. and Ludden, P.W., (2000). Regulation of biological nitrogen fixation. *J. Nutr.*, 130: 1081–1084.

Hall, D.O., Markov, S.A., Watanabe, Y. and Rao, K.K., (1995). The potential applications of cyanobacterial photosynthesis for clean technologies. *Photosyn. Res.*, 46: 159–167.

Hallenbeck, P.C. and Beneman, J.R., (1978). Characterization and partial purification of the reversible hydrogenase of *Anabaena cylindrical*. *FEBS Lett.*, 94: 261–264.

Hallenbeck, P.C. and Benemann, J.R., (1979). Hydrogen from algae. In: *Photosynthesis in Relation to Model Systems*, (Ed.) J. Barber. Elsevier/North–Holland Biomedical Press, Amsterdam.

Hallenbeck, P.C. and Benemann, J.R., (2002). Biological hydrogen production, fundamentals and limiting processes. *International Journal of Hydrogen Energy*, 27: 1185– 1193.

Hallenbeck, P.C., Kochian, L.V. and Benemann, J.R., (1981). Hydrogen evolution catalyzed by hydrogenase in cultures of cyanobacteria. *Z. Naturforsch.*, 36c: 97–92.

Han, S.K. and Shin, H.S., (2004). Biohydrogen production by anaerobic fermentation of food waste. *Int. J. Hydrogen Energy*, 29: 569–77.

Hans, S.K. and Shin, H.S., (2004). Biohydrogen production by mesophilic fermentation of food wastewater. *Water Sci. Technol.*, 49: 223–228.

Hansel, A. and Lindblad, P., (1998). Towards optimization of cyanobacteria as biotechnologically relevant producers of molecular hydrogen, a clean and renewable energy source. *Applied Microbiological Biotechnology*, 50: 153–160.

Hansen, T.A. and Veldkamp, H., (1973). *Rhgodopseudomonas sulfidophila* nov. spec., a new species of the purple non-sulfur bacteria. *Arch. Microbiol.*, 92: 45–48.

Happe, T. and Kaminski, A., (2002). Differential regulation of Fe-hydrogenase during anaerobic adaptation in green algae *Chlamydomonas reinhardtii*. *Eur. J. Biochem.*, 269: 1022–1032.

Happe, T., Hemschemeier, A., Winkler, M. and Kaminski, A., (2002). Hydrogenases in green algae: Do they save the algae's life and solve our energy problems? *Trends Plant Sci.*, 7: 246–250.

Happe, T., Schute, K. and Bohme, H., 2000. Transcriptional and mutational analysis of the uptake hydrogenase of the filamentous cyanobacterium *Anabaena variabilis*. ATCC 29413. *J. Bacteriol.*, 182: 1624–1631.

Harbinson, J. and Hedley, C.L., (1993). Changes in P-700 oxidation during early stages of the induction of photosynthesis. *Plant Physiol.*, 103: 649–660.

Hardy, R.W.F. and Burns, R.C., (1973). Comparative biochemistry of iron-sulfur proteins and dinitrogen fixation. In: *Iron Sulfur Proteins*, (Ed.) W. Lovenberg. Vol. 1, pp. 65–110. Academic Press, New York.

Hardy, R.W.F., (1977). Rate limiting steps in biological productivity. In: *Genetic Engineering for Nitrogen Fixation*, (Eds.) A. Hollaender *et al.* Plenum Press, New York, pp. 369–400.

Harris, R., Book, D., Anderson, P.A. and Edwards, P.P., (2004). Hydrogen storage: The grand challenge. *The Fuel Cell Review*, p. 17–23, June/July.

Harry, M., Westen, V., Stephen, G. M. and Veeger, C., (1978). Separation of hydrogenase from intact cells of *Desulfovibrio vulgaris* purification and properties. *FEBS Lett.*, 86: 122–126.

Hartman, H. and Krasna, A.I., (1963). Studies on the 'Adaptation' of hydrogenase in *Scenedesmus*. *J. Biol. Chem.*, 238: 749.

Haselkorn, R. and Bulkema, W.J., (1992). Nitrogen fixation in cyanobacteria. In: *Biological Nitrogen Fixation*, (Eds.) G. Staccy, R.H. Burris and H. Evans. Chapman and Hall, London, pp. 166–190.

Hatchikian, E.C., Bruschi, M. and LeGall, J., (1978). Characterization of the periplasmic hydrogenase from *Desulfovibrio gigas*. *Biochem. Biophys. Res. Commun.*, 82: 451–461.

Hattori, A., (1963). Microalgae and photosynthetic bacteria. *Special Issue of Plant and Cell Physiology*, University of Tokya Press, pp. 485–492.

Hawkes, F.R., Dindale, R., Hawkes, D.L. and Hussy, I., (2002). Sustainable fermentative biohydrogen, challenges for process optimization. *Int. J. Hydrogen Energy*, 27: 1339–1347.

Haystead, A., Robinson, and R. and Stewart, W.D.P., (1970). Nitrogenase activity in extracts of heterocystous blue-green algae. *Arch. Mikrobiol.*, 82: 325–336.

He, D., Bultel, Y., Magnin, J.P., Roux, C. and Willison, J.C., (2005). Hydrogen photosynthesis by *Rhodobacter capsulatus* and its coupling to PEM fuel cell. *J. Power Sources*, 141: 19–23.

Healey, F.P., (1970). Hydrogen evolution by several algae. *Planta*, 91: 220–226.

Helgi, H.T., Ingolfssonby, H.P. and Jenssona, T., (2008). Optimizing site selection for hydrogen production in Iceland. *International Journal of Hydrogen Energy*, 33: 3632–3643.

Herrero, A., Muro-Pastor, A.M. and Flores, E., (2001). Nitrogen control in cyanobacteria. *J. Bacteriol.*, 183: 411–425.

Herter, S.M., Kortlüke, C.M. and Drews, G., (1998). Complex I of *Rhodobacter capsulatus* and its role in reverted electron transport. *Arch. Microbiol.*, 169: 98–105.

Hillmer, P. and Gest, H., (1977). H2 metabolism in the photosynthetic bacterium *Rhodopseudomonas capsulata*: H2 production by growing cultures. *J. Bacteriol.*, 129: 724–731.

Hirosawa, T. and Wolk, C.P., (1979). Factors controlling the formation of akinetes adjacent to heterocyst in the cyanobacterium *Cylindrospermum licheniforme. J. Gen. Microbiol.*, 114: 423–432.

Hoekeman, S., Bijmans, M., Janssen, M., Tramper, J. and Wijffels, R.H., (2002). A pneumatically agitated flat–panel photobioreactor with gas recirculation: Anaerobic photoheterotrophic cultivation of a purple non–sulfur bacterium. *Int. J. Hydrogen Energy*, 27: 1331–1338.

Hoffmann, D.R., Thauer, R.K. and Trebst, A., (1977). Photosynthetic hydrogen evolution by spinach chloroplast coupled to a *Clostridium hydrogenase*. *Z.Naturforsch. Sect. C, Biosci.*, 32.

Hordeski, M.F., (2006). *Alternative Fuels: The Future of Hydrogen*. Boca Raton: CRC.

Horiuchi, J.I., Shimizu, T., Tada, K., Kanno, T. and Kobayashi, M., (2002). Selective production of organic acids in anaerobic acid reactor by pH control. *Biores. Technol.*, 82: 209–213.

Horner, D.S., Heil, B., Happe, T. and Embley, T.M., Iron hydrogenase–ancient enzymes in modern eukaryotes. *Trends Biomed. Sci.*, 27: 148–153.

Houchins, J. and Burris, R.H., (1981a). Comparative characterization of two distinct hydrogenase from *Anabaena* sp. strain 7120. *J. Bacteriol.* 146: 215–221.

Houchins, J.P. and Burris, R.H., (1981b). Occurrence and localization of two distinct hydrogenases in the heterocystous cyanobacterium *Anabaena* sp. strain 7120. *J. Bacteriol.*, 146: 209–214.

Houchins, J.P. and Burris, R.H., (1981c). Light and dark reactions of the uptake hydrogenase in *Anabaena* 7120. *Plant Physiol.*, 68: 712–716.

Houchins, J.P. and Burris, R.H., (1981d). Physiological reaction of the reversible hydrogenase from *Anabaena* 7120. *Plant Physiol.*, 68: 717–721.

Houchins, J.P., (1984). The physiology and biochemistry of hydrogen metabolism in cyanobacteria. *Biochim Biophys Acta*, 768: 227– 255.

Houchins, J.P., (1984). The physiology and biochemistry of hydrogen metabolism in cyanobacteria. *Biochim. Biophys. Acta*, 768: 227–255.

Howarth, D.C. and Codd, G.A., (1985). The uptake and production of molecular hydrogen by unicellular cyanobacteria. *J. Gen. Microbiol.*, 131: 1561–1570.

Howitt, C.A. and Vermaas, W.F.J., (1997). Analysis of respiratory mutants of *Synechocystis* 6803. In: Book of Abstracts of the *IX International Symposium on Photosynthetic Prokaryotes*, (Eds.) G.A. Peschek, W. Loffelhardt and G.A. Schmetterer. Vienna, Austria, p. 36.

HP-Gas, (2002). Gas Processes. *Hydrocarbon Processing*, 81(5): 61–121.

Hube, M., Blokesch, M. and Bock, A., (2002). Network of hydrogenase expression in *Escherichia coli*. Role of accessory proteins HypA and HybF. *J. Bacteriol.*, 14: 3879–3885.

Hussy, I., Hawkes, F.R., Dinsdale, R. and Hawkes, D.L., (2000). Continuous fermentative hydrogen production from wheat starch co-product by mixed microflora. *Biotechnol. Bioeng.*, 84: 619–26.

Hydrogen Energy and Fuel cells, (2004). A vision of our future. Summary report (June 2003) Science 305. Toward a hydrogen economy pp. 957–976.

Ike, A., Murakawa, T., Kawaguchi, H., Hirata, K. and Miyamoto, K., (1999). Photo-production of hydrogen from raw starch using a halophilic bacterial community. *J. Biosci. Bioeng.*, 88: 72–77.

International Energy Agency (2006). *Hydrogen Production and Storage, R and D Priorities and Gaps*: http: //www.iea.org/Textbase/papers/2006/hydrogen. pdf (accessed October 2006).

Jackson, D.D. and Ellms, J.W., (1896). On odors and tastes of surface waters, with special reference to *Anabaena*, a microscopical organism found in certain water supplies of Massachusetts. In *The 1896 Report of the Massachusetts State Board of Health*, pp. 410–420.

Jameson, D.J., Sawars, R.G., Rugman, P.A., Boxer, D.A. and Higgins, C.F., (1986). Effects of anaerobic regulatory mutations and catabolic repression on regulation of hydrogen metabolism and hydrogenase isoenzyme composition in *Salmonella typhimurium*. *J. Bacteriol.*, 168: 405–411.

Janssen, H., Bringmann, J.C., Emonts, B. and Schroeder, V., (2004). Safety-related studies on hydrogen production in high-pressure electrolysers. *International Journal of Hydrogen Energy*, 29: 759–770.

Jee, H.S, Ohashi. T., Nishizawa, Y. and Nagai, S., (1987). Limiting factro of nitorgenase system mediating hydrogen production of *Rhodobacter sphaeroides*. *J. Ferment. Technol.*, 65: 153–158.

Jeffries, T.W. and Leach, K.L., (1978). Intermittent illumination increases biophotolytic hydrogen yield by *Anabaena cylindrica*. *Appl. Environ. Microbiol.*, 35: 1228–1230.

Jeffries, T.W., Timourian, H. and Ward, R.L., (1978). Hydrogen production by *Anabaena cylindrica*: Effect of varying ammonium and ferric ions, pH and light. *Appl. Environ. Microbiol.*, 35: 704–710.

Jeong-II, Oh. and Bowien, B., (1999). Dual control by regulatory gene fdsR of the fds operon encoding the NAD$^+$-linked formate dehy drogenase of *Ralstonia eutropha*. *Mol. Micro.*, 34: 365–376.

Jo, J.H., Lee, D.S. and Park, J.M., (2008). The effects of pH on carbon material and energy balances in hydrogen-producing *Clostridium tyrobutyricum* JM1. *Bioresource Technology*, 99: 8485–8491.

Johansson, T.B (ed.) (1993). *Renewable Energy: Sources for Fuels and Electricity*. Island Press, Washington.

Jones, L.W. and Bishop, N.I., (1976). Simultaneous measurement of oxygen and hydrogen exchange from the blue-green alga *Anabaena*. *Plant Physiol.*, 57: 659–665.

Jones, R.H.Y. and Thomas, G.J., (2008). *Material for the Hydrogen Economy*. CRC Press, Boca Raton.

Jouanneau, Y., Kelley, B.C., Barlier, Y., Lespinat, P.A. and Vignais, P.M., (1980). Continuous monitoring by mass spectrometry of H2 production and recycling by mass spectrocapsulata. *J. Bacteriol.*, 143: 628–636.

Joyner, A.E., Winter, W.T. and Godbout, D.M., (1977). Studies on some characteristics of hydrogen production by cell free extract of rumen anaerobic bacteria. *Can. J. Microbiol.*, 23: 346–353.

Jung, G.Y. *et al.*, (1999). A new chemoheterotrophic bacterium catalyzing water-gas shift reaction. *Biotechnology Letters*, 21: 869–873.

Jung, G.Y. *et al.*, (1999). Isolation and characterization of *Rhodopseudomonas palustris* P4 which utilizes CO with the production of H2. *Biotechnology Letters* 21: 525–529.

Jungermann, K., Thauer, R.K., Leimenstoll. G. and Decker, K., (1973). Function of reduced pyridine nucleotide–ferredoxin oxidoreductase in *Saccharolytic clostridia*. *Biochim. Biophys. Acta*, 305: 268–280.

Kadar, Z., D. Vrijek, T., van Noorden, G.E., Budde, M.A.W., Szengyel, Z., Reczey, K. and Claassen, P.A.M., (2004). Yields from glucose, xylose, and paper sludge hydrolysate during hydrogen production by the extreme thermophile *Caldicellulosiruptor saccharolyticus*. *Appl. Biochem. Biotechnol.*, 113–116: 497–508.

Kajii, Y., Kobayashi, M., Takahashi, T. and Onodera, K., (1994). A novel type of mutant of *Azotobacter vinelandii* that fixes Nitrogen in the presence of tungsten. *Bioscience Biotechnology and Biochemistry*, 58: 1179–1180.

Kaltwasser, H., Stuart, T.S. and Gaffron, H., (1969). Light dependent hydrogen evolution by *Scenedesmus*. *Planta*, 89: 309–322.

Kanai, T., Imanaka, H., Nakajima, A., Uwamori, K., Omori, Y. and Fukui, T., *et al.*, (2005). Continuous hydrogen production by the hyperthermophilic archaeon, *Thermococcus kodakaraensis* KOD1. *J. Biotechnol.*, 116: 271–282.

Kaneko, T., Nakamura, Y., Wolk, C.P., Kuritz, T., Sasamoto, S., Watanabe, A., Iriguchi, M., Ishikawa, A., Kawashima, K., Kimura, T., Kishida, Y., Kohara, M., Matsumoto, M., Matsumo, A., Muraki, A., Nakazaki, N., Shimpo, S., Sugimoto, M., Takazawa, M., Yamada, M., Yasuda, M. and Tabata, S., (2001). Complete genomic sequence of the filamentous nitrogen-fixing cyanobacterium *Anabaena* sp. strain PCC 7120. *DNA Res.*, 8: 205–213.

Kaneko, T., Sato, S., Kotani, H., Tanaka, A., Asamizu, E. and Nakamura, Y. *et al.*, (1996). Sequence analysis of the genome of the unicellular cyanobacterium

Synechocystis sp. PCC 6803. II. Sequence determination of the entire genome and assignment of potential protein–coding regions. *DNA Res.,* 3: 109–136.

Kaneko, T., Sato, S., Kotani, H., Tanaka, A., Asamizu, E., Nakamura, Y., Miyajima, N., Hirosawa, M., Sugiura, M., Sasamoto, S., Kimura, T., Hosouchi, T., Matsuno, A., Muraki, A., Nakazaki, N., Naruo, K., Okumura, S., Shimpo, S., Takeuchi, C., Wada, T., Watanabe, A., Yamada, M., Yasuda, M. and Tabata, S., (1996). Sequence analysis of the genome of the unicellular cyanobacterium *Synechocystis* sp. strain PCC 6803. II. Sequence determination of the entire genome and assignment of potential protein–coding regions. *DNA Res.,* 3: 185–209.

Kapdan, I.K. and Kargi, F., (2006). Bio-hydrogen production from waste materials. *Enzyme and Microbial Technology,* 38: 569–582.

Karube, I., Matsunaga, T., Otsuga, T., Kayano, H. and suzuki, S., (1981). Hydrogen evolution by co-immobilized chloroplast and *Clostridium butyricum. Biochim. Biophys. Acta,* 637: 490–495.

Karube, I., Otsuga, T., Kayano, H., Matsunaga, T. and Suzuki, T., (1980). Photochemical system for regenerating NADPH from NADP with use of immobilized chloroplasts. *Biotechnol. Bioeng.,* 22: 2655–2665.

Kashyap, A.K. and Singh, Surendra, (1989). Photoevolution of hydrogen during oxygenic photosynthesis of cyanobacterium *Nostoc muscorum.* In: *Hydrogen Energy Workshop cum Review Committee Meeting,* Department of Physics, University of Rajasthan, Jaipur, 5–7 July.

Kasting, J.F. (2004). "When Methane Made Climate". *Scientific American,* 291: 71–75.

Kats, E.Y., Kozlov, Y.N. and Kiselev, B.A., (1979). Photosensitized released of hydrogen in photochemical systems using chlorophyll. *Biofizika,* 24: 801–805.

Kawaguchi, H., Hashimoto, K., Hirata, K. and Miyamoto, K. (2001). H2 production from algal biomass by mixed culture of *Rhodobium marinum* A–501 and *Lactobacillus amylovorus. J. Biosci. Bioeng.,* 91: 277–282.

Kayano, H., Karube, I., Matsunaga, T., Suzuki, S. and Nakayama, O., (1981). A photochemical fuel cell system using *Anabaena* N–7363. *Eur. J. Appl. Microbiol. Biotechnol.,* 12: 1–5.

Keasling, J.D., J.R. Benemann, J. Pramanik, T.A. Carrier, K.L. Jones, and S.J. VanDien, (1998). A toolkit for metabolic engineering of bacteria: Applications to hydrogen production. In: *Bio Hydrogen* (Eds.) O. Zaborsky *et al.* Plenum Press, New York, pp. 87–98.

Kellye, B.C, Meyer, C.M., Gandy, C.G. and Vignias, P.M., (1977). Hydrogen recycling in *Rhodopseudomonas capsulata. FEBS Lett.,* 81: 281–285.

Kellye, B.C. and Nicholas, D.J.D., (1981). Inhibition of nitrogenase activity by metronidazole in *Rhodopseudomonas capsulata. Arch. Microbiol.,* 129: 344–348.

Kentemich, T., Bahnweg, M., Mayer, F. and Bothe, H., (1989). Localization of the reversible hydrogenase in cyanobacteria. *Z Naturforch,* 44c: 384–391.

Kentemich, T., Casper, M. and Bothe, H., (1991). The reversible hydrogenase in *A. nidulans* is a component of the cytoplasmic membrane. *Naturwissenschaften,* 78: 559–560.

Kentemich, T., Dannenberg, G., Hundeshagen, B. and Bothe, H., (1988). Evidence for the occurring of the alternative vanadium-containing nitrogenase in the cyanobacterium *Anabaena variabilis. FEMS Microbiology Letters,* 51: 19–24.

Kentemich, T., Haverkamp, G. and Bothe, H., (1991). The expression of a third nitrogenase system in the cyanobacterium *Anabena variabilis. Zeitschrift fur Naturforschung,* 46: 217–222.

Kerby, N.W., Musgrave, S.C., Rowell, P., Shestakov, S.V. and Stewart, W.D.P., (1986). Photoproduction of ammonium by immobilized mutant strains of *Anabaena variabilis. Appl. Microbiol. Biotechnol.,* 24: 42–46.

Kerfin, W. and Boger, P., (1982). Light–induced hydrogen evolution by blue-green algae (cyanobacteria). *Physiol. Plant.,* 54: 93–98.

Kerfin, W., Spiller, H., Ernst, A. and Boger, P., (1978). Properties of the hydrogen producing system in the blue-green alga *Nostoc muscorum.* In: *Hydrogenases: Their catalytic Activity, Structure and Function,* (Eds.) H.G. Schlegel and K. Schneider. E. Goltze KG, Gottingen, pp. 381–386.

Kessler, E. and Czygan, F.C., (1967). Physiologisches und biochemische Bietrage zur Taxonomic der Gottungen *Ankistrodesmus* and *Scenedesmus.* I. Hydrogenase, sekundar carotenoide und Geletaine–Versflussigung. *Arch. Mikrobiol.,* 55: 320–326.

Kessler, E. and Maiforth, H., (1960). Occurrence and activity of hydrogenase in some green algae. *Arch. Mikrobiol.,* 37: 215.

Kessler, E. and Zweier, I., (1960). Physiologische and biochemische Beitrage zur Taxonomic der Gattrung *Chlorella.* V. Die auxotrophen and mesotrophe. *Arton. Arch. Mikrobiol.,* 79: 44–48.

Kessler, E., (1956). Reduction of nitrite with molecular hydrogen in algae containing hydrogenase. *Arch. Biochem. Biophys.,* 624: 241.

Kessler, E., (1967). Physiologische und biochemische Beitrage zur taxonomic der Gattung (*Chlorella*) III Merkmale von 8 autotrophen. *Arten. Arch. Mikrobiol.,* 55: 346–357.

Kessler, E., (1973). Effect of anaerobiosis on photosynthetic reactions and nitrogen metabolism of algae with and without hydrogenase. *Arch Microbiol.,* 93: 91–100.

Kessler, E., (1973). Effect of anerobiosis on photosynthetic reaction and nitrogen metabolism of algae with and without hydrogenase. *Arch. Microbiol.,* 93: 91–100.

Kessler, E., (1974). Hydrogenase photoreduction and anaerobic growth. In: *Algal Physiology and Biochemistry,* (Ed.) W.D.P. Steward. University of California Press, Barkeley, pp. 456–473.

Khanal, S.K., Chen, W.H., Li, L. and Sung, S., (2004). Biological hydrogen production: effects of pH and intermediate products. *Int. J. Hydrogen Energy,* 29: 1123–1131.

Khanna, S., Vellye, B.C. and Nicholas, D.J.D., (1980). The effect of carbon dioxide on nitrogenase related activities of *Rhodopseudomonas capsulata. J. Gen. Microbiol.*, 119: 261–284.

Kiley, P.J. and Beinert, H., (1999). Oxygen sensing by the global regulator FNR: The role of the iron-sulfur cluster. *FEMS Microbiol. Rev.*, 22: 341–352.

Kim, H.Y., (2003). A low cost production of hydrogen from carbonaceous wastes. *Int. J. Hydrogen Energy*, 28: 1179–1186.

Kim, J. and Rees, D.C., (1994). Nitrogenase and biological nitrogen fixation. *Biochemistry*, 33: 389–397.

Kim, J.O., Kim, Y.H., Ryu, J.Y., Song, B.K., Kim, I.H. and Yeom, S.H., (2004). Immobilization methods for continuous hydrogen gas production biofilm formation versus granulation. *Process Biochem.*, 40: 1331–1337.

Kim, J.O., Kim, Y.H., Ryu, J.Y., Song, B.K., Kim, I.H. and Yeom, S.H., (2005). Immobilization methods for continuous hydrogen gas production biofilm versus granulation. *Process Biochem.*, 40: 1137–331.

Kim, J.S., Ito, K. and Takahashi, H., (1980). The relationship between nitrogenase activity and hydrogen evoluation in *Rhodopseudomonas Palustris. Agric. Biol. Chem.*, 44: 824–834.

Kim, J.S., Toshito, K. and Takahashi, H., (1981). Production of molecular hydrogen by *Rhodopseudomonas* sp. *J. Ferment. Technol.*, 59: 185–190.

Kim, J.S., Yamaushi, H., Ito, K. and Takahashi, H., (1982). Selections of a photosynthetic bacterium suitable for hydrogen production in outdoor cultures among strains isolated in the Seoul. Taegu, Sendal and Bangkok areas. *Agric. Biol. Chemi.*, 46: 1469–1474.

Kim, M.S., Back, J.S., Yun, Y.S., Sim, S.J., Park, S. and Kim, S.C., (2006). Hydrogen production from *Chlaymydomonas reinhardtii* biomass using a two-step conversion process: Anaerobic conversion and photosynthetic fermentation. *Int. J. Hydrogen Energy*, 31: 812–816.

Kim, M.S., Baek, J.S. and Lee, J.K. (2006). Comparison of H2 accumulation by *Rhodobacter sphaeroides* KD131 and its uptake hydrogenase and PHB synthase deficient mutant. *Int. J. Hydrogen Energy*, 31: 121–127.

Kim, S.H., Han, S.K. and Shin, H.S., (2004). Feasibility of biohydrogen production by anaerobic co-digestion of food waste and sewage sludge. *Int. J. Hydrogen Energy*, 29: 1607–1616.

Kim, S.H., Han, S.K. and Shin, H.S., (2006). Effect of substrate concentration on hydrogen production and 16S rDNA-based analysis of the microbial community in a continuous fermenter. *Process Biochem*, 41: 199–207.

King, D.E., David, L. and Gibbs, M., (1977). Inhibition of ferricyanide reduction in chloroplast particles by anaerobiocity. *Biochem. Biophys. Res. Commun.*, 78: 734–738.

Kitajima, M. and Butler, W.L., (1976). Microencapsulation of chloroplast particles. *Plant Physiol.*, 57: 746–750.

Klemme, J.H., (1968). Studies on the photoautotrophic growth of new isolated non-sulfur purple bacteria at the expense of molecular hydrogen. *Arch. Mikrobiol.*, 61: 269–282.

Klibnov, A.K., Kaplan, N.O. and Kamen, M.D., (1978). A rationale for stabilization of oxygen labile enzymes: Application to clostridial hydrogenase. *Proc. Natl. Acad. Sci., USA*, 75: 3640–3643.

Klibnov, A.M. and Puglisi. A.V., (1980). The regeneration of coenzytmes using immobilized hydrogenases. *Biotechnol. Lett.*, 2: 445–450.

Klibnov, A.M., (1983). Immobilized enzymes and cells as practical catalysts. *Science*, 219: 722–727.

Kndratieva, E.N. and Gogotov, I.N., (1969b). Production of hydrogen by green photosynthetic bacteria *Chloropseudomonas Nature*, 221: 83–84.

Kobayashi, M., (1975). Role of photosynthetic bacteria in foul water purification. *Progress in Water Technol.*, 7: 309–315.

Kobayashi, M., (1977). Utilization and disposal of water by photosynthetic bacteria. In: *Microbial Energy Conversion*, (Eds.) H.G. Schlegel and J. Barnea. Pergamon Press, Oxford, pp. 443–453.

Kobayashi. M. and Tchan, Y.T., (1973). Treatment of industrial waste solutions and production of useful byproducts using a photosynthetic bacterial method. *Water Research*, 7: 1219–1224.

Kobayashi. M., Ae, N., Kishimoto, M., and Kinoshita, S., (1976). Purification and use of waste solution discharged from wool washing. *J. Agric. Chemi. Soc. Japan*, 50: 157–161.

Kohlmiller, E.F. and Gest. H., (1951). A comparative study of the light and dark fermentation of organic acids by *Rhodospirillum rubrum*. *J. Bacterial.*, 61: 269–282.

Kojima, N., Fox, J.A., Housinger, R.P., Deniels, I., Orme-Johnson, W.H. and Walsh, C., (1983). Paramagnetic centers in the nickel containing deazaflavin-reducing hydrogenase from *Methanobacterium thermoautotrophicum*. *Proc. Natl. Acad. Sci., USA*, 80: 378–382.

Koku, H., Eroglu, I., Gunduz, U., Yucel, M. and Turker, L., (2002). Aspects of metabolism of hydrogen production by *Rhodobacter sphaeroides*. *Int. J. Hydrogen Energy*, 27: 1315–1329.

Koku, H., Eroglu, I., Gunduz, U., Yucel, M. and Turker, L., (2003). Kinetics of bio-hydrogen production by the photosynthetic bacterium *Rhodobacter spheroids* O.U. 001. *Int. J. Hydrogen Energy*, 28: 381–388.

Kondo, T., Arakawa, M., Hiral, T., Wakayama, T., Hara, M. and Miyake, J., (2002). Enhancement of hydrogen production by a photosynthetic bacterium mutant with reduced pigment. *J. Biosci. Bioeng.*, 93: 145–150.

Kondo, T., Arakawa, M., Wakayama, T. and Miyake, J., (2002). Hydrogen production by combining two types of photosynthetic bacteria with different characteristics. *Int. J. Hydrogen Energy*, 27: 1303–1308.

Kondratieva, E.N. and Gogotov, I.N., (1969a). Evolution and Utilization of molecular hydrogen by *Choropseudomonas. Mikrobiologiya,* 38: 803–809.

Kondratieva, E.N. and Gogotov, I.N., (1976). Microorganisms-hydrogen producers, Izvestija Academii nauk. *USSR, Seria Biol.,* p. 69–85.

Kondratieva, E.N., (1977). Phototrophic microorganisms as source of hydrogenase formation. In: *Microbial Energy Conversion,* (Eds.) H.G. Schlegel and J. Barnea. Pergamon Press, Oxford, New York, pp. 205–216.

Kondratieva, E.N., Gogotov, I.N. and Grazinaskii, I.V., (1979). The effect of nitrogen containing compounds on hydrogen photo-evolution and nitrogen fixation in purple bacteria. *Mikrobiologiya,* 48: 389–395.

Koroneos, A., Dompros, G. and Roumbas, N., (2004). Moussiopoulos. *International Journal of Hydrogen Energy,* 29: 1443–1450.

Kosaric, N. and Lyng, R.P., (1988). Microbial production of hydrogen. In: *Biotechnology,* (Eds.) H.J. Rehm and G. Reed. Weinheim: VCH Verlagsgesellschaft, 6: 101–136.

Kosyak, A.V., Gogotov, I.N. and Kulakova, S.M., (1978). Photoproduction of hydrogen by the cyanobacterium *Anabaena cylindrica. Mikrobiologiya,* 47: 605–610.

Kotani, H., Tanaka, A., Kaneko, T., Sato, S., Sugiura, M. and Tabata, S., (1995). Assignment of 82 known genes and gene clusters on the genome of the unicellular cyanobacterium *Synechocystis* sp. strain PCC 6803. *DNA Res.,* 2: 133–142.

Krasna, A.I. and Rittenberg, D., (1954). The mechanism of action of the enzyme hydrogenase. *J. Am. Chem. Soc.,* 76: 3015–3020.

Krasna, A.I., (1977). Catalytic and structural properties of the enzyme hydrogenase and its role in biophotolysis of water. In: *Biological Solar Energy Conversion,* (Eds.) A. Mitsui, S. Miyachi, A. San Pietro and S. Tamura. Academic Press, New York, p. 53.

Krasna, A.I., (1978). Oxygen-stable hydrogenase and assay. *Methods Enzymol.,* 53: 314–332.

Krasna, A.I., (1979). Hydrogenase: Properties and applications. *Enzyme Microbial. Technol.,* 1: 165–172.

Krasnovskii, A.A., Vanni, C., Nikandrov, V.V. and Brin, G.P., (1980). Study of the efficient photoevolution of hydrogen by chloroplasts in the presence of bacterial hydrogenase. *Mol. Biol. (Mosc.),* 14: 287–298.

Kreith, F.Y.R. West, (2004). Fallacies of the hydrogen economy: A critical analysis of hydrogen production and utilization. *Journal of Energy Resources Technology,* 126: 249–257.

Kruger, H.J., Huynh, B.H., Ljungdahl, P.O., Xavier, A.V., Dervartanian, D.V., Moura, J.J.G. and LeGall, J., (1982). Evidence for nickel and three iron centers in the hydrogenase of *Desulfovubrio desulfuricans. J. Biol. Chem.,* 257: 14620–14623.

Kubicki, A., Funk, E., Westhoff, P. and Steinmüller, K., (1996). Differential expression of plastome-encoded *ndh* genes in mesophyll and bundle-sheath chloroplasts of the C4 plant *Sorghum bicolor* indicates that the complex I–homologous NAD(P) H–plasto-quinone oxidoreductase is involved in cyclic electron transport. *Planta*, 199: 276–281.

Kuhl, S.A., Niz, D.W. and Yoch, D.C., (1983). Characterization of a *Rhodospirillum rubrum* plasmid: Loss of photosynthetic growth in plasmidless strains. *J. Bacteriol.*, 156: 737–742.

Kulda, J., (1999). Trichomonads, hydrogenosomes and drug resistance. *Int. J. Parasitol.*, 29: 199–222.

Kulundaivelu, G., Ravi, V. and Thangraj, A., (1988). Hydrogen photoproduction by algae and higher plants: Optimal conditions, polarographic determinations. *Bull. Mater. Sci.*, 9: 393–400.

Kumar, D. and Kumar, H.D., (1988). Light-dependent acetylene reduction and hydrogen production by the cyanobacterium, *Anabaena* sp. (strain CA). *Int. J. Hydrogen Energy*, 13: 573–575.

Kumar, D. and Kumar, H.D., (1989). Effects of glucose and fructose on acetylene reduction, photosynthess, respiration and hydrogen evolution by whole cells of the cyanobacterium *Anabaena* sp. strain CA. In: *Proc. Intl. Phycotalk Symp.*, Dept. of Botany, Banaras Hindu University, Varanasi.

Kumar, D. and Kumar, H.D., (1990). Protection of nitrogenase levels in dark incubated cultures of *Anabaena* sp. strain CA by various carbon sources, and restoration of nitrogenase activity by oxygen. *Br. Phycol. J.* 25 (in press).

Kumar, D. and Kumar, H.D., (1990b). Effects of some inhibitors and carbonsources on acetylene reduction and hydrogen production by isolated heterocysts of *Anabaena* sp. (strain CA). *Birtish Phycol. J.* (Accepted).

Kumar, D. and Kumar, H.D., (1990c). Protection of nitrogenase levels in dark-incubated cultures of *Anabaena sp.* strain CA by various carbon sources, and restoration of nitrogenase activity by oxygen. *Br. Phycol. J.* (in press).

Kumar, D., (1986). Effects of amitrol on cyanobacterium *Nostoc linckia J. Gen. Appl. Microbiol.*, 32: 51–56.

Kumar, D., (1990). The effect of Shinkimate on *Nostoc linckia. World J. Microbiol. Biotechnol.* (In Press).

Kumar, D., Pandey, A.B. and Kumar, H.D., (1990). Effect of some physiological factors on the nitrogenase and hydrogenase of the *Azolla–Anabaena azollae* association. *Int. J. Hydrogen Energy*, 15: 313–317.

Kumar, N. and Das, D., (2000). Enhancement of hydrogen production by *Enterobacter cloacae* IIT–BT 08. *Process Biochem.*, 35: 589–593.

Kumar, N. and Das, D., (2001). Continuous hydrogen production by immobilized *Enterobacter cloacae* IIT–BT 08 using lignocellulosic materials and solid matrices. *Enzyme Microbial Technol.*, 29: 280–287.

Kumar, N. and Das, D., (2001). Continuous Hydrogen production by immobilized *Enterobacter cloacae* IIT–BT 08 using lignocellulosic material as solid matrices. *Enzyme Microbiological Technology*, 29: 280–287.

Kumar, N., Ghosh, A. K. and Das, D., (2001). Redirection of biochemical pathways for the enhancement of H₂ production by *Enterobacter cloacae* IIT–BT 08. *Biotechnol. Lett.*, 23: 537–541.

Kumar, S. and Polasa, H., (1991). Influence of nickel and copper on photobiological hydrogen production and uptake in *Oscillatoria subbrevis* strain III. *Proc. Indian Natn. Sci. Acad.*, 281–285.

Kumazawa, S. and Mitsui, A., (1981). Characterization and optimization of hydrogen photo-production by a saltwater blue-green alga, *Oscillatoria* sp. Miami BG7. I. Enhancement through limiting the supply of nitrogen nutrients. *Int. J. Hydrogen Energy*, 6: 339–348.

Kumazawa, S. and Mitsui, A., (1982). Hydrogen metabolism of photosystehic bacteria and algae. In: *CRC Handbook of Biosolar, Vol. I: Basis Principles*, (Eds.) A. Mitsul and C.C. Black. CRC Press, Inc., Boca Raton, Florida, pp. 299–316.

Kumazawa, S. and Mitsui, A., (1985). Comparative amperometric study of uptake hydrogenase and hydrogen photoproduction activities between heterocystous cyanobacterium *Anabaena cylindrica* B 629 and non heterocystous cyanobacterium *Oscillatoria* sp. strain Miami BG 7. *Appl. Environ. Microbiol.*, 50: 287–291.

Kumazawa, S., Skjodal, H.R. and Mitsui, A., (1987). Dark hydrogen evolution and adenine nucleotides of substreopical marine unicellular green algae during anaerobic incubation. *Plant Cell Physiol.*, 28: 653–662.

Kurokawa, T. and Tanisho, S., (2005). Effects of formate on fermentative hydrogen production by *Enterobacter aerogenes*. *Marine Biotechnology*, 7: 112–118.

Kuwada, Y. and Ohta, Y., (1987). Hydrogen production by an immobilized cyanobacterium, *Lyngbya* sp. *J. Ferment. Technol.*, 65: 597–602.

Kuwada, Y. and Ohta, Y., (1989). Hydrogen production and carbohydrate consumption by *Lyngbya* sp. No. 108. *Agric. Biol. Chem.*, 53: 2847–2851.

Kuwada, Y., Nakatsukasa, M. and Ohta, Y., (1988). Isolation and characterization of anon heterocystous cyanobacterium *Lyngbya* sp. isolate No. 108, for large quantity hydrogen production. *Agric. Biol. Chem.*, 52: 1923–1928.

Laczko, I. and Barabas, K., (1981). Hydrogen evolution by photo bleached *Anabaena cylindrica*. *Planta*, 153: 312–316.

Laczko, I., (1980). Mechanism of hydrogen evolution by nitrogen starved *Anabaena cylindrica*. *Z. Pflanzenphysiol.*, 100: 241–246.

Lalla-Maharajh, W.W., Hall, D.O., Cammack, R., Rao, K.K. and LeGall, J., (1982). Purification and properties of the membrane bound hydrogenase from *Desulfovibrio desulfuricans*. *Biochem. J.*, 209: 445–454.

Lambert, G.R. and Smith, G.D., (1977). Hydrogen formation by marine blue-green algae. *FEBS Letts*. 83: 159–162.

Lambert, G.R. and Smith, G.D., (1980). Hydrogen metabolism by filamentous cyanobacteria. *Arch. Biochem. Biophys.*, 205: 36–50.

Lambert, G.R. and Smith, G.D., (1981). The hydrogen metabolism of cyanobacteria (Blue-green algae). *Biological Reviews*, 56: 589–660.

Lambert, G.R. and Smith, G.D., (1981a). Hydrogen uptake by the nitrogen starved cyanobacterium *Anabaena cylindrica*. *Arch. Biochem. Biophys.*, 211: 360–367.

Lambert, G.R., Daday, A. and Smith, G.D., (1979a). Hydrogen evolution from immobilized cultures of the cyanobacterium *Anabaena cylindrica*. *FEBS Letts.* 101: 125–128.

Lambert, G.R., Daday, A. and Smith, G.D., (1979b). Duration of hydrogen formation by *Anabaena cylindrica* B 629 in atmospheres of argon, air and nitrogen. *Appl. Environ. Microbiol.*, 38–530–536.

Lambert, G.R., Daday, A. and Smith, G.D., (1979c). Effects of ammonium ions, oxygen, carbon monoxide and acetylene on anaerobic and aerobic hydrogen formation by *Anabaena cylindrica* B–629. *Appl. Environ. Microbiol.*, 38: 521–529.

Lappi, D.A., Stolzenbach. F.E., Kaplan, N.O. and Kamen, M.D., (1976). Immobilization of hydrogenase on glass beads. *Biochem. Biophys. Res. Commun.* 69: 878–884.

Larsen, H., (1952). On the culture and general physiology of the green sulfur bacteria. *J. Bacteriol.*, 64: 187–196.

Laurinavichene, T.V., Tolstygina, I.V., Galiulina, R.R., Ghirardi, M.L., Seibert, M., Tsygankov, A.A., (2002). Dilution methods to deprive *Chlamydomonas rein-hardtii* cultures of sulfur for subsequent hydrogen photoproduction. *Int. J. Hydrogen Energy*, 27: 1245–1249.

Lay, J.J., (2000). Modeling and optimization of anaerobic digested sludge converting starch to hydrogen. *Biotechnol. Bioeng.*, 68: 269–278.

Lay, J.J., (2001). Biohydrogen generation by mesophilic anaerobic fermentation of microcrystalline cellulose. *Biotechnol. Bioeng*, 74: 281–287.

Lay, J.J., Fan, K.S., Chang, J. and Ku, C.H., (2003). Influence of chemical nature of organic waste on their conversion of hydrogen by heat-shock digested sludge. *Int. J. Hydrogen Energy*, 28: 1361–1367.

Lay, J.J., Lee, Y.J. and Noike, T., (1999). Feasibility of biological hydrogen production from organic fraction of municipal solid waste. *Water Res.*, 33: 2579–2586.

Lee, C.M., Chen, P.C., Wang, C.C. and Tung, Y.C., (2002). Photohydrogen production using purple nonsulfur bacteria with hydrogen fermentation reactor effluent. *Int. J. Hydrogen Energy*, 27: 1309–1313.

Lee, H., (1986). Characterization of a hydrogen evolving strain of *Rhodopseudomonas sphaeroides*. *Korean J. Microbiol.*, 24: 62–66.

Lee, K.S., Lo, Y.C., Lin, P.J. and Chang, J.S., (2006). Improving biohydrogen production in a carrier- induced granular sludge bed by altering physical configuration and agitation pattern of the bioreactor. *Int. J. Hydrogen Energy*, 31: 1648–1657.

Lee, K.S., Lo, Y.S., Lo, Y.C., Lin, P.J. and Chang, J.S., (2003). Hydrogen production with anaerobic sludge using activated-carbon supported packed-bed bioreactors. *Biotechnol. Lett.*, 25: 133–138.

Lee, K.S., Lo, Y.S., Lo, Y.C., Lin, P.J. and Chang, J.S., (2004). Operating strategies for biohydrogen production with high-rate anaerobic granular sludge bed bioreactor. *Enzyme Microbial. Technol.*, 35: 605–612.

Lee, K.S., Wu, J.F., Lo, Y.S., Lo, Y.C., Lin, P.J. and Chang, J.S., (2004). Anaerobic hydrogen production with an efficient carrier-induced granular sludge bed bioreactor. *Biotechnol. Bioeng.*, 87: 648–657.

Lefèvre, M., Proietti, E., Jaouen, F. and Dodelet, J., (2009). Iron based catalysts with improved oxygen reduction activity in polymer electrolyte fuel cells. *Science*, 324: 71–74, 3 April.

Levin, D.B., Islam, R., Cicek, N. and Sparling, R., (2006). Hydrogen production by *Clostridium thermocellum* 27405 from cellulosic biomass substrates. *International Journal of Hydrogen Energy*, 31: 1496–1503.

Levin, D.B., Pitt, L. and Love, M., (2003). Feasibility of biological hydrogen production from organic fraction of municipal solid waste. *Int. J. Hydrogen Energy*, 29: 173–185.

Levin, D.B., Pitt, L. and Love, M., (2004). Biohydrogen production: Prospects and limitations to practical application. *Int. J. Hydrogen Energy*, 29: 173–185.

Li, C.L. and Fang, H.H.P., (2007). Fermentative hydrogen production from wastewater and solid wastes by mixed cultures. *Crit. Rev. Environ. Sci. Technol.*, 37: 1–39.

Lien, S. and Pietro, A.S., (1981). Effect of unbcouplers on anaerobic adaptation of hydrogenase activity in *Chlamydomonas reinhardtii*. *Biochem. Biophys. Rev. Commun.*, 103: 139–147.

Liessens, J. and Verstraet, W., (1986). Selective Inhibitors for continuous non-axenic hydrogen production by *Phodobacter capsulatus*. *J. Appl. Bacteriol.*, 61: 547–558.

Ligon, J.M. and Nakos, J.P., (1985). Isolation and characterization of the genes from *Frankia* that code for the component I Proteins of nitrogenase. In: *Nitrogen Fixation Research Progress*, (Eds.) H.J. Evans, P.J. Bottomley and W.E. Newton. Dordrecht, Nijhoff, p. 188.

Lin, C.N., Wu, S.Y., Lee, K.S., Lin, P.J., Lin, C.Y. and Chang, J.S., (2007). Integration of fermentive hydrogen process and fuel cell for online electricity generation. *Int. J. Hydrogen Energy*, 32: 802–808.

Lin, C.Y. and Chang, R.C., (1999). Hydrogen production during the anaerobic acidogenic conversion of glucose. *J. Chem. Technol. Biotechnol.*, 74: 498–500.

Lin, C.Y. and Chang, R.C., (2004). Fermentative hydrogen production at ambient temperature. *Int. J. Hydrogen Energy*, 29: 715–720.

Lin, C.Y. and Chou, C.H., (2004). Anaerobic hydrogen production from sucrose using an acid-enriched sewage sludge microflora. *Eng. Life Sci.*, 4: 66–70.

Lin, C.Y. and Jo, C.H., (2003). Hydrogen production from sucrose using an anaerobic sequencing batch reactor process. *J. Chem. Technol. Biotechnol.*, 78: 678–684.

Lin, C.Y. and Lay, C.H., (2004). Carbon/nitrogen ratio effect on fermentative hydrogen production by mixed microflora. *Int. J. Hydrogen Energy*, 29: 41–45.

Lindberg, P., Hansel, A. and Lindblad, P., (2000). hupS and buf)L constitute a transcription unit in the cyanobacterium *Nostoc* sp. PCC 73102. *Arch. Microbiol.*, 174: 129–133.

Lindberg, P., Schutz, K., Happe, T. and Lindblad, P., (2002). A hydrogen producing, hydrogenase-free mutant strain of *Nostoc punctiforme* ATCC 29133. *Int. J. Hydrogen Energ.*, 27: 1291–1296.

Lindblad, P., Christensson, K., Lindberg, P., Fedorov, A., Pinto, F. and Tsvgankov, A., (2002). Photoproducdon of H, by wild type Anabaena PCC 7120 and a hydrogen uptake deficient mutant: from laboratory' experiments to outdoor culture. *Int. J. Hydrogen Energ.* 27: 1271–1281.

Lindstrom, E.S., Lewis, S.M. and Pinsky, M.I., (1951). Nitrogen fixation and hydrogenase in various bacterial species. *J. Bacterial*, 61: 481–487.

Linus Pauling, (1970). *General Chemistry*, Section 15–2. San Francisco.

Lippert, K.L. and Pfenning, N., (1969). Utilization of molecular hydrogen by *Chlorobium thiosulfatophilum* growth and CO_2 fixation. *Arch. Mikrobiol.*, 65: 29–47.

Liu, G. and Shen, J., (2004). Effects of culture medium and medium conditions on hydrogen production from starch using anaerobic bacteria. *J. Biosci. Bioeng.*, 98: 251–256.

Liu, H., Cheng, S. and Logan, B.E., (2005). Power generation in fed-batch microbial fuel cells as a function of ionic strength, temperature and reactor configuration. *Environmental Science and Technology*, 39: 5488–5493.

Liu, H., Grot, S. and Logan, B.E., (2005). Electrochemically assisted microbial production of hydrogen from acetate. *Environmental Science and Technology*, 39: 4317–4320.

Liu, H., Zhang, T. and Fang, H.P.P., (2003). Thermophilic H_2 production from cellulose containing wastewater. *Biotechnol Lett.*, 25: 365–369.

Liu, W., Wang, A., Ren, N., Zhao, X., Liu, L., Yu, Z. and Lee, D., (2007). Electrochemically assisted biohydrogen production from acetate. *Energy Fuels*, 22: 159–163.

Llama, M.J., Serra, J.L., Rao, K.K. and Hall, D.O., (1979). Isolation and characterization of the hydrogenase activity of the non-heterocystous cyanobacterium *Spirulina maxima*. *FEBS Lett.*, 98: 342–346.

Lloyd, D., Ralphs, J.R. and Harris, J.C., (2002). *Giardia intestinalis*, a eukaryote without hydrogenosomes, produces hydrogen. *Microbiology*, 148: 727–733.

Lockau, W., Peteson, R.B., Wolk, C.P. and Burris, R.H., (1978). Modes of reduction of nitrogenase in heterocysts isolated from *Anabaena* species. *Biochim. Biophys. Acta*, 502: 298–308.

Logan, B.E., Oh, S.E. and Ginkel, S.V., (2002). Biological hydrogen production measured in batch anaerobic respirometer. *Environ. Sci. Technol.*, 36: 2530–2535.

Logan, B.E., Oh, S.E., Kim, I.S. and Van Ginkel, S., (2002). Biological hydrogen production measured in batch anaerobic respirometers, *Environ. Sci. Technol.*, 37: 1055–1055.

Logan, B.E., Van Ginkel, S.W. and Oh, S.A., (2003). Green and sustainable energy system built upon biological hydrogen production. *Am. Chem. Soc. Div. Fuel Chem. Prepr.*, 48(1): 227–228.

Lopes Pinto, F.A., Troshina, O. and Lindblad, P., (2002). A brief look at three decades of research on cyanobacterial hydrogen evolution. *International Journal of Hydrogen Energy*, 25: 1209–1215.

Losada, M., Nozaki, M. and Arnon, D.I., (1961). Photoproduction of molecular hydrogen from thiosulfate by *Chromatium* cells. In: *Light and Life*, (Eds.) W. McElroy and B. Class. The John Hopkins Press, Baltimore, pp. 570–575.

Lovins, A.Y. and Lovins, H., (2005). A new age of resource productivity. In: *Environmentalism and the Technologies of Tomorrow: Shaping the Next Industrial Revolution*, (Eds.) R. Olson and D. Rejeski. Washington, D.C: Island Press, pp. 129–145.

Lutz, R.J., Trujillo, M.A., Denham, K.S., Wenger, L. and Sinen-sky, M., (1992). Nucleoplasmic localization of prelamin A, implications for prenylation-dependent lamin A assembly into the nuclear lamina. *Proc. Natl. Acad. Sci. USA*, 89: 3000–3004.

Maacka, M.H., Skulasonb, J. B. (2006). "Implementing the hydrogen economy". *Journal of Cleaner Production* 14: 52–64.

Macker, B.A. and Bassham. J.A., (1988) Carbon allocation in wilds type and GLC–positive Rhodobacter sphearoides under photoheterotrophic conditions. Appl. Environ. Micribiol, 54: 2737–2741.

Macler, B.A., Pelry, R.A. and Bassham, J.A., (1979) Hydrogen formation in nearly stoichiometric amounts from glucose by a *Rhodopseudomonas sphaeroides* mutant. J. Bacteriol. 138: 446–452.

Mactsev, S.V. and Krasnovskii, A.A., (1982). Production and consumption of molecular hydrogen by isolated chloroplasts of higher plants. Fiziol. Rast. (Mosc.), 29: 951–958.

Mactsev, S.V. and Krasnovskii, A.A., (1983). Hydrogen photoproduction by chloroplast during suppression of oxygen evolving system. Fiziol. Rast. (Mosc.), 30: 915–924.

Mactsev, S.V., Ananev, G.M. and Krasnovskii, A.A., (1986). Effect of dark on hydrogen photoproduction by chloroplasts of higher plants. Biofizika, 31 L: 529–530.

Macy, J., Kulla. H. and Gottschalk, G., (1976) H_2–dependent anaerobnic growth of *Escherichia coli* on L–malate: Succinate formation. *J. Bacteriol.* 125: 423–428.

Madawar D, Garg N, and Shah V, Cyanobacterial hydrogen production, *World J Microbiol Biotechnol*, 16 (2000) 757–767.

Madigan. M.T. and Gest, H., (1978). Growth of a Photosythetic bacterium anaerobically in darkenss, suported by "Oxidant–Dependent" Sugar fermentation. *Arch. Microbiol.* 117: 119–122.

Maeda I, Miyasaka H, Umeda F, Kawase M, Yagi K. Maximization of hydrogen production ability in high–density suspension of *Rhodovulum sulfidophilum* cells using intracellular poly(3–hydroxcbutyrate) as sole substrate. *Biotechnol Bioeng* 2003, 81: 474–81.

Mah. R.A., Smith, M.R., Baresi, I., (1978). Studies on Aacetate fermenting strains of *Metanosarcinba Appl. Envrion. Microbiology* 35: 1174–1184.

Mahro, B. and Grimme, L.H., (982a). H2 photoproduction by green algae. The significant of anaerobic preincubation periods and of highlight intensities for H2 photoproductivity of *Chlorella fusca. Arch. Mikrobiol.* 132: 82–86.

Mahyudin, A. R., Furutani, Y., Nakashimada, Y. and Kakizono, T. N., Enhanced hydrogen production in altered mixed acid fermentation of glucose by Enterobacter aerogenes. *J. Ferment. Bioeng.*, 1997, 83, 358–363.

Maier T, Jacobi A, Sauter M, and Bock A. (1993). The product of the hypB gene, which is required for nickel incorporation into hydrogenases, is a novel guanine nucleotide–binding protein. *J Bacteriol* 175: 630–635.

Majizat A, Mitsunori Y, Mitsunori W, Michimasa N and Junichiro M, hydrogen as gass production from glucose and its microbial kinetics in anaerobic systems. *Water Sci Technol.* 36 (1997) 279–286.

Malcolm A. Weiss, John B. Heywood, Andreas Schafer, and Vinod K. Natarajan, Comparative Assessment of Fuel Cell Cars. MIT Report No. LFEE 2003–001, Laboratory For Energy and the Environment Reports, MIT, 2003.

Malda, J., J.C. Radway, R.W. Babcock Jr., Characterization of hydrodynamics and mass transfer in a tubular photobioreactor. U. of Hawaii, Dept. Civil Engineering Research Report No. UHM/CE/99/06.

Malki, S., Saimmaime, I., De Luca, G., Rousset, M. and Belaich, J. P., Characterization of an operon encoding an NADP–reducing hydrogenase in *Desulfovibrio fructosovorans. J. Bacteriol.*, 1995, 177, 2628–2636.

Maness, P.–C. and Weaver, P., *"Production of poly–3–hydroxyalkanoates from CO and H₂ by a Novel Photosynthetic Bacterium"*, Applied Biochemistry and Biotechnology 45/46: 395–406 (1994).

Manish S and Banerjee R, Comparison of biohydrogen production processes, *Int J Hydrogen Energy*, 33 (2008) 279–286.

Maroto–Valer, M. M., Song, C., Soong, Y. Environmental Challenges and Greenhouse Gas Control for Fossil Fuel Utilization in the 21st Century. Kluwer Academic/ Plenum Publishers, New York, 2002, 447 pp.

Marrs, B., (1974) Genetic recombinzation in *Rhodopseudomonas Capsulata Proc. Natl. Acad. Sci.*, USA 71: 971–973.

Marrs. B., (1981). Mobilization of the genes for photosynthesis from *Rhodopseudomonas Capsulata* by promiscuous plasmid. *J. Bacteriol.* 146: 1003–1012.

Martin, W. and Muller, M., The hydrogen hypothesis for the first eukaryote. *Nature,* 1998, 392, 37–41.

Masukawa H, Mochimaru M, and Sakurai H. (2002). Disruption of the uptake hydrogenase gene, but not of the bidirectional hydrogenase gene, leads to enhanced photobiological hydrogen production by the nitrogen–fixing cyanobacterium *Anabaena* sp. PCC 712(1. *Appl Microbiol Biotechnol* 58: 618–624.

Masukawa H, Mochimaru M, Sakurai H. Hydrogenase and photobi–ological hydrogen production utilizing nitrogenase enzyme system in cyanobacteria. *Int. J. Hydrogen Energy* 2002, 27: 1471–4.

Matheron. R. and Baulaique, R., (1983). Photoproduction of hydrogen from sulfru and suflide by Chromatiaceae. *Arch. Microbil* 135: 211.214.

Matsumoto, M. and Nishimura, Y., (2007). Hydrogen production by fermentation using acetic acid and lactic acid. *Journal of Bioscience and Bioengineering,* Vol. 103, pp. 236–241.

Matsunaga T, Hatano T, Yamada A, Matsumato M. Microaerobic hydrogen production by photosynthetic bacteria in a double phase photobioreactor. *Biotechnol Bioeng* 2000, 68: 647–51.

Matsunaga, T. and Mitsul, A., (1982). Seas water based hydrogen production by immobilized marine photosynthetic bacteria. In: Biotechnolgoy and Biolengineering Symposium No. 12, John Wiley, New York, pp. 441–450.

Mayhew, S.G. and O'Connor, M., (1982). Structure and mechanism of bacterial hydrogenase. Trends in Biochem. Sci., 7: 18–21.

Mcbride, C.A., Stephen, C., Togasaki, R.K., San Pietro, A., (1976). Mutational analysis of *Chlamydomonas reinhardtii*: application to biological solar energy conversion. In: Biological Solar Energy Conversion (eds. Mitsui, A., Miyachi, S., San Pietro, A. and Tamura, S.) pp. 77–86. Academic Press, New York, London.

Mctavish, H., Sayavedrra–Soto, L. A. and Arp, D. J., Substitution of Azotobacter vinelandii hydrogenase small–subunit cysteines by serines can create insensitivity to inhibition by O2 and preferentially damages H2 oxidation over H_2 evolution. *J. Bacteriol.,* 1995, 177, 3960–3964.

Meeks JC, Elhai J, Thiel T, Potts M, Larimer F, Lamerdin J, Prcd.ki P, and Atlas R. (2001). An overview of the genome of *Nostoc punctiforme*, a multicellular, symbiotic cyanobacterium. *Photosynth Res* 70: 85–106.

Mehta N, Olson J~ and Maier RJ. (2003). Characterization of lleUwbactcrpxlm nickel metabolism accessory proteins needed for maturation of both urease and hydrogenase. *J Bacteriol* 185: 726–734.

Melis A (1999). Photosystem–II damage and repair cycle in chloro–plasts: what modulates the rate of photodamage *in vivo*? *Trends Plant Sci* 4: 130–135.

Melis A and Melnicki MR, Integrated biological hydrogen production, *Int J Hydrogen Energy*, 31 (2006) 1563–1573. (75).

Melis A, Zhang L, Forestier M, Ghirardi ML, Seibert M. Sustained photohydrogen production upon reversible inactivation of oxygen evo–lution in the green algae *Chlamydomonas reindhardtii. Plant Physiol* 2000, 122: 127–35.

Melis A. Green alga hydrogen production: progress, challenges and prospects. *Int J Hydrogen Energy* 2002, 27: 1217–28.

Melis A., J. Neidhardt, I. Baroli, and J.R. Benemann, "Maximizing photosynthetic productivity and light utilization in microalgae by minimizing the light–harvesting chlorophyll antenna size of the photosystems". In *BioHydrogen*, O. Zaborsky *et al.*, eds., Plenum Press, New York pp. 41 – 52 (1998).

Melis, A. and Happe, T., Hydrogen production, green algae as a source of energy. *Plant Physiol.*, 2001, 127, 740–748.

Melis, A., J. Neidhardt and John R. Benemann, Dunaliella salina(Chlorophyta) with small chlorophyll antenna sizes exhibit higher photosynthetic productivities and photon use effeiciencies than normally pigmented cells. *J. App. Phycol. 10*: 515 – 525 (1999).

Messenbock, R.C., Dugwell, D.R., Kandiyoti, R. CO_2 and steam–gasification in a high–pressure wire–mesh reactor: the reactivity of Daw Mill coal and combustion reactivity of its chars. *Fuel*, 1999, 78, 781–793.

Meyar, J., Kelley, B.C. and Vignais, P.M., (1978b). Effect of light on nitrogenase function and synthesis in *Rhodopseudomonas Capsulata. J. Bacteriol.* 136: 201–208.

Meyer, J. and Gagnon, J., Primary structure of hydrogenase I from *Clostridium pasteurianum*. Biochemistry, 1991, 30, 9697–9704.

Meyer, J. Kelley, B.C. and Vignais, P.M., (1978c). Aerobic nitrogen Fixation by *Rhodopseudomonas Capsulata. FEBS Lett.* 85: 224–228.

Meyer, J., Kellye, B.C. and Vignais, P.M., (1978a). Nitrogen fixation and hydrogen metabolism in photosynthetic bacteria. *Biochimic* 60: 245–260.

Mi H, Endo T, Ogawa T, Asada K (1995). Thylakoid membrane–bound, NADPH–specific pyridine nucleotide dehydrogenase complex mediates cyclic electron transport in the cyanobacterium *Synechocystis* sp. PCC 6803. *Plant Cell Physiol* 36: 661–668.

Mi H, Endo T, Schreiber U, Ogawa T, Asada K (1992a). Donation of electrons from cytosolic components to the intersystem chain in the cyanobacterium *Synechococcus* sp. PCC 7002 as determined by the reduction of P700+. *Plant Cell Physiol* 33: 1099–1105.

Mi H, Endo T, Schreiber U, Ogawa T, Asada K (1992b). Electron donation from cyclic and respiratory electron flow to the pho–tosynthetic intersystem chain is mediated by pyridine nu–cleotide dehydrogenase in the cyanobacterium *Synechocystis* PCC 6803. *Plant Cell Physiol* 33: 1233–1237.

Millbank, J.W., (1981). Nitrogenase and hydrogenase in cyanophilic lichens. New Phytol. 92: 221–228.

Minnan L, Jinli H, Xiaobin W, Huijuan X, Jinzao C, Chuannan L, *et al.*, Isolation and characterization of a high H_2–producing strain *Klebsialle oxytoca* HP1 from a hot spring. *Res Microbiol* 2005, 156: 76–81.

Mishra, J., Khurana, S., Kumar, N., Ghosh, A. K. and Das, D., (2004). Molecular cloning, characterization, and over expression of a novel [Fe]–hydrogenae isolated from a high rate of hydrogen producing *Enterobacter cloacae* IIT–BT 08. *Biochemical and Biophysical Research Communications*, Vol. 324, pp. 679–85.

Mishra, J., Kumar, N., Ghosh, A. K. and Das, D., Isolation and molecular characterization of hydrogenase gene from high rate of hydrogen producing bacterial strain Enterobacter cloacae IIT BT 08. *Int. J. Hydrogen Energy*, 2002, 27, 1475–1479.

Mistui, A, Otha, Y., Frank. J., Kumazawa, Sa., Hill. C., Rosner, D., Barciella, S., Greenbaum, J., Haynes, L., Oliva, L., Dalton, P., Radway, J. and Griffard, P., (1980). Photosynthetic bacteria as an alternative energy sources: overview on hydrogen production research. In: Alternative Energy Resources, Vol. 8, *Hydrogen Energy* (ed. Veziroglu, T.N.). Hemisphere Publishing Company, Washington, D.C., pp. 3483–3510.

Mitsui, A (1975a). Photoproduction of hydrogen via microbial and biochemical processes: Symposium proceedings of Hydrogen Energy Fundamental (ed. veziroglu, T.N.) pp. S2 31–48. University of Miami Press, Miami.

Mitsui, A (1980b). Salt water based biological solar energy conversion for fuel, chemicals, fertilizer, food and feed. In: Proceedings of Bio–Energy 80. World Congress and Exposition, ed.) pp. 486–491. Bio Energy Council, Washington, D.C.

Mitsui, A. (1879a). Biological and biochemical hydrogen production In: *Solar Hydrogen energy systems* (ed. Ohta, T.) pp. 171–191. Pergamon Press, Oxford and New York.

Mitsui, A. and Black, C.C. (eds.) (1982). CRC Handbook of Biosolar Resources. Vl. 1. Basic Principles, Part 1, pp. 643. CRC Press, Boca Raton, Florida.

Mitsui, A. and Kumazawa, s., (1977). Hydrogen production by marine photosynthetic sources as a potential energy Mitsui, A., Miyachi, s., San pietro, A. and Tamura, S.).

Mitsui, A. and Kumazawa, S., (1978). Hydrogen production by marine photosynthetic organisms as a potential energy resource. In: Biological solar energy conversion (eds. Mitsui, A., Miyachi, S., San Pietro, A. and Tamura, S.). Academic Press, london and New York.

Mitsui, A. and Kumazawa, S., (1979). Solar energy biological conversion research. *J. Solar Energy* (in Japanese) 5: 55–67.

Mitsui, A. and Matsunaga, T (1983). Hydrogen production by photosynthetic microorganisms. In: biomass Energy conversion (ed. Suzuki, S.) pp. 194–228. Kodansha Scientific Press, Tokyo.

Mitsui, A., (1974). The association of photosynthetic organisms with debris of macroalgae, sea grasses and mangrove leaves. *Abstract of Annual Meeting of American Society for Microbiology*, p. 24.

Mitsui, A., (1975). Utilization of solar energy for hydrogen production by cells free systems of photosynthetic organism. Hydrogen Energy (ed. Veriroglu, T.N.) Part A, pp. 309–316. *Proceedings of the Hydrogen Economy Miami Energy Conference, University of Miami, Plenum Publishing Corpn.* New York.

Mitsui, A., (1975b). Hydrogen production via photosynthetic processes. In: Key Technologies for hydrogen Energy systems (ed. Ohta, T.) pp. 75–92. Yokohama National University Press (Revised from the Symposium Proceeding *"Hydrogen Energy Fundamentals"*).

Mitsui, A., (1975c). Utilization of solar energy for hydrogen production by cell–free system of photosynthetic organisms. In: Hydrogen Energy, Part A (ed. Veziroglu, T.N.) pp. 309–316. Plenum Publishing Corp., New York.

Mitsui, A., (1976a). Bioconversion of solar energy in salt water photosynthetic hydrogen production system. In: *Proceedings of First World Hydrogen Energy Conferences: Progress in the Worlds Enegery Projects and Flanning* (ed. Veziroglu, T.N.). University of Miami Press. Miami. Vol. 2, 4B, pp. 77–99.

Mitsui, A., (1976b). Long range concepts: Application of photosynthetic hydrogen production and nitrogen fixation research. *Proceedings of Conference on Capturing the Sun through Bioconversion.* The Washington Center for Metropolitan Studies, Washington, D.C., pp. 653–672.

Mitsui, A., (1976c). Progress reports: A survey of hydrogen producing photosynthetic organism in tropical substropical marine environments. *Proceedings of Grantee– Users Conferences on Enzymes Technology* NSF–RANN (ed. Gainer, J.L.). Univ. Virginia. pp. 39–43.

Mitsui, A., 1976 d) A survey of hydrogen producing phtosynthetic organisms in tropical and subtropical marine environmens. *NSF Annual Reports*: 1–68.

Mitsui, A., (1976). Bioconversion of solar energy in salt water photosynthetic hydrogen production jsystem. In: Proceedings of 1st World Hydrogen Energy Conference: Progress in World Hydrogen Energy Projects and Planning (ed. Veziroglu, T.N.) pp. 4B: 77–99. University of Miami Press, Vol. II.

Mitsui, A., (1978a). Bio–solar hydrogen production. In: Hydrogen Energy system, Vol. 3 (eds. Veziroglu, T.N. and Seifritz, W.) pp. 1267–1291. Pergamon Press, Oxford, England.

Mitsui, A., (1978b). Solar energy conversion to hydrogen by bacteria and algae. In: Bio–Energy (ed.). Gotib Dutteweiler–Institute. Zurich, pp. 205–236.

Mitsui, A., (1978c). Solar kenergy conversion in marine biological systems. In: *An international compendium* (ed. Veziroglu, T.N.). Hemisphere Publ. Co., Washingtn, 2: 1013–1017.

Mitsui, A., (1980a). Nitrogen fixation and hydrogen production by marine blue-green algae used as a source of food, feed and chemicals, In: *Proceedings of the international symposium on Biological Application of Solar Energy* (eds. Gnanam, A. and Kahn, J.) pp. 109–115. Viking Publ., New Delhi.

Mitsui, A., (1981). Hydrogen production by marine photosynthetic microorganisms. In: Biomass production and conversion (ed. Shibdata, K. and Kitani, O.) pp. 165–180. *Japan Scientific Society Pres*, Tokyo.

Mitsui, A., (1981). Progress reports: A study of hydrogen production by tropical marine photosynthetic bacteria for applied system.s In: *Proceedings of Solar Hydrogen Production Programme* SERI/SP–624–1095, pp. 1–19.

Mitsui, A., (1987). Photobiological production of hydrogen from water and waste. In: Proceedings of the Hydrogen photo production workshop (eds. Seki, A., Morgan, B. and Takahashi, P.) pp. 32–39.

Mitsui, A., Duerr, E., Kumazawa, S., Philips, E. and Skjoldal, H., (1979). Biological solar energy conversion: Hydrogen production and nitrogen fixation by marine blue-green algae. In: Sun II. *Proceedings of the International Solar Energy Society Silver Jubilee Congress*, Vol. I (eds. Boer, K. and Glenn, B.H.). pp. 31–35. Pergamon Press, New York and Oxford.

Mitsui, A., Kumazawa, S., Philps, E.J., Reddy, K.J., Gill, K., Renuka, B.R., Kusumi, T., Reyes–Vasques, G., Miyazawa, K., Haynes, L., Ikemoto, H., Duerr, E., Leon, C.B., Rosner, D., Sesco, R. and Moffat, E., (1985). Mass cultivation of algae and photosynthetic bacteria: Concepts and application. In: Biotechnology and bioprocess Engineering (ed. Ghosh, T.K.) pp. 119–155. International Union of Pure and Applied Chemistry and Indian National Science Academy, united Inda Press, New Delhi.

Mitsui, A., Kumazawa, S., Phlips, E.J., Reddy, K.J., Gill, K., Renuka, B.R., Kusumi, T., Reyes–Vasquez, G., Miyanawa, K., Kaynes, L., Ikemoto, H., Duerr, E., Leon, c.B., Rosner, D., Sesco, R. and Morrat, E., (1984). Mass cultivation of algae and photosynthesis bacteria: concepts and application. In: *Proceedings VIIth international biotechnology Symposium* (ed. H. Ghose). pp. 1–43.

Mitsui, A., Kumazawa, S., Takahashi, A., Ikemoto, H., Coa, S. and Arai, T., (1987). Relevance of the study of the strategy by which nitrogen fixing unicellular cyanobacteria grow phuotoautotrophically to the hydrogen photoproduction research. *J. Hydrogen Energy systems society of Japan* 12: 39–49 (in Japanese).

Mitsui, A., Matsumaga, T., Ikemoto, H. and Renuka, B.R, (1985). Organic and inorganic waste water treatment and simultaneous photoperduction of hydrogen by immobilized photosynthetic bacteria. Devleopment Indis. *Microbiol.* 26: 209–222.

Mitsui, A., Miyachi, S., San Pietro, A. and Tamura, S. (ed) (1977). Biological Solar Energy conversion. Academic Press, London and New York.

Mitsui, A., Miyachi, S., San–Pietro, A. and Tamura, S., (1977). Biological Solar Energy Conversation. Academic Press, New York, San Fransiscon London.

Mitsui, A., Philips, E.J., Kumazawa, S., Reddy, K.J., Ramachandran, S., Matsunaga, T., Haynes, L. and Ikemoto, H., (1983). Progress in research toward outdoor biological hydrogen production using solar energy, sea water and marine photosynthetic microorganisms. *Ann. N.Y. Acad. Sci.* 413: 514–530.

Mitsui, A., Rosner, D., Kumazawa, S., Barciela, S., Philps, E., Ramachandra, S., Takahashi, A. and Richard, J., (1985). Hydrogen production from salt water by marine blue green algae and solar radiation. In: *Proceedings of Twenty–Second Space congress* (ed. Haise, F.W.) pp. 11.7–11.14. Canaveral Council of Technical Societies, Revised form.

Miura Y, Saitoh C, Matsuoka S and Miyamoto K, Stably sustained hydrogen production with high molar yield through a combination of a marine green alga and a photosynthetic bacterium. *Biosci Biotechnol Biochem*, 56 (1992) 751–754.

Miura, Y., Ohta, Mano, M. and Miyamoto, K., (1987). Isolation and characterization of a unicellular marine green algae exhibiting high activity in dark hydrogen production. *Agric. Biol. Chem.* 50: 2837–2844.

Miura, Y., Yokoyama, H., Kimura, M., Iwamoto, T. and Miyamoto, K., (1981). Effect of acetylene and carbon monoxide on hydrogen uptake and evolution by nitrogen starved cultures of *Mastigocladus laminosus*. *Plant Cell Physiol.* 22: 1375–1383.

Miura, Y., Yokoyama, H., Takahara, K. and Miyamoto, K., (1982). Effects of acetylene and carbon monoxide on long term hydrogen production by *Mastigocladus laminosus*: a thermophilic blue-green alga. *J. Ferment. Technol.* 60: 411–416.

Miura, Y.H., Kanaoka, K., Saito, S., Iwasa, K., Okazaki, M. and Komemushi, S., (1980). Hydrogen evolution by a thermophilic blue-green alga *Mastigocladus laminosus*. *Plant cell physiol.* 21: 149–156.

Miyake J, Miyake M, Asada Y. Biotechnologyical hydrogen produc–tion: research for efficient light conversion. *J Biotechnol* 1999, 70: 89–101.

Miyake J, Wakayama T, Schnackenberg JS, Arai T, Asada Y. Simula–tion of daily sunlight illumination pattern for bacterial photo–hydrogen production. *J Biosci Bioeng* 1999, 88: 659–63.

Miyake, J. and kawamura, S., (1987). Efficiency of light energy conversation to hydrogen by the photsynthetic bacterium *Rhodobacter sphaeroides*. *Int. J. Hyd. Energy* 12: 147–149.

Miyake, J., Mao, X.Y. and Kawamura, S., (1985). Photoproduction of hydrogen from glucose by a coculture of a photosynthetic bacterium and *Clostridium butyricum*. *Rep. Ferment. Reds. Inst.* (Yatabe) 964: 9–18.

Miyake, J., Tomizuka, N. and Kamibayashi, A., (1982). Prolonged photohydrogen production by *Rhodospirillum rubrum*. *J. Ferment. Technol.* 60: 199–203.

Miyamoto, K. and J.R. Benemann, "Vertical Tubular Photobioreactor: Design and Operation" *Biotechnology Letters*, 10: 703 –710 (1988).

Miyamoto, K., Hallenbeck, P.C and Benemann, J.R., (1979). Hydrogen production by the thermophilic alga *Mastigocladus laminosus:* Effects of nitrogen, temperature and inhibitors of photo synthesis. *Appl. Environ. Microbiol.* 38: 440–446.

Miyamoto, K., Ohta, S., Nawa, Y., More, Y. and Miura, Y., (1987). Hydrogen production by mixed culture of green alga *Chlamydomonas reinnhardtil* and a photosynthesis bacterium *Rhodospirillum rubrum. Agric. Biol. Chem.* 51: 1319–1324.

Miyamoto, K., Takahara–Ishikawa, K., Horie, H. and Miura, Y., (1984). Dark period nitrogen fixation by *Mastigocladus laminosus* and its application to a biophotolysis system. *J. Ferment. Technol.* 62: 13–18.

Mizuno O, Ohara T, Shinya M and Noike T, Characteristics of hydrogen production from bean curd manufacturing waste by anaerobic microflora. *Water Sci Technol*, 42 (2000). 345–350.

Modigell M, Holle N. Reactor development for a biosolar hydrogen production process. *Renewable Energy* 1998, 14: 421–6.

Momirlan M, Veziroglu TN. Current status of hydrogen energy. Renew Sust Energ Rev, 2002, 6 (1–2), 141–179.

Moretra, D. and Garcia, L., Symbiosis between methanogenic archaea and proteobacteria as the origin of eukaryotes, the syntrophic hypothesis. *J. Mol. Evol.*, 1998, 47, 517–530.

Morimoto M, Atsuko M, AtifAAY, Ngan MA, Fakhru'l–Razi A, Iyuke SE, *et al.*, Biological production of hydrogen from glucose by natural anaerobic microflora. *Int J Hydrogen Energy* 2004, 29: 709–13.

Morris, P., Nash, G.V. and Hall, D.O., 1982). The stability of electron transport in vitor chloroplast membranes. *Photosynthetic Research*, 3: 227–240.

Mortensen, I.E., (1978). The role of dihydrogen and hydrogenase in nitrogen fixation. *Biochimie* 60: 219–223.

Mortensen, L.E. and Chen, J.S., (1974). Hydrogenase. In: Microbial Iron Metabolism, a Comprehensive Treatise (ed. Neilands, J.B.) p. 231 Academic Press, London and New York.

Mortensen, L.E. and Thorneley, R.N.F., (1979). structure and function of nitrogenase. *Ann. Rev. Biochem.* 48: 387–418.

Mortenson, I.E. and Chen. J.E., (1974). Hydrogenase. In: Microbial Iron Methbolism. Vol. 1, Academic Press, New York. London, pp. 231–282.

Mu Y and Yu HQ, Biological hydrogen production in a UASB reactor with granules. I: Physicochemical characteristics of hydrogen– producing granules, *Biotechnol Bioeng*, 94 (2006) 980–987.

Muallem, A. and Hall, D.O., (1982). Ascorbate as a substrate for photoproduction of photosystem I of chloroplasts. *J. Plant Physiol.* (Bathseda), 69: 1116–1120.

Mulkidjanian AY, Junge W (1996). New photosynthesis or old. *Nature* 379: 304–305.

Najafpour G, Younesi H, Mohammed AR. Effect of organic sub–strate on hydrogen production from synthesis gas using *Rho do spirillum rubrum* in batch culture. *Biochem Eng J* 2004, 21: 123–30.

Nakada E, Asada Y, Arai T, Miyake J. Light penetration into cell suspensions of photosynthetic bacteria and relation to hydrogen pro–duction. *J Ferment Bioeng* 1995, 80: 53–7.

Nakada E, Nishikata S, Asada Y and Miyake J, Photosynthetic bacterial hydrogen production combined with a fuel cell, *Int J Hydrogen Ennergy*, 24 (1999) 1053–1057.(26).

Nakamura, H. (1939). Metabolism of purple bacteria. V. Hydrogen transformations in the purple bacteria including the mutual relationship between *Thio* and athiorhodoceae. *Acta Phytochim.* (Japan) 11: 109–125.

Nakashimada, Y., Rachman, M.A., Kakizono, T. and Nishio, N., (2002). Hydrogen production of Enterobacter aerogenes altered by extracellular and intracellular redox states. *Int. J. Hydrogen Energy*, Vol. 27, pp. 1399–1405.

Nandi, R. and Sen Gupta, S., Microbial production of hydrogen: an overview. *Crit. Rev. Microbiol.*, 1998, 24, 61–84.

Nath, K. and Das, D., (2004). Improvement of fermentative hydrogen production: various approaches, *Appl Microbiol Biotechnol.*, Vol. 65, pp. 520–529.

National Research Council. (2004). *The hydrogen economy: opportunities, costs barriers and r and d, needs.* Washington: National Academy Press.

NaTrompme, T. K., Shia, R.–L., Allen, M., Eiler, J. M., Yung, Y. L. Potential environmental impact of a hydrogen economy on the stratosphere. *Science*, 2003, 300, 1740 – 1742.

Nazri, G.A. Zazri, H. Joung, R. Chen P. eds. (2004). *Materials and technology.* Warrendale: MRS.

Neidhardt J., J.R. Benemann, L. Zhang, and A. Melis, : Photosystem–II repair and chloroplast recovery from irradiance stress: relationship between chronic photoinhibition, light–harvesting chlorophyll antenna size and photosynthetic productivity in *Dunaliella salina* (green algae)." *Photosynth. Res. 56*: 175 – 184, (1998).

Nelson, L.M. and Salminen, S.O., (1982). Uptake hydrogenase activity and ATP formation in *Rhizobium leguminosarum* bacteroids. *J. Bacteriol.*, 151: 989–995.

Newton, J.W., (1976). Photoproduction of molecular hydrogen by a plant algal symbiotic system. *Science* (Wash. D.C.) 191: 559–561.

Newton. J.W. and Wilson, P.W. (1953). Nitrogen fixation and photoproduction of molecular hydrogen by Thiorhodaceae. Antonie van Leeuwenhoek, 19: 71–77.

Nicolella C, van Loosdrecht MCM and Heijen JJ, Wastewater treatment with particulate biofilm reactors, *J Biotechnol*, 80 (2000) 1–33.

Nicolet, Y., Pras, C., Legarand, P. and Fonticella, C. E., *Desul–fovibrio desulfuricans* iron hydrogenase, the structure shows un–usual coordination to an active site Fe binuclear centre. *Struct. Fold. Des.*, 1999, 7, 13–23.

Niel, G,. Nicholas, D.I.D., Bockris, J.O.M., McCann, J.F., (1976). The photosynthetic production of hydrogen. In: proceedings of the 1st World Hydrogen Conference. Vol. I I (ed. Veziroglu, T.N.) pp. 69–76, Florida, University of Miami.

Nieuwenhuis, P y P. Wells. (2003). *The automotive industry and the environment.* Boca Raton: CRC.

Norbeck, J.M., Heffel, J.W., Durbin, T.D., Tabbara, B. Bowden, J, M. and Montani, M.C. (1996). Hydrogen Fuel for Surface Transportation, Society of Automotive Engineers Inc., Warrendale, PA, p. 548.

Norling B, Zak E, Andersson B, Pakrasi H (1998). 2D–isolation of pure plasma and thylakoid membranes from the cyanobacterium *Synechocystis* sp. PCC 6803. *FEBS Lett* 436: 189–192.

Norlund, S. and Eriksson, U., (1979). Nitrogenase from *Rhodospirillum rubum:* Relation between switch effect and membrane component, hydrogen production and acetylene reduction with different nitrogenase component ratios. *Biochim. Biophys.* Acta, 547: 429–437.

Norskov, J.K y C.H. Christensen. (2006). "Toward Efficient Hydrogen Production at Surfaces". *Science* 312: 1131–1326.

Novochinskii, I., Ma, X., Song, C., Lambert, J., Shore, L., Farrauto, R. A ZnO–Based Sulfur Trap for H_2S Removal from Reformate of Hydrocarbons for Fuel Cell Applications.

nPeschek, G.A., (1979a). Anaerobic hydrogenase activity in *Anacystis nidulans:* Hydrogen dependent photoreduction and related reaction. *Biochim. Biophys. Acta* 548: 187–202.

Ochiai, H., Shibata, H., Matsuo, T., Hashinokuchi, K., Inamura, I., (1978a). Immobilization of chloroplast systems with polyvinyl alcohols. *Agri. Biol. Chem.*, 42: 683–685.

Ochiai, H., Shibata, H., Matsuo, T., Hashinokuchi, K., Sawa, Y. and Inaura, I., (1978b). Properties and availability of the immobilized chloroplast particles. *Amino acid Nucl. Acid.*, 37: 53–63.

Ochiai, H., Shibata, H., Sawa, Y. and Katoh, T., (1980). 'Living electrode' as a long–lived photoconverter for biophotolysis of water. proc. *Natl.Acad. Sci.*, USA 77: 2442–2444.

Ochiai, H., Shibata, H., Sawa, Y., Shoga, M. and Ohta, S., (1983). Properties of semiconductor electrodes coated with living filsm of cyanobacteria. *Appl. Biochem. Biotechnol.* 8: 289–303.

Odom, J.M. and Wall, J., (1983). Photoproduction of hydrogen from cellulose by an anerobic bacterial coculture. *J. Appl. Environ. Microbiol.* 45 L: 1300–1305.

Ogden JM. Prospects for building a hydrogen energy infrastructure. *Annu Rev Energ Env*, 1999, 24, 227–279.

Ogden, J. (2002)."Hydrogen: The Fuel of the future? *Physics Today* 55: 69–74.

Oh SE, Lyer P, Bruns MA and Logan BE, Biological hydrogen production using a membrane bioreactor, *Biotechnol Bioeng*, 87 (2004) 119–127.

Oh YK, Kim HJ, Park S, Kim MS and Ryu DYD, Metabolic–flux analysis of hydrogen production pathway in *Citrobacter amalonaticus* Y19, *Int J Hydrogen Energy*, 24 (2004) 173–185.

Oh, Y.K., Kim, S.H., Kim, M.S. and Park, S., Thermophilic biohydrogen production from glucose with trickling biofilter, *Biotechnol Bioeng*, 88 (2004) 690–698.

Oh, Y.K., Park M.S., Seol, E.H., Lee, S.J., Park, S. Isolation of hydrogen–producing bacteria from granular sludge of an upflow anaerobic sludge blanket reactor. *Biotechnol Bioprocess Eng* 2003, 8: 54–7.

Oh, Y.K., Scol, E.H., Kim, M.S., Park, S. Photoproduction of hydrogen from acetate by a chemoheterotrophic bacterium *Rhodopseudomonas Palustris*. *Int J Hydrogen Energy* 2004, 29: 1 115–21.

Oh, Y.K., Seol, E.H., Kim, J.R. and Park, S., Fermentative bihydrogen production by new chemoheterotrophic bacterium *Citrobacter* sp. Y19, *Int J Hydrogen Energy*, 28 (2003) 1353–1359.

Oh, Y.K., Seol, E.H., Lee, E.Y. and Park, S., Fermentative hydrogen production by chemoheterotrophic bacterium *Rhodopseudomonas Palustris*, *Int J Hydrogen Energy*, 27 (2002) 1373–1379.

Ohmiya, K., K. Sakka, T. Kimura and K. Morimoto, Application of microbial genes to recalcitrant biomass utilization and environmental conservation, *J Biosci Bioengg*, 95 (2003) 549–561. (77).

Ohta, Y. and Mitsui, A., (1981). Enhancement of hydrogen photoproduction by marine *Chromatium* sp., Miami PBS 1071 grown in molecular nitrogen. In: Advances in Biotechnology (eds. Moo–young, M. and Robinson, C.W.). Pergamon Press, Toronto, pp. 303–307.

Ohta, Y., Frank, J. and Mitsui, A. (1980). Environmental factors and substrate specificity on the growth of a hydrogen producing marine photosynthetic bacteria, *Chromatium* sp. Miami PBS 1071. IN: *Proceedings of the 3rd World Hydrogen Energy Conference* (ed. Ohta, T. and Veziroglu, N.T.). Pergamon Press, Oxford, and New York.

Okura, I., (1985). Hydrogenase and its application for photoinduced hydrogen evolution. *Coordination Chem. Rev.*, 68: 53–99.

Olson JW, and Maier RJ. (1997). The sequences of hf>F, kyfC and hypD complete the /yp gene cluster required for hydrogenase activity' in *Bradyrhizobium japonicum*. *Gene* 199: 93–99.

Omerod, H.G. and Gest, JH. (1962). Hydrogen photosynthesis and alternative metabolic pathways in Photosynthetic bacteria. *Bacteriol. Rev.* 26: 51–65.

Omerod, J.A., Omerod, K.S. Gest, H. (1961). Light dependent utilization of organic compounds and photoproduction of molecular hydrogen by photosynthetic bacteria: Relationships with nitrogen metabolism. *Arch. Biochem. Biphys.* 94: 449–463.

Onifade AA, Al–Sane NA, Al–Musallam AA, Al–Zarban S (1998). A review: potentials for biotechnological applications of keratin–degrading micro-organisms and their enzymes for nutritional improvement of feathers and other keratins as livestock feed resources. *Bioresour Technol* 66: 1–11.

Orme–Johnson, W.H., (1973). Iron–sulfur proteins: structure and function. *Annual Rev. Biochem.* 42: 159.

Ormeland, R., (1983). Hydrogen metabolism by decomposing cyanobacterial aggregates in Bio–Soda Lake, Nevada (USA). *Appl. Environ. Microbiol.* 45: 1519–1525.

Oschchepkov, V.P., Nikitina, A.A., Gusev, M.V. and Krasnovskii, A.A., (1974). Evolution of molecular hydrogen by cultures of blue-green algae. *Doklady Akademii Nauk,* SSSR 213: 557–560.

Oschchepokov, V.P., Nikitina, K.A., Gusev, M.V. and Krasnovskii, A.A., (1973). Evolution of molecular hydrogen by blue-green algal cultures. *Doklay Nauk SSSR Biol.* 213: 739–742.

Oshchepkov, V.P. and Krasnovskii, A.A., (1972). Study of hydrogen evolution during the illumination of algae (in Russian). *Fiziol. Rast.* (Mosc.) 19: 1090–1097.

Oshchepkov, V.P. and Krasnovskii, A.A., (1974). Photoproduction of molecular hydrogen by *Chlorella* action spectrum (in Russian). *Fiziol. Rast.* (Mosc.) 21: 462–467.

Oshchepkov, V.P. and Krasnovskii, A.A., (1976). Photoproduction of molecular hydrogen by green algae. Izv. Akad. *Nauk. SSSR. SER. Biol.* 1187: 11100.

Oshchepkov, V.P., Dyurian, I., Vorobeva, L.M., Chernyadev, I.I. and Abroskina, L.S., (1970). A comparative study on H_2 and CO_2 photoproduction by the mutants of green algae. *Fiziol. Rast.* (Mosc.) 25: 821–828.

Oxelfelt, F., Tamagnini, P., and Lindblad, P. (1998). Hydrogen uptake in NasiM sp. strain PCC 73102. Cloning and characterization of a hupsl homolgoue. *Arch Microbiol* 169: 267–274.

Oxelfelt F., Tamagnini P., Salema R., Lindblad P. (1995). Hydrogen uptake in *Nostoc* strain PCC 73102: effects of nickel, hydrogen, carbon and nitrogen. *Plant Physiol Biochem* 33: 617–623.

Packer, L. and Cullingford, W., (1978). Stoichiometry of H_2–plroduction by an *in vitro* chloroplast ferredoxin, hydrogenase reconstituted system. Z. *Naturforsch, Sect.* C, *Biosci.,* 33: 113–115.

Padan, E. and Schuldiner, S., (1978). Energy transduction in the photosynthetic membranes of the cyanobacterium (blue-green alga) *Plectonema boryanum. J. Biol. Chem.* 253: 3281–3286.

Paerl, H.W. and Bland, P.T., (1982). Localized tetrazolium reduction in relation to N2 fixation, CO_2 fixation, and H_2 uptake in aquatic filamentous yanobacteria. *Appl. Environ. Microbiol.* 43: 218–226.

Paerl, H.W., (1980). Ecological rationale for hydrogen metabolism during aquatic blooms of the cyanobacterium *Anabaena* Oecologia 47: 43–45.

Paerl, H.W., (1982). I situ hydrogen production and utilization by natural population of nitrogen fixing blue-green algae. *Can. J. Bot.* 60: 2542–2546.

Palazzi E, Perego P, Fabiano B. Mathematical modelling and opti–mization of hydrogen continuous production in a fixed bed bioreactor. *Chem Eng Sci* 2002, 57: 3819–30.

Palazzi, E., Fabiano, B. and Perego, P., (2000). Process development of continuous hydrogen production by Enterobacter aerogenes in a packed column reactor. *Bioprocess Eng,* Vol. 22, pp. 205–213.

Partridge, C.D.F., Walker, C.C., Yates, M.G. and Postage, J.R., (1980). The relationship between hydrogenase and nitrogenase in *Azotobacter chroococcum*. *J. Gen. Microbiol.* 119: 313–320.

Partridge, C.D.P. and Yates, M.G (1982). Effect of chelating agents on gydrogenase in Azotobacter chroococcum: Evidence that nickel is required for hydrogenase synthesis. *Biochem. J.,* 204: 339–344.

Pascal, M.G., Casse, F., Chippaux, M. and Lepelleteir, M., (1975). Genetic analysis of mutants of *Escherichi coli* K12 and *Salmnella typhimurium* LT23 deficient in hydrogenase activity. *Molec. Gen. Gent.* 141: 173–179.

Paschinger, H., (1974). A changed nitrogenase activity in *Rhodospirillum rubrum* after substitution of tungsten for molybdenum. *Arch. Microbiol.* 101: 379–389.

Peck, H. D., San Pietro, A. and Gest, H., On the mechanism of hydrogenase action. *Proc. Natl. Acad. Sci., USA,* 1956, 42, 13–23.

Peck, H.D. Jr. and Gest, H. (1956). A new procedure for assay of bacterial hydrogenase. *J. bacteriol.* 71: 70–80.

Pedro, J. S., Eyke, C. D., Wassink, H., Haaker, H., Castro, F., Robb, T. and Hagen, W. R., Enzymes of hydrogen metabolism in *Pyrococcus furiosus*. *Eur. J. Biochem.,* 2000, 267, 6541–6551.

Pedrosa, F.O. and Yates, M.G., (1983). Effect of chelating agents and nickel ions on hydrogenase activity in *Azospirllum brasilense*, A. lipoferum and Derxia gummosa. *FEMS Microbiol. Lett.,* 17: 101–106.

Peltier G, Thibault P (1988). Oxygen–exchange studies in *Chlamy–domonas* mutants deficient in photosynthetic electron trans–port: evidence for a photosystem II–dependent oxygen uptake *in vivo*. *Biochim Biophys Acta* 936: 319–324.

Pembeston, J.M. and Bowen, A.R.St. G., (1981). High freq2uency chromosome transfer in *Rhodopseudomonas sphaeroides* promoted by borad–host–range plasmid RP1 carrying mercury transposon Tn 501. *J. Bact.* 147: 110–117.

Peng Y., Stevens, P., Devos, P. and Deley, J., (1987). Relation between pH, hydrogenase and nitrogenase activity: NH_4^+ concentration and hydrogen production in cultures of *Rhodobacter sulfidophilus. J. Gen. Microbiol.* 133: 1243–1247.

Perderson, D.M., Daday, A. and Smith, G.D., (1986). The use of nickel to probe the role of hydrogen metabolism in cyanobacterial nitrogen fixation. *Biochimie* 68: 113–120.

Perego, P., Fabiano, B., Ponzano, G.P. and Palazzi, E., Experimental study of hydrogen kinetics from agricultural by- product: optimal conditions for production and fuel cell feeding, *Bioprocess Eng,* 19 (1998) 205–211.

Perei, K., Kovacs, K.L., Takacs, J., Bagi, Z. (2000). Method of decomposing keratin by microorganisms, biomass obtained by the method, keratin–decompos–ing bacterium and its extracellular protease. *Patent No.:* 224974.

Pereira, A. S., Tavares, P., Moura, I., Moura, J. J. and Huynh, B. H., Mossbauer characterization of the iron–sulphur clusters in D. vulgaris hydrogenase. *J. Am. Chem. Soc.,* 2001, 123, 2771–2782.

Perraju, B.T.V.V., Rai, A.N., Kumar, A.P. and Singh, H.N., (1986). *Cycas circinalis– Anabaena cycadeae* symbiosis: Photosynthesis and the enzymes of nitrogen and hydrogen metabolism in symbiotic and cultured *Anabaena cycadeae.* Symbiosis 1: 239–250.

Persanov, V.M., Gogotov, I.N., Gins, V.K. and Mukhin, E.N., (1977). Photoevolution of hydrogen by chloroplasts of different plants. *Fizool. Rast.* (Mosc.), 24: 699–703.

Peschek, G.A., (1978). Reduced sulfur and nitrogen compounds and molecular hydrogen as electron donors for anaerobic CO_2 photoreduction in *Anacystis nidulans. Arch. Microbiol.* 119: 313–322.

Peschek, G.A., (1979). Aerobic hydrogenase activiy in Anacystis nidulans. The oxyhydrogen reaction. *Biochimica et Biophysica Acta,* Vol. 548, pp. 203–215.

Peschek, G.A., (1979c). Evidence for two functionally distinct hydrogenases in *Anacystis nidulans. Arch. Microbiol.* 123: 81–92.

Peschek, G.A., (1979d). Nitrate and nitrite reductase and hydrogenase in *Anacystis nidulans,* grown in Fe– and Mo– deficient media. *FEMS Microbiol. Letts.* 6: 371–374.

Peschek, G.A., (1980). Electron transport reactions in respiratory particles of hydrogenase induced *Anacystis nidulans. Arch. Microbiol.* 125: 123–131.

Peters, G.A., Evans, W.R. and Toia, R.E. Jr., (1976). *Azolla Anabaena azollae* relationship IV. Photosynthetically driven, nitrogenase catalyzed H_2 production. Plant Physiol. 58: 119–126.

Peters, G.A., Toia, R.E. Jr. and Lough, S.M., (1977). *Azolla Anabaena azollae* rlationship. V. 15N2 fixation, acetylene reduction and H_2 production. Plant physiol. 59: 1021–1025.

Peters, J. W., Lanzilotta, W. N., Lemon, B. J. and Seefeldt, L. C., X–ray crystal structure of the Fe–only hydrogenase (CpI) from *Clostridium pasteurianum* to 1.8 Å resolution. *Science*, 1998, 282, 1853–1858.

Peterson, R.B. and Wolk, P.C. (1978). Localization of an uptake hvdrogenase in Anabrnna. *Plant Physiol* 61: 688–691.

Peterson, R.B. and Burris, R.H., (1978a). Hydrogen metabolism in isolated heterocysts of *Anabaena* 7120. *Arch. Microbiol.* 116: 125–132.

Peterson, R.B. and Wolk, C.P., (1978b). Localization of an uptake hydrogenase in *Anabaena*. *Plant Physiol.* 61: 688–691.

Pettersson, B., Ramsey, D. and Harrison., (2006). A review of the latest developments in electrodes for unitised regenerative polymer electrolyte fuel cells. *Journal of Power Sources* Vol. 157, pp. 28–34.

Pfennig, N. (1967). Photosynthetic bacteria. *Ann. Rev. 'Microbiol.* 21: 285–324.

Pfennig, N. (1969), *Rhodospirillum tenue* sp. n., a new species of the purple nonsulfur bacteria *J. Bacteriol.* 99: 619–620.

Pfennig, N. and Truper, H.G., (1974). The phototrophic bacteria. In: Bergey's Manula of Determinative Bacteriology (eds. Buchanan, R.E. and Gibbons, N.E.), 8th Edition, Baltimore, Williams and Wilkins, pp. 24–64.

Pfennig, N., (1977). Phototrophic green and purple bacteria, a comparative systemic survey. *Ann. Rev. Microbiol.* 31: 275–290.

Phillips, D.A., (1980). Efficiency of symbiotic nitrogen fixation in legumes. *Ann. Rev. Plant Physiol.*, 31: 29–49.

Phlips, E.J. and Mitsui, A., (1983a). Role of light intensity8 and temperature in the regulation ofhydrogen photoproduction by the marine cyanobacterium, *Oscillatoria* sp. strain Miami BG 7. *Appl. Environ. Microbiol.* 45: 1212–1220.

Phlips, E.J. and Mitsui, A., (1983b). Role of light intensity and temperature in the regulation of hydrogen potoproduction by the marine cyanobacterium *Oscillatora* sp. strain Miami BG 7. *Appl. Environ. Microbiol.* 45: 1212–1220.

Phlips, E.J. and Mitsui, A., (1984). Development of hydrogen production activity in the marine blue-green alga *Oscillatoria* sp. Miami BG 7 under natural sunlight conditions. In: Advances in Photosynthesis Research (ed. Sybesma, C.) pp. 810–804, Vol. II, Martinus/Dr. W. Junk Publishers, Hague, .

Pienkos, P.T., Bodmer, S. and Tabifa, F.R. (1983). Oxygen inactivation and recovery of nitrogenase activity in cyanobacteria.*J Bacteriol* 153: 182–90.

Pierik, A. J., Hulstein, M., Hagen, W. R. and Albracht, S. P., A low–spin iron with CN and CO as intrinsic ligands forms the core of the active site in Fe–hydrogenases. *Eur. J. Biochem.*, 1998, 258, 572–578.

Pinto FAL, Troshina O, Lindbald P. A brief look at three decades of research on cyanobacterial hydrogen evolution. *Int J Hyrdogen Energy* 2002, 27: 1209–15.

Planchard, A., Mignot, L., Jouenne, T. and Jinter, G.A., (1989). Phtoproduction of molecular hydrogen by *Rhodospirillum rubrum* immobilized in composite agar layer/microprous membrane structures. *Appl. Microbiol. Biotech.* 31: 49–54.

Plasterk, R.H.A., Rao, K.K. and Hall, D.O., (1981), Immobilization of hydrogenases for biophotlytic hydrogen production, stability and kinetics. *Biotechnol. Lett.* 3: 99–104.

Podest´a JJ, Navarro AMG, Estrella CN, Esteso MA. Electrochemical measurements of trace concentrations of biological hydrogen produced by *Enterobacteriaceae*. Inst Pasteur 1997, 148: 87–93.

Pohorelic, B. K., Voordouw, J. K., Lojou, E., Dolla, A., Harder, J. and Voordouw, G., Effects of deletion of genes encoding Fe–only hydrogenase of *Desulfovibrio vulgaris* Hildenborough on hydro–gen and lactate metabolism. *J. Bacteriol.*, 2002, 184, 679–686.

Polle, J., S. Kanakagiri, J.R. Benemann and A. Melis. "Maximizing Photosynthetic Efficiencies and Hydrogen Production in Microalga Cultures". Proc. Annual Rev. Meeting, U.S. DOE May, 1999, and *Proc.Workshop on Biohydrogen 99*, June 22–23, Tsukuba, Japan, in press (2000).

Portman, N., Hurley, I. and Woodward, J. A., A simple method to demonstrate the enzymatic production of hydrogen from sugar. *J. Chem. Educ.*, 1998, 75, 1270–1274.

Powell, J.C., Peters, M.D., Ruddell, A. and Halliday, J. (2002). *Fuel Cells for a Sustainable Future?* Tyndall Working Paper TWP 50. Tyndall Centre for Climate Change: http://www.tyndall.ac.uk/publications/working_papers/wp50.pdf (accessed October 2006).

Prasad, P., Pandey, K.D. and Kashyap, A.K., (1988). Optimization of hydrogen photoproduction by blue-green algae. In: Hydrogen Energy Workshop cum Review Committee Meeting, Department of Physics, Banaras Hindu University, Varansi, January 29–30.

Presses Polytechnique Romandes, Lausanne, pp. 1–48.

Proceedings of Topical Conference on Fuel Cell Technology. AIChE (2002). Spring National Meeting, New Orleans, March 11–14, 2002, pp. 98–105.

Quadri, S.M.H. and Hoare, D.S. (1968). Formic hydrogenlyase and the photoassimilation of formate by a strain of *Rhodopseudomonas palustris*. J. Bacteriol. 95 ¨2344–2357.

Rachman, M.A., Furutani, Y., Nakashimada, Y., Kakizono, T. and Nishio, N. Enhanced hydrogen production in altered mixed acid fermentation of glucose by *Enterobacter aerogenes*, *J Ferment Bioeng*, 83 (1997) 358–363.

Rachman, M.A., Nakashimada, Y., Kakizono, T. and Nishio, N., Hydrogen production with high yield and high evolution rate by self– flocculated cells of *Enterobacter aergenes* in a packed bed reactor, *Appl Microbiol Biotechnol*, 49(1998) 450–454.

Radway, J.C., B.A. Yozua, J.R. Benemann, G. Chini Zitelli, J. Malda, R.W. Babcock Jr., M.R. Tredici "Evaluation of a near–horizontal tubular photobioreactor system in Hawaii", *Abstracts 8th International Conference on Applied Algology*, Montecassini, Italy, September 1999 (1999).

Rainbird, R.M., Atkins, C.A., Pate, J.S. and Sanford, P., (1981). Significance of hydrogen evolution in the carbon and nitrogen economy of nodulated cowpea *Vigna unquiculata* cultivar calcona. Plant Physiol., 71: 122–127.

Ramachandran, S. and Mitsui, A., (1984). Effect of seawater quality on biomass and hydrogen photoproduction by a marine blue-green alga *Oscillatoria* sp. Miami BG 7 In: Advances in Photosynthesis Research (ed. Sybesma, C.) Vol. II, pp. 805–808. Martinus Nijhff/Dr. W.Junk Publishers. Hague.

Ramachandran, S. and Mitsui, A., (1986). Biosystem catalyzed continuous hydrogen photoproduction from sea water using solar energy. In: Proc. 6th International symp. Hydrogen Energy (ed. Veziroglu, T.N.) pp. 1371–1381.

Ramana, Ch. V., (1989). Screening of blue-green algae for biomass and photoproduction of hydrogen with particular reference to *Synechococcus* sp. Ph.D. thesis, Osmani University, Hyderabad.

Ramasubramanian, T.S., Wci, T.F., and Golden, J.W. (1994). Two Anabaena sp. Strain PCC 7120 DNA–binding factors interact with vegetative cell– and hetcrocyst–specific genes. *J Bacteriol* 176: 1214–1223.

Randt, C. and Senger, H., (1985). Participation of the two photosystem in light dependent hydrogen evolution in *Scenedesmus obliquus*. Photochem. Photochem. Photobiol. 42: 553–558.

Randt, C., Strecker, V. and Scenger, H., (1985). Changes in hydrogen metabolism during greening of pigment mutant C–2A from *Scenedesmus obliquus*. *Plant Cell Physiol.* 26: 1111–1118.

Rao, G. and Mutharasan, R., (1988). Altered electron flow in a reducing environment in *Clostridium acetobutylicum*. *Biotechnol. Lett.* 10: 129–132.

Rao, K.K. and Hall, D.O., (1979). Hydrogen production from isolated chloroplasts. Topics in photosynthesis, 3: 299–329.

Rao, K.K. and Hall, D.O., (1983). Photobiological production of fuels and chemicals. In: Photochemical Conversions (ed. Brzun, A.M.).

Rao, K.K. and Hall, D.O., (1983). Photobiological production of fuels and chemicals. In: Photochemical conversions (ed. Braun, A.M.) pp. 1–48, Presses Polytechnique Romandes, Laussane.

Rao, K.K. and Hall, D.O., (1988). Hydrogenases: Isolation and assay. Methods Enzymol. 167: 501–509.

Rao, K.K., Bruce, D.L., Gisby, P.E., Muallem, A. and Hall, D.O., (1982). Biophotolysis of water for hydrogen production via natural and artificial catalytic systems. In: *Photochemical, Photoelectrochemical and Photobiological Processes*, Vol. I (eds.

Hall, D.O. and Palz, W.) pp. 195–212, Reidel Publishing Company, Dordrecht, The Netherlands.

Rao, K.K., Gogotov, I.N. and Hall, D.O., (1978). Hydrogen evolution by chloroplast hydrogenase systems: Improvement and additional observations. *Biochemie* 60: 291–296.

Rao, K.K., Rosa, L. and Hall, D.O., (1976). Prolonged production of hydrogen gas by a chloroplast– bioeatalytic system. *Biochem. Biophys. Res. Commun.*, 68: 21–28.

Rao, R., Banerjee, M., Kumar, A. and Kumar, H.D., (1987). Hydrogen supported nitrogenase activity in two cyanobacteria. *Current. Microbiol.* 16: 75–78.

Reddy, K.J. and Mitsui, A., (1984). Simultaneous production of hydrogen and oxygen as affected by light intensity in unicelluar aerobic nitrogen fixing blue-green alga *Synechococcus* sp. Miami BG 043511. In: Advances in Photosynthetic Research (ed. Sybesma, C.) Vol. II, pp. 785–788. Martinus Nijhoff/Dr W Junk Publishers, hague.

Reeves, M. and Greenbaum, E., (1985). Long term endurance and selection studies in hydrogen and oxygen photoproduction by *Chlamydomonas reinhardtii*. *Arch. Biochem. Biophys.* 213: 37–44.

Reeves, S.G., Rao, K.K., Rosa, L. and Hall, D.O., (1977). Biocatalytic production of hydrogen. In: Microbial Energey conversion (eds. Schlegel, H.G. and Barnea, J.). Pergamon Press, Oxford, New York, pp. 235–243.

Reissniann, S., Hochleitner E, Wang H, Paschos A, Lottspeich F, Glass RS, and Bock A. (2003). Taming of a Poison: Biosynthesis of the NiFe–Hydrogenase Cyanide Ligands. Science 299: 1067–1070.

Resnik, D. (2007). *The price of truth.* London: Oxford University Press.

Rey L, Imperial J, Palacios JM, and Ruiz–Argueso T. (1994). Purification of *Rhizobium leguminosarum* Ipewninosarum HypB, a nickel–binding protein required for hydrogenase synthesis. *J Bacteriol* 176: 6066–6073.

Rey L, Murillo J, Hernando Y, Hidalgo E, Cabrera E, Imperial], and Ruiz–Argueso T. (1993). Molecular analysis of a microaei–obically induced opeton required for hydrogenase synthesis in Rhi~pbium lenrtpnosairtm biovar viciae. *Mol. Microbiol* 8: 471–481.

Rieder, R., Cammack, R. and Hall, D.O., (1985). Purification and properties of cytoplasmic hydrogenase from Desulfovibrio desulfuricans (strain Norway 4). *Eur. J. BioChe.*, 145: 637–643.

Rifkin, J. (2002). *The hydrogen econom: the creation of the worldwide energy and the redistribution of power on earth.* New York: Putnam.

Rippka R, Deruelles J, Waterbury JB, Herdman M, Stanier R (1979). Generic assignments, strain histories and properties of pure cultures of cyanobacteria. *J Gen Microbiol* 111: 1–61.

Robinson, P.K., Mak, A.L., Trevan, M.D., (1986). Immobilized algae: a review. *Process Biochem* (Aug.) 122–127.

Robson, R.L. and Postgate, J.R., (1980). Oxygen and hydrogen in biological nitrogen fixation. *Annual Review of Microbiology*, Vol. 34, pp.183–207.

Rocheleau, R., S. Turn, Y. Nemoto, O.Zaborsky and J. Radway, Sustainable Bioreactor Systems for Hydrogen Production. Annual DOE Hydrogen R and D Program Review Meeting, 1999.

Rohrback, G.H., Scott, W.R. and Canfield, J.H. (1962). Biochemical fuel cell. Proc of the 16th Annual power sources conferences, pp. 18–21.

Romm, J.J. (2004). *The Hype about hydrogen, fact and fiction in the race to save the climate.* Washington: Island Press.

Rossler, P. and Lien, S., (1982). Anionic modulation of the catalytic activity of hydrogenase from *Chlamydomonas reinhardtii. Arch. Biochem. Biophys.* 213: 37–44.

Rossler, P. and Lien, S., (1984). Effect of electron mediator charge properties on the reaction kinetics of hydrogenase from *Chlamydomonas reinhardtii. Arch. Biochem. Biophys.* 230: 103–109.

Rostrup–Nielsen, J. R., Sehested, J., Udengaard, N. Paper presented at Am. Chem. Soc. Symp. H_2 Energy for the 21st Century, March 23–27, 2003, New Orleans, LA, Gunardson, H. Industrial Gases in Petrochemical Processing. Marcel Dekker, New York, 1998, 283 pp, .

Rostrup–Nielsen, J.R. and Rostrup–Nielsen, T. Large–scale Hydrogen Production., Cattech., (2002), 6, 150–159.

Roualt, T. A. and Klausner, R. D., Iron–sulfur clusters as biosensors of oxidants and iron. *Trends Biochem. Sci.*, 1996, 21, 174– 177.

Roy Chowdhury, S., Cox, D. and Levandowsky, M., Production of hydrogen by microbial fermentation. *Int. J. Hydrogen Energy*, 1988, 13, 407–410.

Rozendal, R. A., Hamelers, H. V. M., Euyerink, G.J.W. , Metz, S.J. and Buisman, C. J. N.(2006). Principles and perspectives of hydrogen production through biocatalyzed electrolysis. *Int. J. of Hydrogen Energy*, Vol. 31, pp. 1632–1640.

Rozendal, R., Jeremiasse, A., Hamelers, H. and Buisman, C., (2008). Hydrogen production with a microbial biocathode. *Environmental Science and Technology*, Vol. 42, pp. 629–634 Sawers, R.G., 2005). Formate and its role in hydrogen production in *Escherichia coli. Biochemistry Society Transactions*, Vol. 33, pp. 42–46.

RTD info Magazine on European Research 42. The hydrogen fairy pp 1–11 (August 2004).

Ruiz–Argueso, T., Hanus, J. and Evans, H.J., (1978). Hydrogen production and uptake by pea nodules as affected by strains of Rhizobium leguminosarum. *Arch. Microbiol.*, 116: 113–118.

Russel M. (2003). "Liability and the hydrogen economy". *Science* 301: 47.

Salih, F.M., (1989). Improvement of hydrogen photproduction form *E. coli* pretreated cheese whey. *Int. J. Hydrogen Energy* 14: 661–663.

San Pietro, A., (1977). Hydrogen formation from water by photosynthesis and artificial systems. In: Microbial Energy Conversion (eds. Schlegel, H.G. and Barnea, J.). Pergamon Press, Oxford, New York, pp. 217–233.

Santos, C.P. and Hall, D.O., (1982). Thylakoid polypeptides of light and dark aged chloroplasts. *Plant Physiol.*, 79: 795–802.

Sasikala, K. and Ramana, Ch. V., (1989). Photoproduction of hydrogen by photosynthetic purple non–sulphur bacteria: I. Isolation, characterization, identification and growth of *Rhodobacter sphaeroides* OU 001. *Ind. Natl. Acad. Sci.*, B (in press).

Scherer, S., Kerfin, W. and Boger, P., (1980a). Regulatory effect of hydrogen on nitrogenase activity8 of the blue-green alga (cyanobacterium) *Nostoc muscorum J. Bacteriol.* 141: 1037–1040.

Scherer, S., Kerfin, W. and Boger, P., (1980b). Increase of nitrogenase activity in the blue green alga *Nostoc muscorum* (cyanobacterium) *J. Bacteriol.* 144: 1017–1023.

Schick, H.J., (1971a). Substrate and light dependent fixation of molecular hydrogen in *Rhodospirillum rubrum. Arch. Mikrobiol.* 75: 89–101.

Schick, H.J., (1971b). Interrelationship of nitrogen fixation hydrogen evolution and photoreduction in *Rhodospirillum rubrum. Arch. Mikrobiol.* 75: 102–109.

Schink, B. and Schlegel, H. G., (1979). The membrane bound hydrogenase of Alkaligenes eutrophus. I. Solubilization, purification and biochemical properties. *Biochim. Biophys. Acta,* 567: 315–324.

Schink, B. and Schlegel, H. G., (1979). The membrane bound hydrogenase of Alkaligenes eutrophus. I. Solubilization, purification and biochemical properties. *Biochim. Biophys. Acta,* 567: 315–324.

Schlegel, H.G. and Barnea, H. (ed) (1976). Introductory report: distribution and physiological role of hydrogenases in microorganisms. In: Hydrogenases: their catalytic Activity8 Structure and Function (eds. Schlegel, H.G. and Schneider, K.). p. 15. Goltze, Gottingen.

Schlegel, H.G. and Maria, M., (1985). Isolation of hydrogenase regulatory mutants of hydrogen oxidizing bacteria by a colony–screening method. *Arch. Microbiol.* 141: 377–383.

Schlegel, H.G. and Schneider, K. (eds.) (1978). Hydrogenases and their Catalytic Activity, Structure and Function. Erich Galtze, Gottingen, K.G.

Schlegel, H.G. and Schneider, K.(1978). Introductory report: distribution and physiological role of hydrogenases in microorganisms. In *Hydrogenases:* The Shin, J.H., Yoon, J.H., Sim, S.J., Kim, M.S. and Park, T.H., 2007). Fermentative hydrogen production by the newly isolated Enterobacter asburiae SNU–1. *Int. J. Hydrogen Energy,* Vol. 32, pp.192–199.

Schlegel, H.G., (1974). Production, modification and consumption of atmospheric trace gases by microorganisms. Tellus 26: 11–20.

Schmetterer G (1994). Cyanobacterial respiration. In: Bryant DA (ed) The molecular biology of cyanobacteria. Kluwer, Dor–drecht, pp. 409–435.

Schmitz O, and Bothe H. (1996). NAD(P)+–dependent hydrogenase activlt~ m extracts from the cyanobacterium *Anabaena inaequalis*. *FEMS Microbiol Lett* 135: 97–101.

Schmitz, O., Boison G, Hilscher R, Hundeshagen B, Zimmer W, Lottspeich F, Bothe H (1995). Molecular biological analysis of a bidirectional hydrogenase from cyanobacteria. *Eur. J. Bio–chem.* 233: 266–276.

Schmitz, O., Boison G, HUscher R, Hundeshagen B, Ziinmcr W, Lottspeich F, and Rothe H. (1995). Molecular biological analysis of a bidirectional hyd–rogenase from cyanobacteria. *Eur J Biochem.* 233: 266–276.

Schmitz, O., Boison, G., and Botlie 11. (2001). Quantitative analysis of expression. of two ciTcadian dock–coatroUed gene dusters coding for the bi–directional bydrogenase in the cyanobacterium *Synechocuccus* sp. PCC 7942. *Mol Microbiol.* 41: 1409–1417.

Schmitz, O., Bothe H (1996). The diaphorase subunit HoxU of the bidirectional hydrogenase as electron transferring protein in cyanobacterial respiration. *Naturwissenschaften* 83: 525–527.

Schmitz, O., Boison, G., Salxtnan, H., Bothe, H. Schiite, K., and Happe, T. (2002). HoxE–a subanit specific for the petameric bidirection hydrogenase coplex (HosEI, ~TI'H)) of cyanobacteria. *Biochem BiophysActa* 1554: 66–74.

Schnackenberg, J., Ikemoto, H., Miyachi, S. Photosynthesis and hydrogen evolution under stress conditions in a CO_2–tolerant marine green alga, *Chlorococcum littorale. J Photochem Photobiol B: Biol* 1996, 34: 59–62.

Schneider, K. and Schlegel, H.G., (1976). Purification and Properties of soluble hydrogenase from *Alkaligenes eutrophus* H 16. *Biochim. Biophys. Acta* 452: 66–80.

Schneider, K. and Schlegel, H.G., (1977). Localization and stability of hydrogenase from aerobgic hydrogen bacteria. *Arch. Microbiol.* 112: 229–238.

Schneider, K., Cammack, R. and Schlegel, H.G., (1984b). Content and localization of FMN, Fe–S clusters and nickel in the NAD linked hydrogenase in Nocardia opaca 1b. *Eur. J. Biochem.,* 142: 75–84.

Schneider, K., Cammack, R., Schlegel, H.G. and Hall, D.O., (1979). The iron sulfur cluster of soluble hydrogenase from Alkaligenes eutrophus. *Biochim. Biophys. Acta,* 578: 445–461.

Schneider, K., Johim, K. and Schlegel, H.G., (1984a). Effect of nickel on activity and subunit composition of the purified hydrogenase from Nocardia opaca 1b. *Eur. J. Biochem.,* 138: 533–541.

Schobert, H.H., Song, C. Chemicals and Materials from Coal in the 21st Century. *Fuel,* (2002). 81 (1), 15–32.

Schoenmaker, G.S., Oltman, L.F. and Stouthamer, A.H., (1979). Purification and properties of the membrane– bound hydrogenase from Proteus mirabilis. *Biochim. Biophys. Acta,* 567: 511–521.

Schon, G. (1968). Fructoseverwertung and Bacteriochlorophyll synthese in anaeroben Dunkel and Lichtkulturen von *Rhodospirillum rubrum. Arch. Microbiol.* 63: 362–375.

Schon, G. and Biedermann, M., 1973). Growth and adaptive hydrogen production of *Rhodospirillum rubrum* (F) in anaerobic dark cultures. *Biochim. Biophys. Acta* 304: 65–75.

Schroder, C., Selig, M. and Schonheit, P., Glucose fermentation to acetate, CO_2 and H_2 in the hyperthermophilic eubacterium *Thrmotoga –maritima–* involvement of the Embden –Meyerhof pathway, *Arc Microbiol*, 161(1994) 460–470.

Schubert, K.R., Jennings, N.T. and Evans, H.J., (1978). Hydrogen reactions of nodulated plants. II. Effects on dry matter accumulation and nitrogen fixation. *Plant Physiol.*, 61: 398–401.

Schultsz, J.E. and Weaver, P.F., (1982). Fermentation and aerobic respiration by *Rhodospirillum capsulata. J. Bacteriol.* 149: 181–190.

Scientific American 42. Questions about a hydrogen economy pp 41–47 (May 2004).

Scolnik, P.A. and Haselkorn, R., (1984). Activation of extra copies of genes coding for nitrogenase in *Rhodopseudomonas capsulata. Nature*, London 307: 289–292.

Scolnik, P.A., Virosco, J. and Haselkorn, R., (1983). The Wild–type gene for glutamine synthetase restores ammonia control of nitrogen fixation to Gln (gln A) mutants of *Rhodopseudomonas capsulata. J. Bacteriol.* 155: 180–185.

Scrantn, M.I, Novelli, P.C., Michaels, A. Horrigan, S.G. and Carpenter, E.J., (1987). Hydrogen roduction and nitrogen fixation by *Oscillatoria thiebautii* iduring *in situ* incubations. *Limnol. Oceanogr.* 32: 998–1006.

Sellstedt, A., (1986). Acetylene reduction, hydrogen evolution and nitrogen15 fixation in the *Alnus incana* and *Frankia* symbiosis. *Planta*, 167: 382–386.

Sellstedt, A., (1988). Nitrogenase activity, hydrogen evolution and biomass production in different Casuarina symbioses. *Plant and Soil*, 105: 33–40.

Sellstedt, A., Hoss–Dannel, K. and Ahlquist, A.S., (1986). Nitrogen fixation and biomass production in symbioses between *Alnus incana* and *Frankia* strains with different hydrogen metabolism. *Physiol. Plant.*, 66: 99–107.

Seol, W.G. and Kho, Y.H., (1986). Relationship between hydrogenase and nitrogenase for hydrogen evolution in *Rhodopseudomonas* sp. KCTC 1437. *Korean J. Apl. Microbiol. Bioeng.* 14: 385–390.

Serebryakova, L.T., Zorin, N.A. and Lindblad, P. (1994). Reversible hydrogenase in *Anabaena variables* ATCC 29413: presence and localization in non–N., fixing cells. *Arch Microbiol* 161: 140–144.

Serebryakova, L.T., Zorin, N.A., Lindblad, P. (1994). Reversible hydro–genase in *A. variabilis* ATCC 29413. *Arch Microbiol* 161: 140– 144.

Serebryakova, L.T., Teslya, E.A., Gogotov, I.N. and Kondratieva, E.N., (1980). Nitrogenase and hydrogenase activity of non–sulfur purple bacteria *Rhodopseudomonas capsulata. Microbiologiya* 49: 401–407.

Serehryakova, L.T., Sheremetieva, M., and Lindblad, P. (2000). H, –uptake and evolution in the unicellular cyanobacterium *Chroococciduipsis thermus* CALTJ 758. *Plant Physiol Biochem* 38: 525–530.

Serra, J., Llama, M.J. and Hall, D.O., (1984). Comparison of the properties of two hydrogenases from the photosynthetic bacterium *Chromatium vinosum*. *Arch. Biochem. Biophys.* 234: 73–81.

Shah, V., Garg, N., Madamwar, D. Ultrastructure of the fresh water cyanobacterium *Anabaena variabilis* SPU 003 and its applica–tion for oxygen–free hydrogen production. *FEMS Microbiol Lett* 2001, 194: 71–5.

Sheremetieva, M., Troshina, O., Serebryakova, L.T. and Lindblad, P. (2002). Identification of bnrgenes and analysis of their transcription in the unicellular cyanobacte.aum Ghewapsa alptcila CALU 743 growing under nitrate–limited conditions. *FEMS Microbiol Lett* 214: 229–233.

Shi X and Yu H, Continuous Production of hydrogen from mixed volatile fatty acids with *Rhodopseudomonas capsulata*, *Int J Hydrogen Energy*, 31 (2006) 1641–1647.

Shi XY, Yu HQ. Hydrogen production from propionate by *Rhodopseu–domonas capsulate*. *Appl Biochem Biotechnol* 2004, 1 17: 143–54.

Shi XY, Yu HQ. Response surface analysis on the effect of cell con–centration and light intensity on hydrogen production by *Rhodopseu–domonas capsulate*. *Process Biochem* 2005, 40: 2475–81.

Shi, D–J., Brouers, M., Hall, D.O. and Robins, R.J., (1987). The effects of immobilization on the biochemical, physiolgical and morphological features of *Anabaena azollae*, *Planta* 172: 298–308.

Shin, H.S., Youn, J.H., Kim, S.H. Hydrogen production from food waste in anaerobic mesophilic and thermophilic acidogenesis. *Int. J. Hydrogen Energy* 2004, 29: 1355–63.

Shinnar, R. (2005). "The mirage of the H2 economy". In *Water Encyclopedia*, J.Lehr y J. Keeley, eds., 4, pp. 477–480. New York: Wiley.

Show, K,Y., Zhang, Z.P., Tay, J.H., Liang, D.T., Lee, D.J., Jiang, W.J., Production of hydrogen in a granular sludge– based anaerobic stirred tank reactor, *Int. J. Hydrogen Energy*, 32 (2007) 185–191.

Siebert, M., Lien, S. and Weaver, P.S., (1980). Photobiological production of hydrogen – A solar energy conversion option. SERI/TR–33–133, Golden Colorado.

Siefert, E. and Pfennig, N., (1978). Hydrogen metabolism and nitrogen fixation in wild type and nif mutants of *Rhodopseudomonas acidophilla*. *Biochimie* 60: 261–265.

Silver, W.S., (1971). Physiological chemistry of non–legume symbiosis. In: The Chemistry and Biochemistry of Nitrogen Fixation (ed. Postage, J.R.) pp. 245–281. Plenum Press, London–New York.

Simon, H., Egerer, P. and Gunther, H., (1978). Some mechanistic aspects and immobilization of soluble hydrogenase from *Alkaligenes eutrophus*. In: Hydrogenases: Their Catalytic Activity, Structure and Function (eds. Schlegel, H.G. and Schneider, K.). Eric Goltze KG, Gottingen, p. 235–251.

Sindhu, S.S., Dadarwal, K.R. and Dahiya, B.S., (1986). Hydrogen evolution and relative efficiency in chick pea (*Cicer arietinum*): Effect of *Rhizobium* strain, host cultivar and temperature. *Ind. J. Exp. Biol.*, 24: 416–420.

Singh, S.P., Srivastava, S.C., Pandey, K.D. Hydrogen–production by *Rhodopseudomonas* at the expense of vegetable starch, sugar juice and whey. *Int. J. Hydrogen Energy* 1994, 19: 437–40.

Singh, S. and Kashyap, A.K., (1988). Photoevolution of hydrogen during oxygenic photosynthesis of cyanobacterium *Nostoc muscorum. Biotech Lett.* 10: 921–925.

Singh, S.P., Srivastava, S.C. and Pandey, K.D., (1990). Photoproduction of hydrogen by a non–sulphur bacterium isolated from root zones of water ferm *Azolla pinnata. Int. J. Hydrogen Energy*, 15: 403–406.

Singh, Surendra and Kashyap, A.K., (1988). Photoevolution of hydrogen during oxygenic photosynthesis of cyanobacterium *Nostoc muscorum. Biotechnol. Lett.* 10: 921–925.

Smith R.L, Maguire ME. (1998). Microbia! magnesium transport: unusual transporters searching for identity, *Mol. Microbiol* 28: 217–26.

Smith, B.E., Campbell, F., Eady, R.R., Eldridge, M., Ford, C.M., Hill, S., Kavangh, E.P., Lowe, D.J. Miller, R.W., Richardson, T. H., Robson, R.L., Thornley, R.N.F. and yates, M.G., (1987). Biochemistry of nitrogenase and the physiology of related metabolism. *Phil. Trans. Roy. Soc. Lond.* B 317: 146.

Smith, G.D. and Lambert, G.R., (1981). An outdoor biophotolytic system using the cyanobacterium, *Anabaena cylindrica* B 629. *Biotechnol. Bioeng.* 23: 213–220.

Smith, G.D., Mualiem, A. and Hall, D.O., (1982). Hydrogenase catalyzed photoproduction of hydrogen by photosystem I of the thermophilic blue-green algae *Mastigocladus laminosus* and Phormidium laminosum. Photobiochem. Photobiophys. 4: 307–320.

Smith, R.L., Kumar, D., Xiankong, Z., Tabita, F.R. and Van Baalen, C., (1985). H2, N2 and O2 metabolism by isolated heterocysts from *Anabaena* sp. strain CA. *J. Bacteriol.* 162: 565–570.

Smith, R.V., Noy, R.J. and Evans, M.C.W., (1971). Physiological electron donors to the nitrogenase of the blue-green alga Anabaena cylindrica. *Biochim. Biophys. Acta*, 253: 104–109.

Smith, W.N. y J.D. Santangelo, eds. (1980). *Hydrogen production and marketing* Washington D.C.: American Chemical Society.

Sodimkova, M. (1969). Physiologische und biochemische charakterisierung einiger *Coelastrum Arten. Arch. Protistenkd.* 111: 223.

Solioz, M., Yen, H.C. and Marrs, B., (1975). Release and uptake of gene transfer agent by *Rhodopseudomonas capsulata. J. Bacteriol.* 123: 651–657.

Song, C. Catalytic Research for Clean Energy and Ultra–Clean Fuels in the 21st Century. Future Perspectives. Acta Petrolei Sinica (Petroleum Processing Section, China), 2002d, 18 (6), 1–19.

Song, C. CO$_2$ Conversion and Utilization: An Overview. In "CO$_2$ Conversion and Utilization". ACS Symposium Series, Vol. 809, Edited by C. Song, A. M. Gaffney, K. Fujimoto. American Chemical Society, Washington DC, 2002b, Chapter 1, pp. 2–30.

Song, C. Fuel Processing for Low–Temperature and High–Temperature Fuel Cells. Challenges, and Opportunities for Sustainable Development in the 21st Century. Catalysis Today, 2002, 77 (1), 17–50.

Song, C. Future Perspectives of Catalytic Research for Clean Energy and Ultra–Clean Fuels in the 21st Century. *Energy and Resources* (Japan), 2001, 22 (1), 77–81.

Song, C. Recent Advances in Selective Conversion of Polycyclic Hydrocarbons into Specialty Chemicals over Zeolite Catalysts. Catalysis, Royal Society of Chemistry, 2002c, 16, 272–321.

Song, C. Tri–reforming: A New Process for Reducing CO$_2$ Emissions. Chemical Innovation (formerly Chemtech, ACS), 2001, 31, (1), 21–26.

Song, C., Gaffney, A. M., Fujimoto, K. CO$_2$ Conversion and Utilization. American Chemical Society (ACS), Washington DC, ACS Symposium Series, Vol. 809, 2002, 448 pp.

Song, C., Pan, W., Srimat, S. T. Tri–reforming of Natural Gas Using CO$_2$ in Flue Gas of Power Plants without CO$_2$ Pre–separation for Production of Synthesis Gas with Desired H$_2$/CO Ratios. In "Environmental Challenges and Greenhouse Gas Control for Fossil Fuel Utilization in the 21st Century". Edited by M. M. Maroto–Valer, C. Song, and Y. Soong.

Song, C., Saini, A. K., Yoneyama, Y. A New Process for Catalytic Liquefaction of Coal Using Dispersed MoS2 Catalyst Generated *In–Situ* with Added H$_2$O. Fuel, 2000, 79 (3), 249–261.

Song, C., Schmitz, A. D. Zeolite–Supported Pd and Pt Catalysts for Low–Temperature Hydrogenation of Naphthalene in the Absence and Presence of Benzothiophene. Energy and Fuels, 1997, 11 (3), 656–661.

Song, C., Schobert, H. H. Non–fuel Uses of Coals and Synthesis of Chemicals and Materials. *Fuel*, 1996, 75 (6), 724–736.

Song, C., Schobert, H. H. Opportunities for Developing Specialty Chemicals and Advanced Materials from Coals. *Fuel Processing Technology*, 1993, 34 (2), 157–196.

Song, H.Y., Hanchi, C. Morgan, W., Binging, C. and Baolin, Y., (1980). Relationship between hydrogenase and nitrogenase in hydrogen metabolism of photosynthetic bacterium *Rhodopseudomonas capsulata*. *Plant Physiol*. (Acad. Scinica 23: 252–260.

Sorensen, B. (2005). *Hydrogen and fuel cells*. Burlington: Elsevier Academic Press.

Sparling R, Risbey D and Poggi– Varaldo HM, Hydrogen production from inhibited anaerobic composters, *Int. J. Hydrogen Energy*, 22 (1997) 563–566.

Sperling, D. y J. Ogden. (2004). "The Hope for Hydrogen". *Science and Technology*, 2: 82–86.

Sperling, D. y J.S. Cannon. (2004). *The Hydrogen energy transition. Morning toward the post petroleum age in transportation*. Burlington: Elsevier Academic Press.

Spiller, H., Bookjans, G. and Shanmugam, K.T., (1983). Regulation of hydrogenase activity in vegetative cells of *Anabaena variabilis. J. Bacteriol.* 155: 129–137.

Spiller, H., Ernst, E., Kerfin, W. and Boger, P., (1978). Increase and stabilization of photoproduction of hydrogen in *Nostoc muscorum* by photosynthetic electron transport inhibitors. Z. Naturforsch. 33c: 541–547.

Spruit, C.J.P. (1958). Simultaneous photoproduction of hydrogen and oxygen by *Chlorella*. Meded van de Landlougschool., Waganingen 58: 1–17.

Spruit, C.J.P. (1962). Photoreduction and anaerobiosis in physiology and biochemistry of algae (ed. Lewin, R.A.). Academic press, New York, p. 47.

Srivastava, M., (1990). Regulation of nitrogen fixation and hydrogen production. Ph.D. Thesis, Department of Botany, Banaras Hindu University.

Srivastava, M., Sharma, A., Kumar, A. and Kumar, H.D., (1989). Hydrogen production by nitrogen fixing cyanobacteria. In: *Proc. Intl. Phycotalk Symposium*, department of Botany, Banaras Hindu University, Varanasi 221005, pp. 26–29.

Stal, L,J, and Moczelaar, R. (1997). Fermentation in cyanobacteria. *FEMS Microbiol Rev.* 21: 179–211.

Stanier, R.Y., Pfennig, N. and Truper, H.G., (1981). Introduction to the phototrophic prokaryotes. In: The Prokaryotes: A Handbook on Habitats, Isolation and Identification of of Bacteria (eds. Starr, M.P., Stolp, H., Truper, H.G. Balows, A. and Schlegel, H.G.). Springer Verlag, Berlin, Heidelberg, New York, pp. 197–211.

Stevens, P., Plovie, N., De Vos., P. and De Ley, J., (1986). Photoproduction of molecular hydrogen by *Rhodobacter sulfidophilus*. Syst. *Appl. Microbiol.* 8: 19–23.

Stevens, P., Van der Sypt, H., De Vos, P. and De Ley, J., (1983). Comparative study on H2 evolution from DL–lactate, acetate and butyrate by different strains of *Rhodopseudomonas capsulata* in a new type of reactor. *Biotechnol. Lett.* 5: 369–374.

Stevens, P., Vertonghen, C., Vos, D.P. and De Ley, J., (1984). The effect of temperature and light intensity on hydrogen gas production by different *Rhodopseudomonas capsulata* strains. *Biotech. Lett.* 6: 277–282.

Stewart, W.D.P., 1973). Nitrogen fixation by photosynthetic microorganisms. *Annual Review of Micro biology*, Vol. 27, pp. 283–316.

Stiegel, G. J., Maxwell, R. C. Gasification Technologies: the Path to Clean, Affordable Energy in the 21st Century. *Fuel Processing Technology*, 2001, 71, 79–97.

Stiffer, H.J. and Gest, H. (1954). Effects of light intensity and nitrogen growth source on hydrogen metabolism in *Rhodospirillum rubrum. Science*, 120: 1024–1026.

Stoker, K., Oltmann, L. F. and Stouthamer, A. H., Partial charac–terization of an electrophoretically labile hydrogenase activity of *Escherichia coli* K–12. *J. Bacteriol.*, 1988, 170, 1220–1226.

Stuart, T.S. and Gaffron, H., (1971). The kinetics of hydrogen photoproduction by adopted *Scenedesmus. Planta* 100: 228–243.

Stuart, T.S. and Gaffron, H., (1972a). The gas exchange of hydrogen–adopted algae as followed by mars spectrometry. *Plant Physiol* 50: 136–140.

Stuart, T.S. and Gaffron, H., (1972b). The mechanism of hydrogen production by several algae. I. The effect of inhibitors of photophosphorylation. *Planta*, 106: 91–100.

Stuart, T.S. and Gaffron, H., (1972c). The mechanism of hydrogen photoproduction.

Stuart, T.S. and Kaltwasser, H., (1970). Photoproduction of hydrogen by photosystem I of *Scenedesmus*. *Planta* 91: 302–313.

Stuart, T.S. Herold, E.W. Jr. and Gaffron, H., (1972) A simple combination mass spectrometer inlet and oxygen electrode chamber for sampling gases dissolved in liquids. *Anal Biochem.* 46: 91–100.

Stwart, W.D.P., (1980). Some aspects of structure and function in N2 fixing cyanobacteria. *Ann. Rev. Microbiol.* 34: 497–536.

Taguchi F, Chang J.D., Takiguchi S and Morimoto M, Efficient Hydrogen–Production from starch by a Bacterium Isolated From Termites, *J Ferment Bioeng* 73 (1992) 244–245.

Taguchi F, Mizukami N, Yamada K, Hasegawa K, Saito–Taki T. Direct conversion of cellulosic materials to hydrogen by *Clostridium* sp. strain no. 2. *Enzyme Microbial Technol* 1995, 17: 147–50.

Taguchi F, Yamada K, Hasegawa K, Saito–Taki T, Hara K. Con–tinuous hydrogen production by *Clostridium* sp. strain no. 2 from cellulose hydrolysate in an aqueous two–phase system. *J Ferment Bio–eng* 1996, 82: 180–3.

Takabatake H, Suzuki K, Ko IB, Noike T. Characteristics of anaerobic ammonia removal by mixed culture of hydrogen producing photosyn–thetic bacteria. *Bioresour Technol* 2004, 95: 151–8.

Takahashi, P.K., Hydrfogen Energy From Renewable Resources". Final Report to the Solar Energy Research Inst., Subcontract XK–5–05097–1 Volume I., 1986).

Takakuwa, S. and Wall, J.D., (1981). Enhancement of gydrogenase activity in Rhodopseudomonas capsulata. *FEMS Microbiol. Lerr.*, 12: 359–363.

Takakuwa, S., Odom, J.M. and Wall, J.D., (1983). Hydrogen Uptake Deficient mutants of *Rhodopseudomonas capsulata*. *Arch. Microbiol.* 136: 20–25.

Tamagnini P, Costa, L, Almeid.a L, Oliveria M–J, Salema R, and. Undblad P. 2000. Diversity of cyanobacterial hydrogenases, a molecular approach. *Curr Microbiol* 40: 356–360.

Tamagnini P, Troshina O., Oxelfelt F, Salema R, and Lindblad P. (1997). Hydrogenases in Nnstoc sp. PCC 73102, a strain lacking a bi–directional hydrogenase. *Appi Environ Microbiol* 63: 1801–1807.

Tamagnini, P., Axelsson, R., Lindberg, P., Oxelfelt, F., Wunschiers, R. and Lindblad, P., Hydrogenases and hydrogen me–tabolism of cyanobacteria. *Microb. Mol. Biol. Rev.*, 2002, 66, 1– 20.

Tamagnini, P., Troshina, P., Oxelfelt, F., Salema, R. and Lindblad, P., (1997). Hydrogenase in Nostoc sp. strain PCC 73120, a strain lacking a bi–directional enzyme. *Applied and Environmental Microbiology*, Vol. 63, pp. 1801–1807.

Tamiya, N., Kondo, Y., Kameyama, T. and Akabori, S. (1955). Determination of hydrogenase by the hydrogen evolution from reduced methyl viologen. *J. Biochem.* 42: 613–614.

Tanisho, S., Ishiwata, Y. Continuous hydrogen production from molasses by bacterium *Enterobacter aerogenes*. *Int. J. Hydrogen Energy* 1994, 19: 807–12.

Tanisho, S., Ishiwata, Y. Continuous hydrogen production from molasses by fermentation using urethane foam as a support of flocks. *Int. J. Hydrogen Energy* 1995, 20: 541–4.

Tao, Y., Chen, Y., Wu, Y., He, Y. and Zhou, Z., High hydrogen yield from a two–step process of dark– and photo– fermentation of sucrose, *Int. J. Hydrogen Energy*, 33 (2007) 83–92.

Taqui Khan, M.M. and Bhatt, J.P., (1989). Light dependent hydrogen production by *Halobacterium halobium* coupled to *Escherichia coli*. *Int. J. Hydrogen Energy* 14: 643–645.

Taylor, L.A., Rose, R.E. (1988). A correction in the nucleotide se–quence of the Tn903 kanamycin resistance determinant in pUC4K. *Nucleic Acids Res* 16: 358.

Tel–Or, E., Luijk, L.W and Packer, L., (1977). An inducible hydrogenase incyanobacteria enhances N2 fixation. *FEBS Lett.* 78: 49–52.

Tel–Or, E., Luijk, L.W. and Packer, L., (1978). Hydrogenase in N2 fixing cyanobacteria. *Arch. Biochem. Biophys.* 185: 185–194.

Tetley, R.M. and Bishop, N.I (1979). The differential action of metronidazole an nitrogen fixation, hydrogen metabolism, photosynthesis and respiration in *Anabaena and Scenedesmus. Biochim. Biophys. Acta* 546: 43–53.

Thauer, R., (1977). Limitation of microbial H_2–formation via fermentation. IN: Microbial Energy Conversion (eds. Schlegel, H.G. and Barnea, J.). Pergamon Press, Oxford, New York, pp. 201–204.

The Scientist 17. Hydrogen–the new fuel? pp. 28–29 (April 2003).

Theil, T.(1994). Genetic analysis of cyanobacteria, In: Bryant DA(ed) The molecular biology of cyanobacteria. Kluwer academic Publishers, Dordrecht, The Netherlands. pp. 581–611.

Thiel, T., and Pratte, B. (2001). Effect on hcterocyst diffrentiation of nitrogen fixation in vegetative cells of the cyanobacterium.AMbaena iwriabilif ATCC 2941.3. *J Bacteriol* 183: 280–286.

Thiel, T., Lyons, E.M., Erker, C., and Ernst, A. (1995). A second nitrogenase in vegetative cells of a heterocyst.–forming cyanobacterium. *Proc Nad Acad Sci* 92: 9358–9362.

Thiele, H.H. (1968). Sulfur metabolism in Thiorhodaceae, IV. Assimilatory reduction of sulfate by *Thiocapsa floridana* and *Chromatium* sp. Ant. Van Leeuwenhoek *J. Microbiol. Serol.* 34: 341–349.

Tjepkema, J.D., Ormorod, W. and Torrey, J.C., (1980). Vesicle formation and acetylene reduction activity in Frankia sp. CP 11 cultured in defined nutrient media. *Nature* (London), 287: 633–635.

Torrey, J.C., 1985). The site of nitrogenase in Frankia in free–living culture and symbiosis. In: Nitrogen Fixation Research Progress (eds. Evans, H.J., Bottomley, P.J. and Newton, W.W.) pp. 293–299. Dordrecht, Nijhoff.

Tragut, V., J. Xiao, E.J. Bylina, and D. Borthakur, Characterization of DNA Restriction–modification systems in Spirulina platensis pacifica. *J. App. Phycol.*, 7–561–564 (1995).

Traore, A.S., Hatehikian, C.E., Belaich, J.P. and Gall, J.L., (1981). Microcolorimetric studies of the growth of sulfate reducing bacteria: Energetics of *Desulfovibrio vulgaris* growth. *J. Bacteriol.* 145 191–199.

Tredici, M.R., G.C. Zittelli and J.R. Benemann, "A tubular internal gas exchange photobioreactor for biological hydrogen production: Preliminary cost analysis". *Bio-Hydrogen*, O. Zaborsky *et al.*, eds., Plenum Press, New York pp. 391 – 402 (1998).

Tromp, T.K. (2003)."Potential environmental impact of the hydrogen economy on the stratosphere". *Science* 300: 1470–1472.

Troshina O, Serebryakova L, Sheremetieva M, and Lindblad P. (2002). Production of H\ bv the unicellular cyanobacteriuni gleocapsa nipwola CALU 743 during fermentation. *Int. J. Hydrogen Energ* 27: 1283–1289.

Troshina OY, Screbryakova LT, and Lindblad P. (1996). Induction of H, –uptake and nitrogcnasc actrvitics in the cyanobacterium *Anabaena variables* ATCC 29413: effects of hydrogen and organic substrate, *Curr Microbiol* 33: 1 1–15.

Truper, H.G. (1968). *Ecotohiorhodospira mobilis* Pelsh, a photosynthetic sulfur bacterium depositing sulfur outside the cells. *J. Bacteriol.* 95: 191–1920.

Truper, H.G. and Pfennig, N., (1981). Characterization and identification of the anoxygenic phtotrophic bacteria. In: The Prokaryotes – A Handbook of Habitats, Isolation and Identifixation of Bacteria (eds. Starr, M.P., Stolp, H., Truper, H.G., Balows, A. and Schlegel, H.G.) Springer Verlag, Berlin and New York, pp. 299–312.

Tsygankov AA, Federov AS, Kosourov SN, Rao KK. Hydrogen pro–duction by cyanobacteria in an automated outdoor photobioreactor under anaerobic condition. *Biotechnol Bioeng* 2002, 80: 777–83.

Tsygankov AA, Serebryakova LT, Rao KK, Hall DO. Acetylene reduc–tion and hydrogen photoproduction by wild type and mutant strains of *Anabaena* at different CO_2 and O_2 concentrations. *FEMS Microbiol Lett* 1998, 167: 13–7.

Tsygankov, A.A., Yakunin, A.F. and Gogotov, I.N., (1982) Hydrogenase activity of *Rhodopseudomonas capsulata* growing in organic media. *Mikrobiologiya* 51: 533–537.

Turner, J. (2004). "Sustainable Hydrogen Production". *Science* 305: 972–974.

Turner, J. G., Sverdrup, M.K., Mann, P.C., Maness, B., Kroposki, M., Ghirardi, R.J. and Evans, D., (2008). Renewable hydrogen production. *Int. J. of Hydrogen Energy*, Vol. 32, pp. 379–407.

Ueno Y, Haruta S, Ishii M, Igarashi Y. Characterization of a microor–ganism isolated from the effluent of hydrogen fermentation by microflora. *J Biosci Bioeng* 2001, 92: 397–400.

Ueno Y, Kurano N, Miyachi S. Purification and characterization of hydrogenase from the marine green alga, *Chlorococcum littorale. FEBS Lett* 1999, 443: 144–8.

Uffen, R.L., *"Anaerobic Growth of a Rhodopseudomonas species in the dark with carbon mon–oxide as sole carbon and energy substrate"*, Proceedings of the National Academy of Sciences of the United States of America, 73(9): 3298–3302 (1976).

Uffen, R.L., (1973). Growth properties of *Rhodospirillum rubrum* mutants and fermentation of pyruvate in anaerobic dark conditions. *J. Bacteriol.* 116: 874–884.

Umbereit, W.W., Burris, R.H. and Stauffer, J.F., (1972). Manometric and biochemical techniques. 5th ed. Burgens, Minneapolies, 387.

Umbreit, W.W., Burris, R.H. and Stauffer, J.F., (1972). Manometric and biochemical techniques. 5th Edition, Burgess Publ. Co., Minneapolis, *Minnesota*, pp. 387.

United States Department of Energy. An integrated research, development and demonstration plan. Hydrogen Pasteur Plan, 2004.

US Department of Energy (2003). *Basic Research Needs for the Hydrogen Economy.* Report of the Basic Energy Sciences Workshop on Hydrogen Production, Storage and Use. Office of Basic Energy Sciences: *http: //www.sc.doe.gov/bes/ reports/list.html* (accessed October 2006).

US Department of Energy (2004). *Hydrogen Posture Plan: http: //www.hydrogen.energy. gov/* (accessed October 2006).

Valdez–Vaquez I, Rios–Leal E, Esparza–Garica F, Cecchi F and Poggi–Varaldo HA, Semi–continuous solid substrate anaerobic reactors for hydrogen production from organic waste: Mesophilic versus thermophilic regime, *Int. J. Hydrogen Energy*, 30 (2005) 1383–1391.

Valero–Matas, J.A. (2010). *El espejismo de una energía social: La economía del hidrógeno* in Revista Internacional de Sociología (RIS), Vol.68, n° 2, Mayo–Agosto, pages 429–452.

Van der Oost J, Butlhuis A, Feitz S, Krab K, Kraayenhof R (1989). Fermentation metabolism of the unicellular cyanobacterium *Cyanothece* PCC 7822. Arch Microbiol 152: 415–419.

Van Der Oost, J., Kanneworff. WA., Krab, K. and Kraayenhof, R., (1987). Hydrogen metabolism of three unicellular nitrogen fixing cyanobacteria. *FEMS Microbiol. Lett.* 48: 41–45.

Van Ginkel SW and Logan B, Increased biological hydrogen production with reduced organic loading, *Water Res*, 39 (2005) 3819–3826.

Van Kessel, C. and Burris, R.H., 1983). Effect of hydrogen evolution on nitrogen (15N) fixation, acetylene reduction and relative efficiency of leguminous symbionts. *Physiol. Plant.*, 59: 329–334.

Van Neil EWJ, Budde MAW, de Haas GG, van der Wal FJ, Classen PAM and Stams AJM, Distinctive properties of high hydrogen producing extreme thermophiles, *Caldicellulosiruptor saccharolyticus* and *Thermotoga elfi, Int. J. Hydrogen Energy* 27 (2002) 1391–1398.

Van Niel, C.B. (1932). On the morphology and physiology the purple and green sulphur bacteria. *Arch. Mikrobiol.* 3: 1–112.

Van Niel, C.B. (1944). The culture, general physiology, morphology and classification of the non–sulfur purple and brown bacteria. *Bacteriol. Rev.* 8: 1–118.

Van Niel, C.B. (1962). The present status of the comparative study of photosynthesis. *Ann. Rev. Plant Physiol.* 13: 1–26.

Van Ooteghem SA, Beer SK and Yue PC, Hydrogen production by the thermophilic bacterium *Thermotoga neapolitana, Appl Biochem Biotechnol,* 98 (2002) 1391–1398.

Van Ooteghem SA, Jones A, van der Lelie D, Dong B and Mahajan D, Hydrogen production by the thermophilic bacterium *Thermotoga neapolitana* under anaerobic and microaerobic growth conditions, *Biotechnol Lett,* 26 (2004) 1223–1232.

Van Soom C, de Wilde P, and Vanderleyden J. (1997). HoxA is a transcriptional regulator for expression of the hup structural genes in. free–living *Bradirhizobium japoricum. Mol. Microbiol* 23: 967–977.

Vanderhaegen B, Ysebaert E, Favere K, Vanwambeke M, Peters T, Panic V, Vandenlangenbergh V and Verstraete W, Acidogenesis in relation to in– reactor granule yeid, *Water Sci. Technol.*, 25 (1992) 21–30.

Vardar–Schara, G., Maeda, T. and Wood, T.K., (2008). Metabolically engineered bacteria for producing hydrogen via fermentation. *Mcrobial Biotech* 1 (2) 107-125

Vasileva, L.G., Tsygankov, A.A. and Gogotov, I.N., (1988) Regulation of hydrogenase biosynthesis in *Rhodobacter sphaeroides. Mikrobiologiya* 57: 5–10.

Vas–quez–Bermuder MF, Flores E, HeiTei–o A. (2002). Analysis of binding sites fur the nitrogen–control transcription factor NtcA in the promoters of Sirnrthocnceus nitrogen–regulated genes. *Biochim Biophys Acta* 1578: 95–98.

Vatsala, T.M. and Balaji, V., (1989). Hydrogen production from cellulose by phototrophic organisms. In: Alternative Energy Sources, VIII, Vol. 2, Research and Development (ed. Veziroglu, T.N.). Hemisphere, Washington, pp. 539–546.

Vatsala, T.M. and Ramaswamy, V., (1989). Photohydrogen from cane molasses distillery waste. In Alternative Energy Sources, VIII, Vol. 2. Research and Development (ed. Veziroglu, T.N.). Hemisphere, Washington, pp. 519–528.

Vatsala, T.M. and Ramesh Kumar, G., (1989). Biohydrogen production from industrial wastes using locally isolated *Citrobacter freundii*. In: Alternative Energy Sources, VIII, Vol. 2. Research and Development (ed. Veziroglu, T.N.). Hemisphere, Washington, pp. 529–538.

Vatsala, T.M. and Ramesh Kumar, G., (1989). Hydrogen production from glucose by locally isolated *Citrobacter freundii. Ind. J. Expt. Biol.* 27: 824–825.

Vatsala, T.M. and Seshadri, C.V., (1989). Breakdown of silk cotton to hydrogen and volatile fatty acids by *Rhodospirillum rubrum. Curr. Sci.* 58: 173–175.

Vatsala, T.M., (1987). Influence of nutritional factors on hydrogen production by *Rhodospirillum rubrum* ATCC 11170. *J. Microb. Biotechnol.* 2: 127–131.

Vatsala, T.M., (1989). Degradation of cellulose by the phototrophic bacterium *Rhodospirillum rubrum. Ind. J. Expt. Biol.* 27: 963–966.

Vatsala, T.M., (1989). Microbial production of energy: I. Photoproduction of hydrogen. Monograph Series on the Engineering of Photosynthetic Systems. Vol. 128, Murugappa Chettiar Research Centre, Madras.

Velagapudi, V., Pourpoint, T., Zheng, Y., Fishr, T.S., Mudawar, I., Smith, K., Meyer, S., Anderson, W., Gore, J., (2006) "Onboard Metal Hydride Storage System, " Hydrogen Initiative Symposium, Purdue University, West Lafayette, IN, April 5–6.

Velagapudi, Varsha, Mudawar, I (2007). "Thermal Management of Metal Hydride Systems for Hydrogen Storage Applications, ' MSME, December.

Vignais, P. M., Billoud, B. and Meyer, J., Classification and phy–logeny of hydrogenases. *FEMS Micro. Rev.,* 2001, 25, 455–501.

Vignais, P.M. Colbeau, A., Willison, J.C. and Jouanneau, Y., (1985). Hydrogenase, nitrogenase and hydrogen metabolism in photosynthetic bacteria. *Adv. Microbial. Physiology* 26: 156–234.

Vignias, P.M., Colbeau, A., Willison, J.C. and Jouanneau, Y., (1985). Hydrogenase, nitrogenase and hydrogen metabolism in the photosynthetic bacteria. *Adv. Microbial Physiol.,* 26: 155–234.

Vincenzini, M., Materrasi, R., Sili, C. and Balloni, W., (1985). Evidences for an hydrogenase dependent hydrogen producing activity in *Rhodopseudomonas palustris. Ann. Microbiol. Enzymol.* 35: 155–164.

Vincenzini, M., Materrasi, R., Tredeci, M.R. and Florenzano, G., (1982). Hydrogen Production by immobilized cells. II. H2 photoevolution and waste water treatment by agar entrapped cells of *Rhodopseudomonas palustris* and Rhodospirillum molischianum. *Int. J. Hydr. Energy* 9: 725–728.

Voelskov, H. and Schon, G., (1978). Pyruvate fermentation in light grown cells of *Rhodospirillu rubrum* during adaptation to anaerobic dark conditions. *Arch. Microbiol.* 119: 129–133.

Volbeda, A., Charon, M. H. and Hatchikian, P. C., Crystal struc–ture of the nickel-iron hydrogenase from *Desulfovibrio gigas. Nature,* 1999, 373, 580–587.

Wachi, Y., "Hydrogen Production of Nostoc 1083 in Hawaii Culture Collection. Unpublished (2000).

Wagner R. (2000). The master–bender integration host factor (IHF) and its family of DNA–binding proteins. In: Transcription reflation in prokaryotes pp 227–232. Oxford University Press.

Wakayama T, Miyake J. Light shade bands for the improvement of solar hydrogen production efficiency by *Rhodobacter sphaeroides* RV. *Int J Hydrogen Energy* 2002, 27: 1495–500.

Wakayama T, Nakada E, Asada Y, Miyake J. Effect of light/dark cycle on bacterial hydrogen production by *Rhodobacter sphaeroides Rv. Appl Biochem Biotechnol* 2000, 84–86: 43 1–40.

Walker, C.C., Partridge, C.D.P. and Yates, M.G., (1981). The effect of nutrient limitation on hydrogen production by nitrogenase in continuous cultures of *Azobacter chroococcum. J. Gen. Microbiol.* 124: 317–328.

Wang C C, Chang CW, C P, Lee D J, Chang B V and Liao C S, Producing hydrogen from waste water sludge by *Clostiridium bifermentans, J. Biotechnol,* 102 (2003) 83–92.

Wang CC, Chang CW, Chu CP, Lee DJ, Chang BV, Liao CS and Tay JH, Using filtrate of waste biosolids to effectively produce bio– hydrogen by anaerobic fermentation, *Water Res,* 37 (2003) 1–39.

Wang CC, Chang CW, Chu CP, Lee DJ, Chang BV, Liao CS, *et al.,* Using filtrate of waste biosolids to effectively produce bio–hydrogen by anaerobic fermentation. Water Res 2003, 37: 2789–93.

Wang CC, Chang CW, Chu CP, Lee DJ, Chang BV, Liao CS. Effi–cient production of hydrogen from wastewater sludge. *J Chem Technol Biotechnol* 2004, 79: 426–7.

Wang CC, Chang CW, Chu CP, Lee DJ, Chang BV, Liao CS. Produc–ing hydrogen from wastewater sludge by *Clostridum bifermentans. J Biotechnol* 2003, 102: 83–92.

Wang R, Healey FP, Myers J (1971). Amperometric measurement of hydrogen evolution in *Chlamydomonas. Plant Physiol* 48: 108–110.

Wang, R., Healey, F.P. and Myers, J., (1971). Amperometric measurement of hydrogen evolution in *Chlamydomonas. Plant Physiol.* 48: 108–110.

Ward, H.B., Reeves, M.E. and Greenbaum, E., (1985). Stress–selected *Chlamydomonas reinhardtii* for photoproduction of hydrogen. *Biotech. Bioeng.* Symp. 15 501–507.

Ward, M.A., 1970). Whole cell and cell free hydrogenase of algae. Photochemistry 3: 259.

Watanabe, K.I., Kim, J.S., Ito, K., Buranakari, I., Kampee, T. and Takahashi, H., 1981) Thermostable nature of hydrogen production by non–sulfur purple photosynthetic bacteria isolated in Thialand. *Agric. Biol. Chem.* 45: 217–222.

Waugh, R., Mandrand–Besthold, M. and Boxer, D., (1985). Genetic analysis of hydrogenase in *Escherichia coli.* In: Microbial Gas Metabolism Mechanistic, Metabolic and Biotechnological Aspects (eds. Poole, R.K. and Dow, C.S.). Academic Press, New York, pp. 109–111.

Weare, N. M. and Shanmugam, K. T., Photoproduction of ammo–nium ion from N2 in *Rhodospirillum rubrum. Arch. Microbiol.,* 1976, 110, 207–213.

Weare, N.M. and J.R. Benemann. "Nitrogenase Activity and Photosynthesis by Plectonema boryanum 594." *J. Bacteriol.* 119, 258–268 (1974).

Weare, N.M., (1978). The Photoproduction of H2 and NH+4 fixed from N2 by a derepressed mutant of *Rhodospirillum rubrum. Biochim. Biophys. Acta* 502: 486–494.

Weaver, P. Lien, S., and Seibert, M., "Photobiological Production of Hydrogen", Solar Energy, 24(10): 3–45 (1980).

Weaver, P.E., Lien, S. and Siebert, M., (1980). Photobiological production of hydrogen. *Solar Energy* 24: 3–45.

Weaver, P.E., Wall, J.D. and Gest, H., (1975). Characterization of *Rhodopsedomonas capsulata. Arch. Microbiol.* 105 107–216.

Weaver, P.F., Lien, S. and Siebert, M., (1980). Photobiological production of hydrogen. *Solar Energy*, 24: 3–45.

Weetal, H.H. and Krampitz, L.O., (1980). Production of hydrogen using photolytic methods. *J. Solid Phase Biochem.*, 5: 115–124.

Wei TF, Ramasubramanian TS, and Golden JW (1994). *Anabaena* sp. strain PCC 7120 nfcA gene required for growth on nitrate and hererocyst development. *J Bacteriol* 176: 4473–4482.

Weiss, A.R., Schneider, K. and Schlegel, H.G., (1980). Purification and properties of the mmbrane bound hydrogenase of Pseudomonas pseudoflava. *Current Microbiol.*, 3: 317–320.

Weisshaar, H. and Boger, P., (1983). Sulfide stimulation of light induced hydrogen evolution by the cyanobacterium *Nostoc muscorum.* Z. Naturforsch. 38: 237–242.

Weissman, J.C., R.P. Goebel, and J.R. Benemann, "Photobioreactor Design: Comparison of Open Ponds and Tubular Reactors", *Bioeng. Biotech 31:* 336–344 (1988).

Williams JGK (1988). Construction of specific mutations in photo–system II photosynthetic reaction center by genetic engineering methods in *Synechocystis* 6803. *Methods Enzymol* 167: 766– 778.

Willison, J.C. Madern, D. and Vignais, P.M., (1984). Increased photoproduction of hydrogen by nonautotrophic mutants of *Rhodopsudomonas capsulata. J. Biochem.* 219: 593–600.

Winkler M, Heil B, Heil B, Happe T. Isolation and molecular char–acterization of the [Fe]–hydrogenase from the unicellular green alga *Chlorella fusca. Biochim Biophys Acta* (BBA): Gene Struct Express 2002, 1576: 330–4.

Winkler M, Hemschemeier A, Gotor C, Melis A, Happer T. [Fe]–hydrogenases in green algae: photo–fermentation and hydrogen evolu–tion under sulfur deprivation. *Int. J. Hydrogen Energy* 2002, 27: 1431–9.

Winter CJ. Into the hydrogen energy economy–milestones. *Int. J. Hydrogen Energy* 2005, 30: 681–5.

Winter, C.–J., Nitsch, J. Hydrogen As an Energy Carrier. Springer Verlag, Berlin, 1988, 377 pp.

Winter, H.C. and Burris, R.H., (1976). Nitrogenase. *Ann. Rev. Biochem.*, 45: 409–426.

Winzeler, E. A., Liang, H., Shoemaker, D. D. and Davis, R. W., Functional analysis of the yeast genome by precise deletion and parallel phenotypic characterization. Novartis Found. *Symp.*, 2000, 229, 105–109.

Wittnerberger, C.L. and Repaske, R. (1958). Studies on the elctrion transport system in *Hydrogenomonas eutropha. Bacterioloical Proceedings* 1958: 106.

Wolf I, Buhrke T, DemeddeJ, Pohimann A, and Friedrich B. (1998). Duplication of f}fp genes involved in maturation of [NiFe] hydrogenases in A.lcaliffnts entrophus H16. *Arch Microbiol* 170: 451–459.

Wolfe, R.S., (1971). Microbial formation of methane. *Adv. Microbial. Physiol.* 6: 107–146.

Wolfrum, E. and Watt, A., "*Bioreactor Design Studies for a Hydrogen–Producing Bacterium*", *Applied Biotechnology and Bioengineering* 98–100: 611–625 (2002).

Wolk CP. (1996). Heterocyst. foi–madon. *Annu Rev Genet* 30: 59–78.

Wu JH and Lin CY, Biohydrogen production by mesophilic fermentation of food wastewater, *Water Sci Technol*, 49 (2004) 223–228.

Wu SY, Hung CH, Lin CN, Chen HW, Lee AS and Chang JS, Fermentative hydrogen production and bacterial community structure in high– rate anaerobic bioreactors containing silicone– immobilized and self – flocculated sludge, *Biotechnol Bioeng*, 93 (2006) 934–946.

Wu SY, Lin CN and Chang JS, Biohydrogen production with anaerobic sludge immobilized by ethylene– vinyl acetate copolymer. *Int. J. Hydrogen Energy*, 30(2005)1375–1381.

Wu SY, Lin CN and Chang JS, Hydrogen production with immobilized sewage sludge in three– phase fluidized – bed bioreactors, *Biotechnol Progr*, 19(2003) 828–832.

Wu SY, Lin CN, Chang JS, Lee KS and Lin PJ, Microbial hydrogen production with immobilized sewage sludge, *Biotechnol Progr*, 18 (2002) 921–926.

Wunschicrs R, Scngcr H, and Schuiz R. (2001). Electron pathways involved in.H, –rnctabolisiTi in the green alga Scenedesnuu obliquns. *Biochim Biophys Acta* 1503: 271–278.

Wunschiers R and Lindblad P. (2003). Light–dependent hydrogen uptake and generation by cyanobacteria. Handbook of Photochemistry and Photobiology, in press.

Wunschiers R, Batur M, and Lindblad P. (2003). Presence and expression of hydrogenase specific C–terminal endopeptidases in cyanobacteria. *BMC Microbiol* 3: 8.

Wunschiers, R., Stangier, K., Senger, H. and Schulz, R., Molecu–lar evidence for a Fe–hydrogenase in the green alga *Scenedesmus obliquus*. *Curr. Microbiol.*, 2001, 42, 353–360.

Wynn–Wllliams DD. (2000). Cyanobacteria in deserts — Life at the limit? In: The Ecology of Cyanobacteria (Whitton BA and Potts M, eds) pp 341–366. Kluver Academic Publishers, Dordrecht.

Xiankong Z, Tabita, and van Baalen C. (1984). Nickel control of hydrogen production and uptake in *Anabaena* spp. strains CA a.nd I F. *J Gen. Microbiol* 130: 1815–1818.

Xu, X., Song, C., Andresen, J. M., Miller, B. G., Scaroni, A. W. Novel Polyethyleneimine–Modified Mesoporous Molecular Sieve of MCM–41 Type as Adsorbent for CO_2 Capture. Energy and Fuels, 2002, 16, 1463–1469.

Xu, X.C., Song, C., Andresen, J.M., Miller, B.G., Scaroni, Selective Capture of CO_2 from Boiler Flue Gases by a Novel CO_2 "Molecular Basket" Adsorbent. Proceeding of the Nineteenth Annual International Pittsburgh Coal Conference, September 15–18, 2003, Pittsburgh, PA USA.

Y HQ and Mu Y, Biological hydrogen production in UASB reactor with granules. II: Reactor performance in 3– year operation, *Biotechnol Bioeng*, 94 (2006) 988–995.

Yagi, T., 1976). Properties of purified hydrogenase from the particulate fraction of Desulfovibrio vulgaris Miyazaki. *J. Biochem.* (Tokyo), 79: 661–671.

Yagi, T., (1976). Separaction of hydrogenase catalysed hydrogen evolution system from electron donating system by means of enzymic electric cell technique. *Proc. Natl. Acad. Sci.*, U.S.A. 73: 2947–2949.

Yagi, T., (1977). Use of enzymic electric cell and immobilized hydrogenase in the study of biophotolysis of water to (eds. Mitsui, A., Miyachi, S., San Pietro, A. and Tamura, S.) pp. 61–68, Academic Press, New York.

Yagi, T., (1981). Function and structure of hydrogenases. *Natural Science*, Vol. 32, pp. 29–83 Yumurtaci, Z. and Bilgen, E., 2004). Hydrogen production from excess power in small hydroelectric Installations. *Int. J. Hydrogen Energy*, Vol. 29, pp. 687–693.

Yamazaki, S., (1982). A selenium containing hydrogenase from Methanococcus vanneilu: Identification of selenium moiety as a selenocysteine residiue. *J. Biol. Chem.*, 257: 7926–7929.

Yanyshin, M.F., (1981). Effect of electron transfer inhibitors on high induced hydrogen evolution by *Chlamydomionas reinhardtii* cells in synchronous culture. *Fiziol., Rast.* (Mosc.) 29: 749–755.

Yanyshin, M.F., (1982). Activation of hydrogenase and light–induced hydrogen evolution in synchronous culture Chlamydomonas reinhardtii, during anaerobic adaptation in the light. *Fiziol. Rast.* (Mosc.) 29: 1126–1133.

Yanyshin, M.F., (1982). Hydrogen evolution and hydrogenase activity in a synchronous culture of Chlamydomonas reinhardtii with respect to anaerobic starch degradation. *Fiziol. Rast.* (Mosc.) 29: 121–126.

Yetis M, G¨und¨uz U, Ero~glu ~I, Y¨ucel M, T¨urker L. Photoproduction of hydrogen from sugar refinery wastewater by *Rhodobacter sphaeroides* O.U.001. *Int. J. Hydrogen Energy* 2000, 25: 1035–41.

Yoch, D.C. and Arnon, D.I., (1974). Biological nitrogen fixation of Photosynthetic bacteria. In: The Biology of Nitrogen Fixation (ed. Quispel, A.). Elsevier, New York, pp. 686–687.

Yokoi H, Maeda Y, Hirose J, Hayashi S and Takasaki Y, H_2 production by immobilized cells of *Cloestridium butyricum* on porous glass beds, *Biotechnol Tech*, 11 (1997) 431–433.

Yokoi H, Maki R, Hirose J, Hayashi S. Microbial production of hydrogen from starch manufacturing wastes. *Biomass Bioenergy* 2002, 22: 89–395.

Yokoi H, Mori S, Hirose J, Hayashi S, Takasaki Y. H_2 production from starch by mixed culture of *Clostridium butyricum* and *Rhodobacter* sp M–19. *Biotechnol Lett* 1998, 20: 895–9.

Yokoi H, Saitsu A, Uchida H, Hirose J, Hayashi S and Takasaki Y, Microbial hydrogen production from wet potato starch residue, *J. Biosci Bioeng*, 91 (2001) 58–63.

Yokoi H, Saitsu AS, Uchida H, Hirose J, Hayashi S, Takasaki Y. Microbial hydrogen production from sweet potato starch residue. J Biosci Bioeng 2001, 91: 58–63.

Yokoi H, Tokushi T, Hirose J, Hayashi S, Takahashi Y. (1997b). Hydrogen production by immobilized cells of aciduric *Enterobacter aerogenes* strain HO–39, *J. Ferment Bioeng*, 83 (1997) 481–484.

Yokoi H, Tokushige T, Hirose J, Hayashi S, Takasaki Y. H_2 production from starch by mixed culture of *Clostridium buytricum* and *Enterobac–ter aerogenes. Biotechnol Lett* 1998, 20: 143–7.

Yokoi H, Tokushige T, Hirose J, Hayashi S, Takasaki Y. Hydrogen Production by immobilized cells of aciduric *Eneterobacter aerogenes* strain HO–39. *J. Ferment Bioeng* 1997, 83: 481–4.

Yoon JH, Sim SJ, Kim M, Park TH. High cell density culture of *Anabaena variabilis* using repeated injection of carbon dioxide for the production of hydrogen. *Int. J. Hydrogen Energy* 2002, 27: 1265–70.

Yoshida, A., Nishimura, T., Kawaguchi, H., Inui, M. and Yukawa, H.(2005). Enhanced hydrogen production from formic acid by formate hydrogen lyase overexpressing *Escherichia coli* strain. *Applied Environmental Microbiology*, Vol. 71, pp. 6762– 6768.

Yu H, Zhu Z, Hu W and Zhang H, Hydrogen production from rice winery wastewater in an upflow anaerobic reactor by using mixed anaerobic cultures. *Int. J. Hydrogen Energy*, 27 (2002) 1359–1365.

Yu H, Zhu Z, Hu W, Zhang H. Hydrogen production from rice winery wastewater in an uplow anaerobic reactor by mixed anaerobic cultures. *Int. J. Hydrogen Energy* 2002, 27: 1359–65.

Yu HQ, Hu ZH and Hong TQ, Hydrogen production from rice winery wastewater by using a continuously – stirred reactor, *J. Chem Eng Japan*, 36 (2003) 1147–1151.

Yu, L. and Wolin, M.J. (1969). Hydrogenase measurement with photochemically reduced methyl viologen. *J. Bacteriol.* 98: 51–55.

Zaborsky, O., "International Marine Biotechnology Culture Collection". Presented at the Annual Review of the U.S. Dept. Energy Hydrogen Program, Miami, Florida, April 29 –May 3, 1996.

Zaborsky, O.R., J.C. Radway, B.A. Yoza, J.R. Benemann, and M.R. Tredici, Sustainable Bioreactor Systems for Producing Hydrogen. Annual DOE Hydrogen R and D Program Review Meeting, 1998.

Zaijic, J.E., Kosaric, N. and Brosseau, J.D., (1978). Microbial production of hydrogen. *Adv. Biochem. Enginner.* 9: 57–109.

Zexing, C. and Michael, B. H., Modeling the active sites in metal–loenzymes.3. density functional calculations on models for Fe–hydrogenase: structures and vibrational frequencies of the observed redox forms and the reaction mechanism at the diiron active cen–tre. *J. Am. Chem. Soc.*, 2001, 123, 3734–3742.

Zhan, Z, and Barnett, S.A., (2006). "Solid oxide fuel cells operated by internal partial oxidation reforming of *iso*–octane", *Journal of Power Sources*, Vol. 155, Issues 2, pages 353–357.

Zhang, T., Liu, H., Fang, H.H.P. Biohydrogen production from starch in wastewater under thermophilic conditions. *J Environ Manag* 2003, 69: 149–56.

Zhang, Y.F., Liu, G.Z. and Shen, J.Q., Hydrogen production in batch culture of mixed bacteria with sucrose under different iron concentrations, *Int. J. Hydrogen Energy*, 30 (2005) 855–860.

Zhang, Z.P., Show, K.Y., Tay, J.H., Liang, D.T. and Lee, D.J., Biohydrogen production with anaerobic fluidized bed reactors– A comparison of biofilm – based and granule – based systems, *Int. J. Hydrogen Energy*, 33 (2008) 1559–1564.

Zhang, Z.P., Show, K.Y., Tay, J.H., Liang, D.T. and Lee, D.J., Enhanced continuous bihydrogen production by immobilized anaerobic microflora, Energy and Fuels, 22 (2008) 87–92.

Zhang, Z.P., Show, K.Y., Tay, J.H., Liang, D.T., Lee, D.J. and Jiang, W.J., Biohydrogen production in a granular activated carbon anaerobic fluidized bed reactor, *Int. J. Hydrogen Energy*, 32 (2007) 185–191.

Zhang, Z.P., Show, K.Y., Tay, J.H., Liang, D.T., Lee, D.J. and Jiang, W.J., Effect of hydraulic retention time on biohydrogen production and anaerobic microbial community, *Process Biochem*, 41 (2006) 2118–2123.

Zhang, Z.P., Show, K.Y., Tay, J.H., Liang, D.T., Lee, D.J. and Jiang, W.J., Rapid formation of hydrogen– producing granules in an anaerobic continuous stirred tank reactor induced by acid incubation, *Biotechnol Bioeng*, 96 (2007) 1040–1050.

Zhang. Z.P., Show, K.Y., Tay, J.H., Liang, T.D., Jiang, W.J. and Lee, D.J., A comparison between suspended sludge and granular sludge hydrogen production in a CSTR system, in *Proc 4th IWA Leading– Edge Conf Water and Wastewater Technol* (IWA Press, London) 2007.

Zhang, C., Xing, X.H. and Lou, K., (2005). Rapid detection of a gfp–marked Enterobacter aerogenes under anaerobic conditions by aerobic fluorescence recovery. *FEMS Microbiology Letters*, Vol. 249, pp. 211–281r Catalytic Activity, Structure and Function. pp. 15–44. Goltze KG, Gottingen.

Zhang, X.K., Haskell, J.B., Tabita, F.R. and Baalen, C.V., (1983). Aerobic hydrogen production by the heterocystous cyanobacteria *Anabaena* spp. strains CA and 1F. *J. Bacteriol.*, 156: 1118–1122.

Zheng, M. and Storz, G., (2000). Redox sensing by prokaryotic transcription factors. *Biochem. Pharmacol.*, 59: 1–6.

Zhilina, T.N., (1976). Biotypes of *Methanosarcina*. *Mikrobiologiya* 40: 674–680.

Zhu, C., Binjian, C. and Hongyu, S., (1987). Enhancement of hydrogenase and nitrogenase activities in *Rhodopseudomonas capsulata* by nickel. *Acta Microbiol. Sin.*, 27: 52–56.

Zhu, H., Suzuki, T., Tsygankov, A.A., Asada, Y. and Miyake, J., (1999). Hydrogen production from tofu wastewater by *Rhodobacter sphaeroides* immobilized agar gels. *Int. J. Hydrogen Energy*, 24: 305–310.

Zhu, H., Ueda, S., Asada, Y. and Miyake, J., (2002). Hydrogen production as a novel process of wastewater treatment-studies on tofu wastewater with entrapped *R. sphaeroides* and mutagenesis. *Int. J. Hydrogen Energy*, 27: 1349–1357.

Zhu, H., Wakayama, T., Asada, Y. and Miyake, J., (2001). Hydrogen production by four cultures with participation by anoxygenic photobacterium and anaerobic bacterium in the presence of NH4+. *Int. J. Hydrogen Energy*, 26: 1149–1154.

Ziomek, E., Martin, W.G., Veliki, I.A. and Williams, R.E., (1982). Immobilization of *Desulfovibrio desulfuricans:* Cell associated hydrogenase in beaded matrices. *Enz. Microbial Technol.*, 4: 405–408.

Zolotukhinsa, I.M., Davydova, M.N. and Belyaeva, M.I., (1987). Influence of growth conditions on hydrogen production and hydrogenase activity of *Desulfovibrio desulfuricans*. *Izv. Akad. Nauk. SSSR Ser Biol.* 30: 607–611.

Zorin, N.A., 1986. Inhibition of *Thiocapsa roseopersicina* hydrogenase by various compounds. *Biochimiya*, 51: 770–774.

Zorin, N.A., Serebryakova, L.T. and Gogotov, I.N., (1984). Effect of redox potential on the activity of purple bacteria hydrogenase. *Biochimiya*, 42: 1316–1319.

Zurrer, H. and Bachofen, R., 1979. Hydrogen production by the photosynthetic bacterium *Rhodospirillum rubrum*. *Appl. Environm. Microbiol.* 37: 789–793.

Zyper, J., Yoza, B.A., Benemann, J.R., Tredici, M.R. and Zaborsky, O.R., (1998). Internal gas exchange photobioreactor development and testing in Hawaii. In: Biohydrogen, (Ed.) O. Zaborsky. Plenum Press, New York, pp. 441–446.

Index

www.ingramcontent.com/pod-product-compliance
Lightning Source LLC
Chambersburg PA
CBHW050523190326
41458CB00005B/1638